Mechanisms and Machine Science

Volume 33

Series editor

Marco Ceccarelli, Cassino, Italy

More information about this series at http://www.springer.com/series/8779

Shaoping Bai · Marco Ceccarelli
Editors

Recent Advances in Mechanism Design for Robotics

Proceedings of the 3rd IFToMM Symposium on Mechanism Design for Robotics

Springer

Editors
Shaoping Bai
Aalborg University
Aalborg
Denmark

Marco Ceccarelli
Laboratory of Robotics and Mechatronics (LARM)
University of Cassino and South Latium
Cassino
Italy

ISSN 2211-0984 ISSN 2211-0992 (electronic)
Mechanisms and Machine Science
ISBN 978-3-319-18125-7 ISBN 978-3-319-18126-4 (eBook)
DOI 10.1007/978-3-319-18126-4

Library of Congress Control Number: 2015937532

Springer Cham Heidelberg New York Dordrecht London

Printed on acid-free paper

Springer International Publishing AG Switzerland is part of Springer Science+Business Media (www.springer.com)

Preface

MEDER 2015, IFToMM International Symposium on Mechanism Design for Robotics, is the third event of a series that started in 2010 as a specific conference activity on mechanisms for robots. The first event was held at Universidad Panamericana de Ciudad de Mexico, Mexico in September 2010 and the second was held at Beihang University, Beijing, China in October 2012.

The aim of the MEDER Symposium is to bring together researchers, industry professionals and students from broad ranges of disciplines dealing with mechanism for robots, in an intimate, collegial and stimulating environment. Again, in the 2015 MEDER the event received increased attention, since the proceedings contain contributions by authors from all over the world.

The proceedings of MEDER 2015 Symposium contains 42 papers, which were selected after review for oral presentation. These papers cover several aspects of the wide field of Mechanism Design for Robotics, from theoretical studies up to practical applications through new robot designs and prototypes. They are authored mainly from the IFToMM community coming from China, Denmark, France, Germany, Italy, Japan, Mexico, The Netherlands, Norway, Russia, Singapore, Spain, UK and USA.

We express grateful thanks to the MEDER Symposium International Scientific Committee members including Marco Ceccarelli, Chair (University of Cassino, Italy), Juan Carretero (University of New Brunswick, Canada), Lu Zhen (Beihang University, China), Pierre Larochelle (Florida Institute of Technology, USA), Ding Xilun (Beihang University, China), Grigore Gogu (French Institute for Advanced Mechanics, France), I-Ming Chen (Nanyang Technological University, Singapore), Mario Acevedo (Panamerican University, Mexico), Teresa Zielinska (Warsaw University of Technology, Poland), Joseph Rooney (Open University, UK), Atsuo Takanishi (Waseda University, Japan), Alfonso Hernandez (Bilbao University, Spain) for cooperating enthusiastically for the success of the MEDER 2015 event.

We thank the authors who contributed with very interesting papers on several subjects, covering many fields of Mechanism Design for Robotics and additionally for their cooperation in revising papers in due time in agreement with the reviewers' comments. We are grateful to the reviewers for the time and efforts they spent in evaluating the papers within a given schedule that has permitted the publication of this proceedings volume. These reviewers are Zheng-Hua Tan (Aalborg University, Denmark), Xuping Zhang (Aarhus University, Denmark), Marco Ceccarelli and Giuseppe Carbone (University of Cassino and South Latium, Italy), Juan Antonio Carretero (University of New Brunswick, Canada), Grigore GOGU (French Institute of Advanced Mechanics in Clermont-Ferrand, France), John Hayes (Carleton University, Canada), Ronghua Li (Dalian Jiaotong University, China), Carl Nelson (University of Nebraska–Lincoln, USA), Latifah Nurahmi (IRCCyN, France), Alba Perez Gracia (Idaho State University, USA), Teresa Zielinsks (Warsaw University of Technology, Poland), Alfonso Hernandez (University of the Basque Country, Spain), Volkert van der Wijk (University of Twente, The Netherlands) and Tao Li (Institute of Advanced Manufacturing Technology, China).

We thank Aalborg University (AAU), in particular, the Department of Mechanical and Manufacturing Engineering and Aalborg U Robotics, for having hosted the MEDER 2015 event. We would like to thank our colleagues at AAU for their efforts and support in the symposium organizing. We thank the local organizing committee members including Ole Madsen, Jørgen Kepler, Guanglei Wu, Ewa Kolakowska and Christoffer Eg Sloth, who put their prestigious time into the event to make it successful. We also extend our thanks for help by the LARM Laboratory of Robotics and Mechatronics of University of Cassino.

We thank The IFToMM (International Federation for the Promotion of Mechanism and Machine Science) for sponsorship of two Young Delegation Travel grants. The symposium received generous support from local sponsors, namely the Thomas B. Thriges Fund and the Danish RoboCluster, which were critical to make this symposium of low registration cost possible.

We thank the publisher Springer and its Editorial staff for accepting and helping in the publication of this Proceedings volume, since its early steps in 2014.

We are grateful to our families, as without their patience and comprehension it would not have been possible for us to organize MEDER 2015, IFToMM International Symposium on Mechanism Design for Robotics and this Proceedings volume.

Aalborg
Cassino
March 2015

Shaoping Bai
Marco Ceccarelli

Contents

Part VI Parallel Manipulators

Part I
Linkage and Manipulators

Finger Mechanisms for Robotic Hands

M. Ceccarelli

Abstract The problem of grasping with robots is solved by using suitable finger mechanisms that are inspired from structures in nature. A variety of solutions have been experienced and are used in a variety of designs all around the world. This paper discusses a survey of possibilities by addressing attention to characteristics and problems in the design and operation of those finger mechanisms. The author's experience with LARM hand is reported to show practical results in attaching the problem of improving efficient solutions with better finger mechanisms.

Keywords Artificial hands · Finger anatomy · Finger mechanisms · Kinematic design

1 Introduction

Manipulation of objects with fingered robotic hands is an aspect which can be used in many applications, [6], also in industry and service contexts and it attracts still great interest as indicated in [1, 4, 16].

Since a recent past, in order to develop anthropomorphic finger mechanisms researchers have used two different approaches: complex mechanisms in order to perform manipulation tasks with high dexterity, or design of mechanisms with a reduced number of degrees of freedoms (DOFs) and actuators with less performances but a fairly simplified device operation.

Using underactuated mechanisms it is possible to achieve an adaptive grasp that mimics the human grasping action for which it is possible to consider two kinds of structures, namely using flexibility of links or designing underactuated mechanisms as pointed out in [8]. A mechanism is defined underactuated when its number of

M. Ceccarelli (✉)
LARM: Laboratory of Robotics and Mechatronics, University of Cassino and South Latium, Cassino (Fr), Italy
e-mail: ceccarelli@unicas.it

S. Bai and M. Ceccarelli (eds.), *Recent Advances in Mechanism Design for Robotics*, Mechanisms and Machine Science 33,
DOI 10.1007/978-3-319-18126-4_1

actuators is smaller than the number of degrees of freedom of the mechanism. It is possible to identify two types of underactuated finger mechanisms, depending on whatever a tendon or a link transmission is used. Example of tendon finger mechanisms is presented in [2], and a pulley-cable solution is described in [3].

When large grasping forces are required, underactuated linkage mechanisms are usually preferred, like in [2] where a 1-DOF mechanism with suitable four-bar linkages and flexible elements is used to move all phalanxes of fingers, in [17] where an underactuated linkage with force control is studied or in [19] where a 5 fingered underactuated prosthetic hand is reported.

Since the end of 1990s at LARM in Cassino design and research activities have been carried out in order to design a low-cost easy-operation robotic hand with anthropomorphic fingers, denominated LARM Hand [5].

In this paper the design of a new underactuated finger mechanism has been proposed for LARM Hand, as focused on requirements referring to 1-DOF, anthropomorphic grasp, and mechanism's compact size.

2 Fingers in Nature

Fingers in nature are related to hands and feet when considering a grasp in a broad sense as a task by which interaction is with an object to be manipulated or with the ground to be in contact with, respectively.

Such a variety of task configurations has generated a variety of finger anatomy but with solutions that still show a common structure. The anatomy of fingers in hands both in humans and primates is illustrated in the example in Fig. 1, [11]. The anatomy is characterized by a bone structure with a serial kinematic configuration and a muscle complex system by which each bone is moved through two or more counteracting muscular tendons. The main terminology of human anatomy is reported in Fig. 1 and it is usually used also for artificial hands.

In Fig. 2, examples of finger configurations are reported as referring to horses and cats with a structure that still shows sequential bones and parallel architecture muscle system. The similarity with human anatomy is recalled by the similar terminology of the parts in those other fingers.

Summarizing, the anatomy of fingers in nature shows the following design characteristics:

- A serial bone-link structure whose main purpose is rigidity and load capability
- An actuation muscle system aiming to rotate each bone-link independently but with coordinated movements with other finger links

The operation of fingers in nature is devoted to grasp and interaction with objects by exerting proper force during a proper motion for pure mechanical purposes.

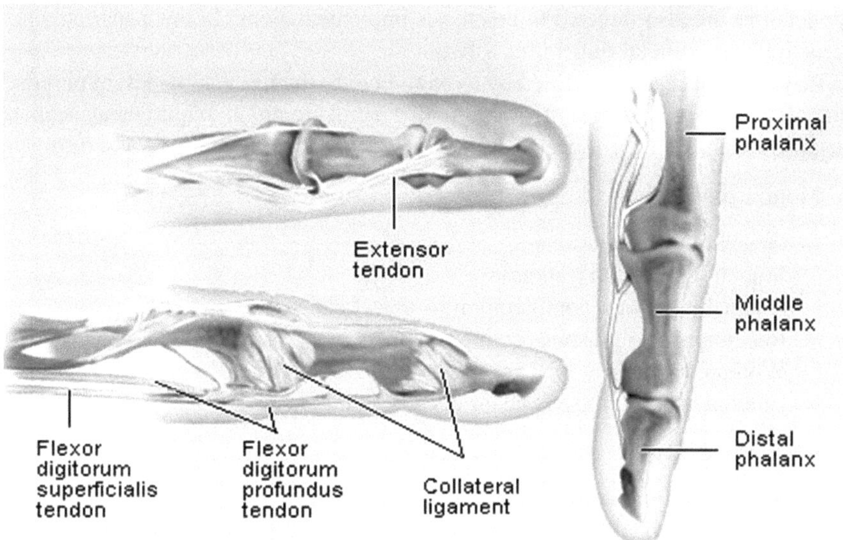

Fig. 1 Anatomy of human fingers, [11]

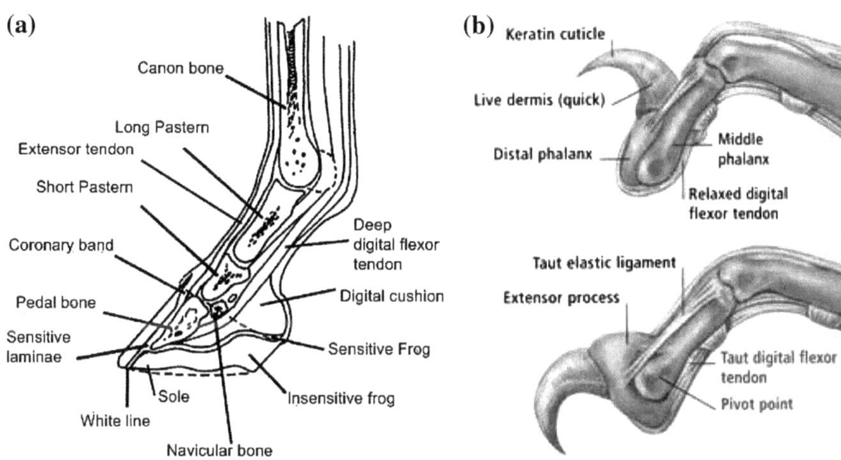

Fig. 2 Anatomy structures in: **a** horse fingers, [9]; **b** cat fingers, [12]

3 Requirements for Artificial Fingers

Artificial fingers are conceived to replicate the design and operation of fingers in nature for performing grasping actions at the most. Artificial fingers are developed for automatic machinery, and recently mainly for robotic systems, [6]. Some applications are also directed to prosthesis implementations. In this paper attention is focused on artificial fingers for robots.

Requirements can be outlined by looking both to the mimicking design purposes and operation peculiarities. However, in general common requirements can be identified mainly in the aspects for:

1. Motion properties in:
   - Grasping configurations
   - Smooth approaching motion
   - Adaptable motion configuration to object shapes
   - Reconfigurable grasping configurations
   - Workspace ranges
   - Limited motion impacts again objects to be grasped
2. Force capability in:
   - Stable grasping configurations
   - Efficient transmission of input power to grasping forces
   - Distribution of grasping forces among several contacts with grasped object
   - Positions of application points of the grasping forces
   - Adjustable grasping forces
3. Mechanical design in:
   - Stiff or compliant structure at grasp
   - Phalanx shape for adaptability to object shape
   - Room for sensors
   - Compact design versus human-like solutions
   - Lightweight solution with smart or traditional materials
   - Low-friction joints
   - Location of power source

In general, the size of grasping devices for robots is designed as function of prescribed grasps with a given set of objects. In particular, for humanoid robots artificial hands are usually of hum-like size but different smaller or larger solutions can be required in other robotic systems and manipulators.

The above-mentioned aspects can be considered with proper numerical ranges of requirements after a careful analysis of peculiarities and aims in design procedures for the adopted finger solution structure. A general procedure can be outlined as in the block diagram of Fig. 3 where the core procedure application of mechanism design procedures with specific algorithms taking into account the grasping purposes and features.

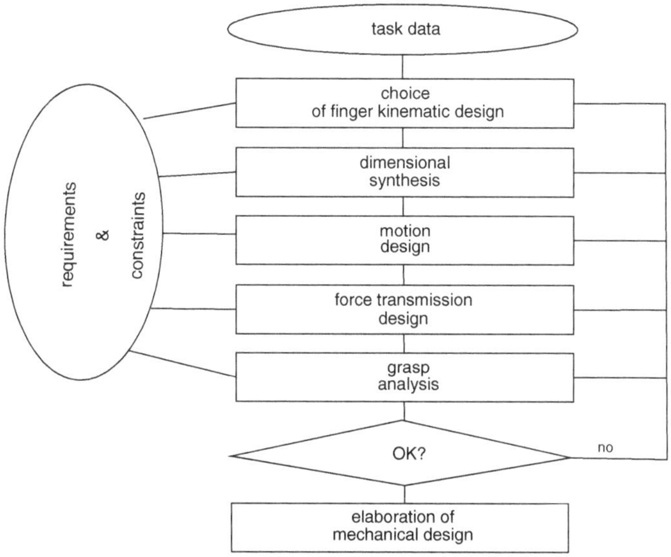

Fig. 3 A flowchart for design procedure of finger mechanisms

4 Mechanisms for Fingers

Artificial fingers are conceived as inspired by natural solutions so that in general they show structures with phalanx bodies with serial chain configuration. Figure 4 summarizes such an inspiration by looking also at the phalanx actuation with motion joint angles θi (i = 1, …, 3) and joint torque τi (i = 1, …, 3) with serial design, Fig. 4a or parallel architectures, Fig. 4b. Indeed a variety of actuating mechanisms are used to obtain actuation solutions with phalanx bodies that are even part of them.

Examples of the most used mechanisms for finger design are reported in Figs. 5, 6, 7 as referring to design finger structures which are based on linkages, tendons, or direct drives, respectively.

In Fig. 5 an artificial human hand is shown with the fingers that are designed with linkages as a product that is available in the market even at very low cost and fairly easy robot implementation, [15]. Phalanxes are links of the linkage and the finger mechanism shows a kinematic chain that can have 1 or more DOFs. In general linkages are used in finger structures to obtain 1 DOF finger mechanism whose actuator can be active on the first joint like in the example in Fig. 5.

Figure 6 shows a general scheme of tendon driven finger design where each phalanx body is actuated by two antagonist cables whose pulling gives motion and force to the phalanxes, [14]. This solution is clearly inspired from the human anatomy and it is the most common used structure in robotic hand for humanoid

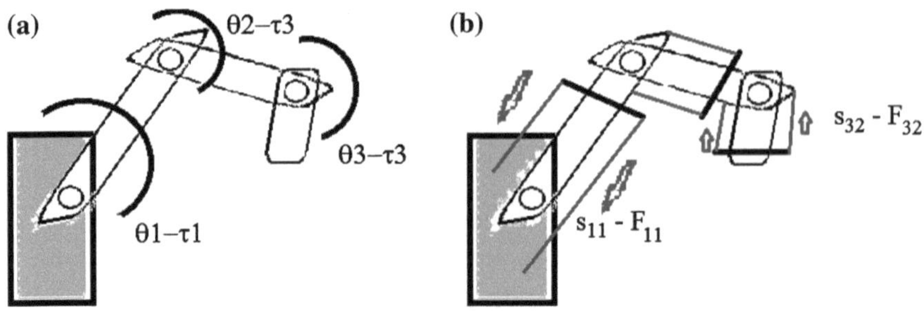

Fig. 4 Kinematic main structures for finger mechanisms: **a** serial chain as inspired by bone structure; **b** parallel architecture as inspired by muscle actuators

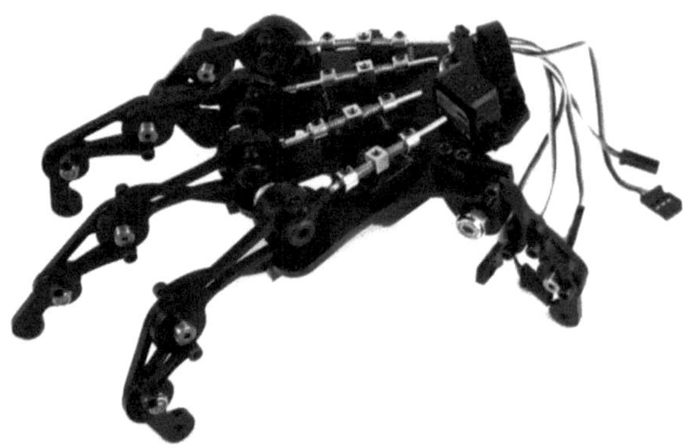

Fig. 5 Finger linkage mechanisms with rigid links in an artificial hand, [15]

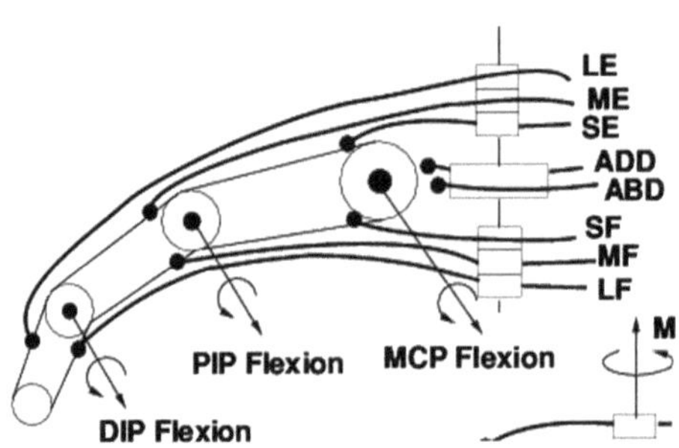

Fig. 6 A scheme of finger mechanism with tendon/cable actuation of the phalanxes, [14]

robots. Complexity refers to the multiple actuators that are needed to act in coordination for the antagonism operations.

Figure 7 shows examples in which the phalanx body is the main issue to obtain grasping adaptability to object characteristics. In Fig. 7a, [13], deformable material

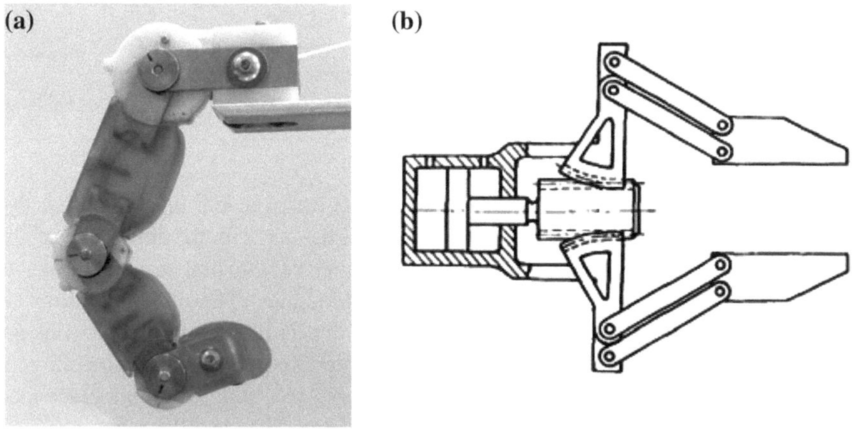

Fig. 7 Examples of finger mechanisms with phalanx bodies: **a** as deformable body, [13]; **b** as rigid link, [10]

is used as active cover of phalanx bodies where as in Fig. 7b, [10], rigid link fingertip is used in a gripper for grasping rigid bodies with possibility to shape its geometry to facilitate multiple contacts with grasped objects.

5 LARM Solutions and New Designs

A last version of LARM Hand is reported in Fig. 8a, with three 1-DOF fingers, a palm, and a standard flange for connection with robots [5]. The size of this prototype is 1.2 times larger than an average human hand as summarized in Table 1 [7]. The actuation system consists of three DC motors with a reduction gear train on each axis. The 1-DOF human-like finger mechanism for LARM Hand is designed according to the scheme in Fig. 8b. Each finger is composed of two cross two

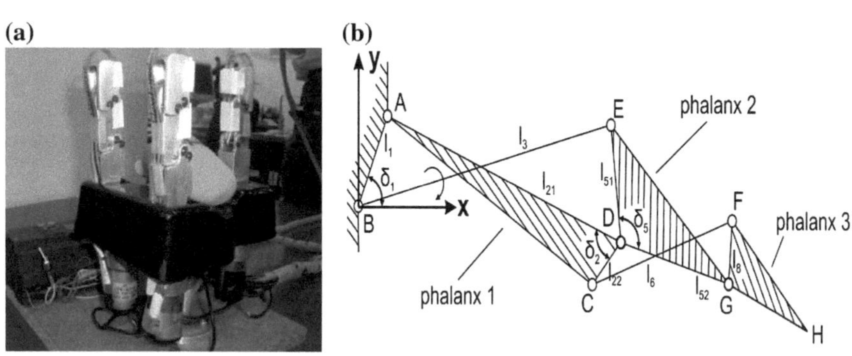

Fig. 8 LARM Hand IV: **a** prototype built in 2007; **b** finger mechanism's scheme [5]

Table 1 Structural parameters of the LARM Hand IV in Fig. 8

(mm)								(deg)		
l_1	l_{21}	l_{22}	l_3	l_{51}	l_{52}	l_6	l_8	δ_1	δ_2	δ_5
8.8	24.1	3.9	28.5	6	19.9	25	6.9	83.5	51	129

four-bar linkages. Phalanx 1 is the input bar of the first four-bar linkage and is also the base frame of the second four-bar linkage. Phalanx 2 is the input bar of the second four-bar linkage and it is also the coupler of the first four-bar linkage. Phalanx 3 is the coupler of the second four-linkage. The linkage design is characterized by a limited grasping adaptability that is determined by the linkage proportions for the finger configurations. In order to improve the capability of grasping objects with different sizes and shapes, solutions with underactuated mechanisms have been considered in previous works, [17, 18, 20].

Underactuation with spring elements as passive joints can be considered as convenient solution for artificial finger designs, [2]. Thus, at LARM attempts have been worked out to define suitable solutions with slight modifications of the original LARM finger design with the aim to improve the grasping adaptability to object shapes but by preserving the original features of 1 DOF actuation and linkage configurations within the finger body.

In particular, Fig. 9 shows a solution with torsional springs in order to obtain underactuated behavior of the four-bar linkages, [18]. Figure 10 refers to a first design for underactuating the LARM finger mechanism by using a linear spring within the body of phalanx 2, [17].

In order to design a new 1-DOF underactuated finger for LARM Hand, new kinematic solutions have been developed by using considerations in [8]. In [20] in order to obtain a new underactuated finger mechanism for LARM Hand a mechanism search has been worked out to identify several possible solutions.

The selected solution for a new finger mechanism is shown in Fig. 11. It is composed by 8 links and 9 revolute joints. Phalanxes are respectively links 2, 6 and 8. This mechanism has a limited manufacturing complexity because of the reduced number of bodies and linkage structure. Because of underactuation this mechanism is able to grasp objects with different shapes remaining within the finger body during its movement.

The mechanism operation can be described according to characteristic configurations for specific contacts by using suitable virtual equivalent mechanisms that have been used in [17, 18] to characterize the underactuated behavior of a finger

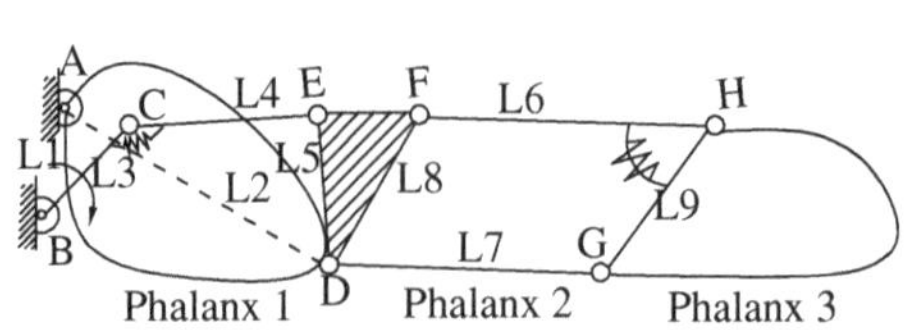

Fig. 9 Underactuated LARM finger mechanism with torsional springs, [18]

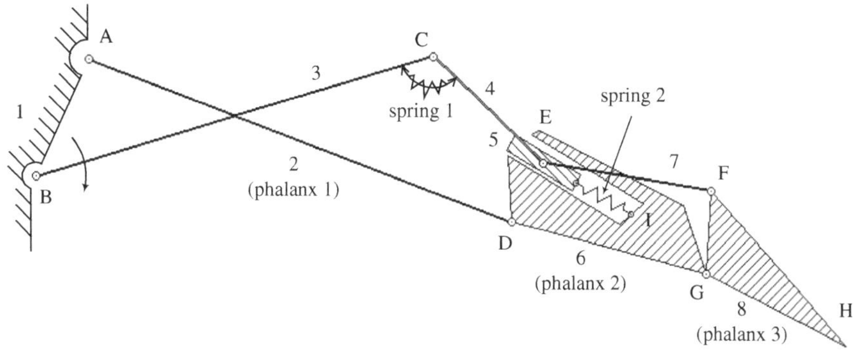

Fig. 10 Underactuated LARM finger mechanism with a linear spring, [17]

mechanism. Namely, equivalent mechanisms can be identified for the cases: no phalanxes in contact; only phalanx 1 in contact with an object; phalanxes 1 and 2 in contact. Referring to the first situation, a phalanx is free when there is no contact force and a torque acts on it. Generally a phalanx is free before it will touch an object. In this case, links that are connected by spring can be considered as a single virtual link. Here links 3 and 4 can be considered as acting as one virtual link 9 as shown by dashed line segment BD in Fig. 11b. Link 6 and 7 can be considered as another virtual link 10 through segment FI. Therefore, the proposed finger mechanism can be simplified as the equivalent mechanism of Fig. 11b, which recall the original linkage in LARM Hand, shown in Fig. 8b. In the second situation, phalanx 1 is stopped while phalanxes 2 and 3 are free. In this case, link 2 and joints E and F are fixed and they act as a virtual base as shown in Fig. 11b. Spring 1 will start to be deformed because of motor push. But spring 2 will be not activated because phalanxes 2 and 3 are free. Links 6 and 7 can still be considered as one single virtual link 10. Thus, the finger mechanism can be simplified as the equivalent mechanism in Fig. 11b. When phalanx 2 is stopped because in contact with object, link 6 and joints E, F and G are fixed and phalanxes 1 and 2 act as a virtual base. Thus, also spring 2 will start to be deformed.

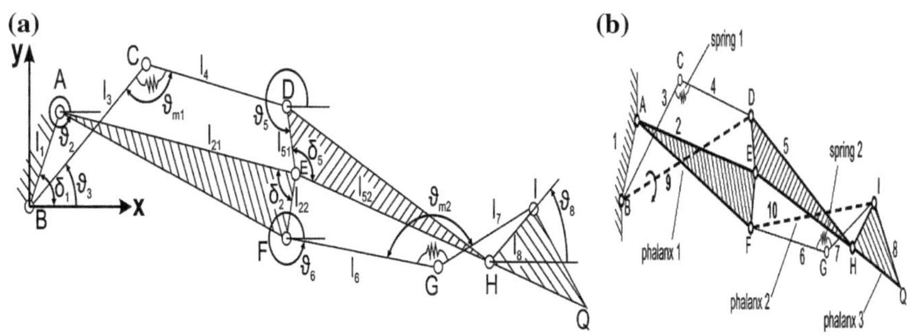

Fig. 11 A new underactuated solution for LARM Hand fingers: **a** a scheme with structural parameters; **b** an equivalent mechanisms during functioning

The experience of LARM finger designs with underactuated linkages has been developed satisfactorily by considering requirements in a procedure like in Fig. 3 and by preserving the peculiarities of the original LARM design. Nevertheless, although underactuated linkages with springs shows feasible operations, practical mechanical design with convenient handsome features is still a problem and further design research is under development.

6 Conclusions

In this paper finger design has been approached by looking at the mechanism design for finger structures that can be also the driving systems as inspired from anatomy of fingers in nature. The role of the finger mechanism is outlined by discussing requirements and peculiarities for the grasping tasks and by referring to the common general topologies that are used in artificial robot hands. In particular the experiences with LARM Hand are reported to outline future developments with solutions that can be based on underactuated linkages.

References

1. Bicchi, A.: Hands for dexterous manipulation and robust grasping: A difficult road toward simplicity. IEEE Trans. Robot. Autom. **16**, 652–662 (2000)
2. Birglen, L., Lalibertè, T., Gosselin, C.: Underactuated Robotic Hands. Springer, Berlin (2008)
3. Cabas, R., Cabas, L.M., Balaguer, C.: Optimized design of the underactuated robotic hand. In: Proceedings of the 2006 IEEE International Conference on Robotics and Automation (ICRA), Orlando, Florida, pp. 982–987, May 2006
4. Carbone, G.: Grasping in Robotics. Springer, Dordrecht (2013)
5. Carbone, G., Ceccarelli, M.: Design of LARM hand: problems and solutions. J. Control Eng. Appl. Inform. **10**(2), 39–46 (2008)
6. Ceccarelli, M.: Notes for a history of grasping devices. In: Carbone, G. (Ed.) Grasping in Robotics, pp. 3–17. Springer, Dordrecht (2013)
7. Ceccarelli, M., Rodriguez, N.E., Carbone, G.: Optimal design of driving mechanism in a 1-DOF anthropomorphic finger. Mech. Mach. Theory **41**(8), 897–911 (2005)
8. Ceccarelli, M., Tavolieri, C., Lu, Z.: Design considerations for underactuated grasp with a one D.O.F. Anthropomorphic Finger Mechanism. In: International Conference on Intelligent Robots and Systems IROS 2006, Beijing, pp. 1611–1616, Oct 2006
9. Fresh, H.: Anatomy of hoof. www.freshhoooves.co.uk. Accessed 12 Aug 2014
10. Lundstrom, G.: Industrial robots—gripper review. International Fluidics Services, Bedford (1977)
11. MedicineNet: Picture of finger anatomy, medicineNet.com. Accessed 12 Aug 2014
12. Pinteterest: Cat fingers. wwww.pinterest.com. Accessed 12 Aug 2014
13. Pisatauro, C.: FlexiBone Robot Finger 2008. http://carlpisaturo.com/Finger_MAIN.html. Accessed 12 Aug 2014
14. Pollard, N.S., Gilbert, R.C.: Tendon arrangement and muscle force requirements for humanlike force capabilities in a robotic finger. In: Proceedings of the IEEE International Conference on Robotics and Automation, Washington, D.C., May 2002

15. Robotshop: MechaTE Robot Right Hand- Product Code : RB-Cus-01. http://www.robotshop.com. Accessed 12 Aug 2014
16. Siciliano, B., Kathib, O.: Springer Handbook of Robotics, Part D, Chapter 28. Springer, Heidelberg (2008)
17. Wu, L., Carbone, G., Ceccarelli, M.: Designing an underactuated mechanism For A 1 active DOF finger operation. Mech. Mach. Theory **44**, 336–348 (2008)
18. Yao, S., Ceccarelli, M., Carbone, G., Zhan, Q., Lu, Z.: Analysis and optimal design of an underactuated finger mechanism for LARM hand. Front. Mech. Eng. **6**, 332–343 (2011)
19. Zhao, J., Jiang, L., Shi, S., Cai, H., Liu, H., Hirzinger, G.: A five-fingered underactuated prosthetic hand system. In: Proceedings of the 2006 IEEE International Conference on Mechatronics and Automation, Luoyang, pp. 1453–1458, June 2006
20. Zottola, M.: Master's Thesis—Design of an underactuated LARM finger mechanism for robotic hand. University of Cassino and South Latium, Cassino (2013)

Dimensional Synthesis of One-Jointed Multi-fingered Hands

Alba Perez-Gracia

Abstract Wristed, multi-fingered hands can be designed for specific tasks, leading to an optimized performance and simplicity. In this work we present the design of the simplest family of multi-fingered hands, with one revolute joint at the wrist and a set of fingers attached to the palm with a single revolute joint each. It is shown that hands with two to five fingers can be designed for meaningful tasks, and that two arbitrary positions can be defined at most for these hands, yielding a good tool for pick-and-place, or grasp-and-release, tasks. For each of those possible designs, dimensional synthesis is performed and an algebraic solution is derived. It is proved that two solutions exist for the general case of this family of hands. Coupled actuation for the grasp-and-release task can be easily implemented for these hands, to create an underactuated design able to be driven with a single actuator. Some examples are presented.

Keywords Robotic hands · Kinematic synthesis

1 Introduction

Kinematic chains with a tree topology consist of several common joints that branch to a number of serial chains, each of them corresponding to a different end-effector. A typical example of a kinematic chain with a tree topology is a wristed, multi-fingered hand.

Compared to other topologies, the tree topologies have not been so widely studied. Kinematic analysis for applications in modular robots and robotic hands can be found in [1, 10, 11], and dynamic analysis is found in [2, 3]. Structural

A. Perez-Gracia (✉)
Department of Mechanical Engineering, Idaho State University, Pocatello, USA
e-mail: perealba@isu.edu

S. Bai and M. Ceccarelli (eds.), *Recent Advances in Mechanism Design for Robotics*, Mechanisms and Machine Science 33,
DOI 10.1007/978-3-319-18126-4_2

synthesis for multiple fingers with no wrist, considering grasping and manipulation requirements, are found in [4]. The first reference to kinematic design of tree topologies is found in [7].

The kinematic synthesis of these topologies presents particular challenges that are different of those that appear in single serial chains or in closed-loop systems. In particular, the kinematic synthesis of multi-fingered hands has been explored also in [9] and more extensively in [8].

When dealing with exact kinematic synthesis, one of the first steps is to define the task in order to size the workspace of the system. In particular, we consider revolute joints without joint limits. In the case of tree topologies, consider a task having the same number of positions for each of the multiple end-effectors; this means that we are dealing with a coordinated action of all those fingertips, denoted as a *simultaneous task*. It has been proved [5] that not all tasks are possible for all fingers, as the relative motion between fingers needs to be considered.

In [5], a chart of solvable multi-fingered hands with identical fingers was presented. Here we focus on the simplest case of that chart, a family of hands with one revolute joint at the wrist and a series of fingers, joined to a single palm by one revolute joint each. For this family of hands, it is possible to obtain a closed, algebraic solution for the design equations. This provides a fast calculation method, as well as information on the number of solutions and the effect of pre-defined constraints. The method is demonstrated with an example.

2 Multi-fingered Hands

A tree topology for a kinematic chain has a set of common joints splitting on several chains, possibly in several stages, and ending in multiple end-effectors [6]. The tree topology is represented as a rooted tree graph; for this we follow the approach of Tsai [13], the root vertex being fixed with respect to a reference system. In tree topologies, a vertex can be connected to several edges defining several branches.

Wristed, multi-fingered hands are kinematic chains with a tree or hybrid topology. For our synthesis formulation, the internal loops in the hand structure are substituted using a reduction process [8], so that the hand has a tree topology with links that are ternary or above.

Tree topologies are denoted as $W - (B_1, B_2, \ldots, B_b)$, where W are the common joints and the dash indicates a branching or splitting stage, with the branches contained in the parenthesis, each branch B_i characterized by its type and number of joints. In the case of using just revolute joints, the joint type is dropped and only the number of joints is indicated. Figure 1 shows the compacted graph for the $R - (R, R, R, R, R)$, or $1 - (1, 1, 1, 1, 1)$ chain. This is one of the member of the one-jointed hand family, in particular the five-fingered hand. The root vertex is indicated with a double circle.

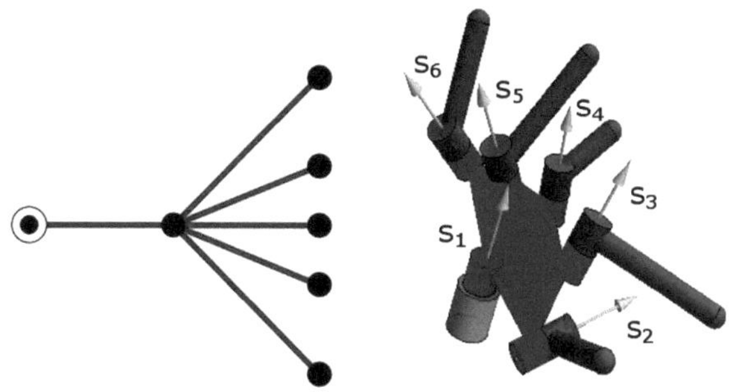

Fig. 1 The tree graph of the $1-(1,1,1,1,1)$ hand, *left*; the kinematic sketch, *right*

3 Dimensional Synthesis of Multi-fingered Hands

Given a desired hand topology with b branches and joint axes S_i, and a simultaneous task for each fingertip, characterized by a set of m_p finite positions $\hat{P}^b_{1k}$ and m_v velocities $\dot{\mathsf{P}}^b_k$, kinematic synthesis is applied by equating the forward kinematics equations of each branch to the relative displacement of the corresponding fingertip. Similarly, velocities can be defined for some of those task positions,

$$\mathbf{F}\left(\mathsf{S}, \Delta\theta, \dot{\theta}\right) = \begin{cases} \hat{P}^c_{1k} - \prod_{i=1}^{n_c} e^{\frac{\Delta\theta^k_{i,c}}{2}\mathsf{S}_{i,c}}, & k = 2, \ldots, m_p \\ & c = 1, \ldots, b \\ \dot{\mathsf{P}}^c_k - \sum_{i=1}^{n_c} \dot{\theta}^k_{i,c}\mathsf{S}^k_{i,c}, & k = 1, \ldots, m_v \\ & c = 1, \ldots, b \end{cases} \tag{1}$$

In most of the cases, this set of equations is solved using numerical methods to obtain a kinematic design.

4 Single-Jointed, Single-Branching Hands

The solvability of this family of hands was studied in [5]. In Table 1, all possible one-jointed trees consisting of revolute joints and able to perform a meaningful task are presented. The first row in the table contains the serial $R-R$ robot, which is known [12] to be solvable for $m=3$ positions, yielding two real solutions. The last row presents the $1-(1,1,1,1,1)$ hand, with five fingers. Notice that any general hand of these characteristics with more than five fingers will not be solvable for a useful task, as it will not be able to reach an initial and final positions. For all the cases in the middle, from two to four fingers, the hand will be able to reach an initial and a final position, as their solvability calculations show two plus one incompletely specified position.

Table 1 Topologies with 1 common joint and 1-jointed branches, single branching

Topology	Graph	Graph solvability
$1-(1)$		Solvable, $m = 3$
$1-(1,1)$		Solvable, $m = 2.33$
$1-(1,1,1)$	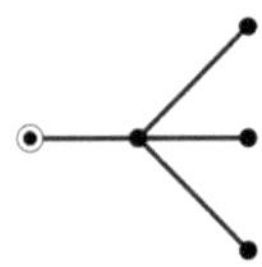	Solvable, $m = 2.14$
$1-(1,1,1,1)$	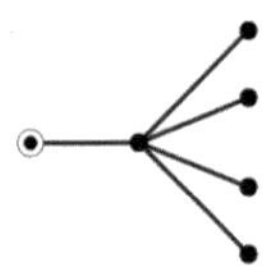	Solvable, $m = 2.05$
$1-(1,1,1,1,1)$	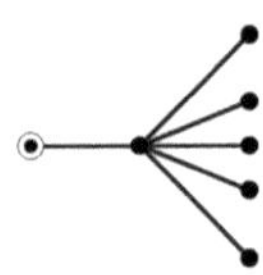	Solvable, $m = 2$

5 Dimensional Synthesis for the 1 − (1, 1, 1, 1, 1) Hand

The $1 - (1, 1, 1, 1, 1)$ hand consists of a single revolute joint at the wrist and a palm spanning five fingers, each of them being a fingertip link connected to the palm by a single revolute joint. Figure 1 shows the reduced tree graph for the hand and the kinematic sketch with the notation used for the axes.

The six joints have Plucker coordinates $\mathsf{S}_i = \mathbf{s}_i + \varepsilon \mathbf{s}_{i0}$, for $i = 0, \ldots, 5$. The rotation angle about each joint is θ_i.

As stated in Table 1, this kinematic system can be solved for exact dimensional synthesis for $m = 2$ positions. In what follows, the algebraic solution is derived.

5.1 Design Equations

Let $\hat{P}_{ij}$ be the ith displacement assigned to finger j. For this robot, $i = 1, 2$ and $j = 1, \ldots, 5$. Construct the relative displacements from first to second position for each finger j, $\hat{P}_j = \hat{P}_{2j}\hat{P}_{1j}^{-1}$, $j = 1, \ldots, 5$, where the composition of displacements is used.

The forward kinematics equations for relative displacements are constructed, for each finger, as the composition of screw displacements about each joint axes along the serial chain. The screw displacements about the joint axes are easily computed as the exponential of the unit twist, and so the product of exponentials is

$$e^{\hat{S}_0\frac{\Delta\theta_0}{2}} e^{\hat{S}_j\frac{\Delta\theta_j}{2}}, \quad j = 1, \ldots, 5, \tag{2}$$

where $\Delta\theta_j = \theta_j^2 - \theta_j^1$ is the relative angle to transform from first to second position, for each finger j.

Impose now that each of the fingers has to be able to perform the desired relative displacement $\hat{P}_j$ from first to second position, to create a set of equations to solve for the design parameters of the robotic hand,

$$e^{\hat{S}_0\frac{\Delta\theta_0}{2}} e^{\hat{S}_j\frac{\Delta\theta_j}{2}} = \hat{P}_j, \quad j = 1, \ldots, 5. \tag{3}$$

This is a set of six independent equations per finger times five fingers, for a total of 30 independent equations. When using unit dual quaternions to express the displacements, the total set has 40 equations. The parameters to solve for are the Plucker coordinates of the axes and the relative joint angles to reach the relative transformation.

5.2 Algebraic Solution

A revolute joint has five independent parameters, and is unable to perform a general displacement. Each finger of the $1-(1,1,1,1,1)$ hand is a serial $R-R$ chain in which the first joint cannot be fully arbitrary. This hand can be seen as five $R-R$ chains in which each chain needs to fully define the second joint axis and a single parameter of the common first joint axes, in order to perform a general displacement.

Using this rationale, solve for the second joint axes of each finger serial chain,

$$e^{\hat{S}_j\frac{\Delta\theta_j}{2}} = \left(e^{\hat{S}_0\frac{\Delta\theta_0}{2}}\right) * \hat{P}_j, \quad j = 1, \ldots, 5, \tag{4}$$

where $*$ denotes the conjugate unit dual quaternion. The expansion of this equation in dual quaternion form yields:

$$\begin{aligned}&\left(\cos\frac{\Delta\theta_j}{2}+\varepsilon\sin\frac{\Delta\theta_j}{2}\right)\hat{S}_j\\&=\left(\cos\frac{\Delta\theta_0}{2}-\sin\frac{\Delta\theta_0}{2}S_0\right)\left(p_{j0}+_{j7}+\mathsf{P}_j\right),\quad j=1,\ldots,5,\end{aligned}\tag{5}$$

where,

$$\hat{S}_j=0+\varepsilon 0+\mathsf{S}_j=0+s_{jx}i+s_{jy}j+s_{jz}k+\varepsilon\left(s^0_{jx}i+s^0_{jy}j+s^0_{jz}k+0\right)\tag{6}$$

is the pure dual quaternion that represents a geometric line in the Clifford algebra. The solution for all S_j, $j=1,\ldots,5$, is obtained as a function of the first, common joint axis, S_0. These equations are also used to fully define this first axes.

Notice that the last component of the dual quaternion must be equal to zero for each joint solution, according to Eq. (5). This forces the product $\left(e^{\hat{S}_0\frac{\Delta\theta_0}{2}}\right)*\hat{P}_j$ to have also last component equal to zero, and creates one equation on the parameters of the first axis S_0 from each finger equations,

$$\cos\frac{\Delta\theta_0}{2}p_{j7}+\sin\frac{\Delta\theta_0}{2}\left(\mathbf{s}_0\cdot\mathbf{p}^0_j+\mathbf{s}^0_0\cdot\mathbf{p}_j\right)=0,\quad j=1,\ldots,5.\tag{7}$$

These are five equations in the five independent unknowns of the rotation about S_0 by $\Delta\theta_0$. Imposing that the joint angle has to be the same for the simultaneous task,

$$\cot\frac{\Delta\theta_0}{2}=\frac{-(\mathbf{s}_0\cdot\mathbf{p}^0_j+\mathbf{s}^0_0\cdot\mathbf{p}_j)}{p_{j7}},\quad j=1,\ldots,5,\tag{8}$$

we end up with four linear equations in the Plucker coordinates of the axes. Those are six parameters subject to two Plucker constraints plus the four linear equations from (8),

$$\begin{aligned}&\frac{\mathbf{s}_0\cdot\mathbf{p}^0_j+\mathbf{s}^0_0\cdot\mathbf{p}_j}{p_{j7}}=\frac{\mathbf{s}_0\cdot\mathbf{p}^0_{j+1}+\mathbf{s}^0_0\cdot\mathbf{p}_{j+1}}{p_{(j+1),7}},\quad j=1,\ldots,4,\\&\mathbf{s}_0\cdot\mathbf{s}_0=1,\quad \mathbf{s}_0\cdot\mathbf{s}^0_0=0,\end{aligned}\tag{9}$$

with at most four solutions. However two of the solutions correspond to the double covering of $SO(3)$, and the final number of different solutions is two.

The two solutions for the Plucker coordinates of the first joint S_0 are used to compute a single rotation angle about this first joint for each solution, using Eq. (8). These values allow us to create the screw transformation about the first joint, and the rest of the joint axes and angles can be calculated using Eq. (4).

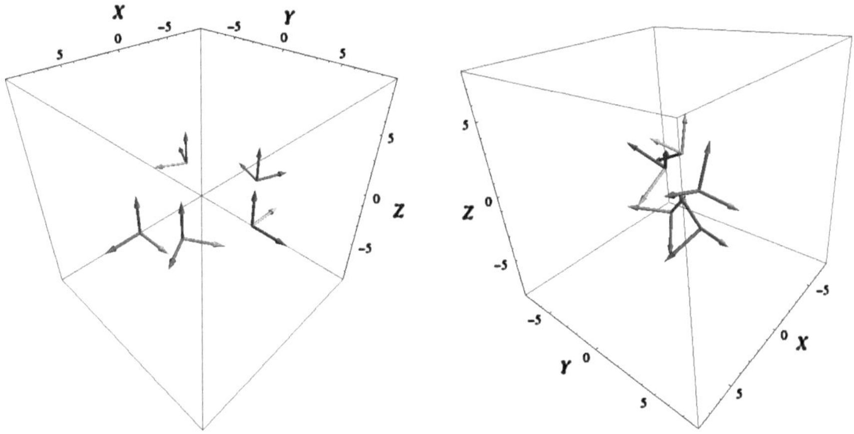

Fig. 2 Initial (*left*) and final (*right*) positions for the five fingers

Table 2 Initial and final positions for each finger in dual quaternion form

Finger	Initial position	Final position
Finger 1	$1.0 + 3.0\varepsilon i$	$-0.96 - 0.11i - 0.19j + 0.16k + \varepsilon(-0.47i - 1.58j + 0.91k + 0.52)$
Finger 2	$0.97 + 0.26k + \varepsilon(2.41i + 0.65j)$	$-0.99 + 0.04i + 0.15j + 0.02k + \varepsilon(-0.63i - 1.62j - 0.38k - 0.29)$
Finger 3	$0.71 + 0.71k + \varepsilon(1.77i + 1.77j)$	$-0.62 - 0.38i + 0.08j + 0.68k + \varepsilon(0.37i + 0.75j - 0.53k - 0.70)$
Finger 4	$0.17 + 0.98k + \varepsilon(0.43i + 2.46j)$	$-0.68 + 0.39i - 0.005j + 0.62k + \varepsilon(-0.15i - 0.05j - 0.48k - 0.52)$
Finger 5	$0.50 - 0.87k + \varepsilon(1.25i - 2.16j)$	$0.81 + 0.41i + 0.42j + 0.07k + \varepsilon(0.63i + -0.33j - 0.75k - 0.08)$

5.3 Example

Consider the simultaneous five-finger task composed of an initial position in which all fingers are spread and more or less evenly spaced in a circular space, to a final randomly-generated position. Figure 2 and Table 2 show the initial and final positions for this task.

The two solutions obtained are shown in Table 3 and Fig. 3. The joint axes are expressed in the reference frame and when reaching the initial position. The design presented in Fig. 4 is one of the infinitely many implementations that are possible for the kinematic solutions without changing the kinematic task.

6 Dimensional Synthesis for the 1 − (1, 1) Hand

Similarly, we can perform dimensional synthesis for the two-fingered to four-fingered hand topologies, according to the solvability results in Table 1. Specifying incomplete positions is not always a very practical approach. A better option could be to add some geometric constraints on the design so that this can be synthesized

Table 3 Joint axes for the two solutions corresponding to the task in Fig. 2

Joint	Solution 1	Solution 2
Wrist (S_0)	$0.68i + 0.66j + 0.31k + \varepsilon(2.23i - 1.31j - 2.14k)$	$0.13i + 0.30j + 0.95k + \varepsilon(-1.41i + 1.53j - 0.29k)$
Finger 1 (S_1)	$0.50i + 0.80j + 0.31k\varepsilon(1.22i - 1.29j + 1.35k)$	$-0.18i + 0.22j + 0.96k + \varepsilon(-1.37i - 2.66j + 0.35k)$
Finger 2 (S_2)	$0.54i + 0.83j + 0.13k + \varepsilon(1.47i - 0.95j - 0.10k)$	$0.15i + 0.37j + 0.91k + \varepsilon(-0.76i - 1.62j + 0.79k)$
Finger 3 (S_3)	$0.62i + 0.73j - 0.28k + \varepsilon(1.63i + 2.18j + 2.04k)$	$-0.72i + 0.22j + 0.66k + \varepsilon(0.98i - 2.09j + 1.76k)$
Finger 4 (S_4)	$-0.81i + 0.11j - 0.58k + \varepsilon(-0.30i + 1.74j + 0.76k)$	$-0.68 + 0.39i - 0.005j + 0.62k + \varepsilon(7.33i - 7.33j + 3.36k)$
Finger 5 (S_5)	$-0.96i + 0.14j + 0.23k + \varepsilon(0.28i + 4.30j - 1.48k)$	$0.12i - 0.79j + 0.60k + \varepsilon(-1.18i + 0.68j + 1.14k)$

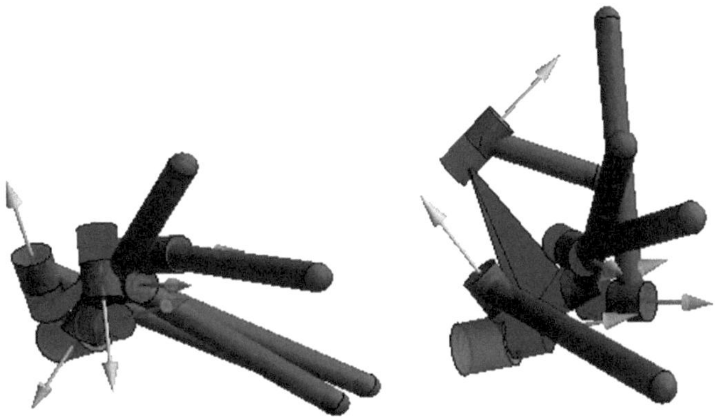

Fig. 3 The two solutions corresponding to the task in Fig. 2

Fig. 4 The CAD model of the second solution

for $m = 2$ positions. Those geometric constraints could be pre-selecting one or more parameters on the joint axes, selecting the desired relative joint angles, or adding angle or distance constraints between consecutive axes. Depending on the nature of these additional constraints, the system of equations to solve will yield more ore less candidate designs. The example of the $1 - (1, 1)$ hand is developed here.

This hand has one joint at the wrist and a palm spanning two one-jointed fingers. Notice that this hand design is used in many simple hooks. This hand is solvable for

Table 4 Joint axes for the $1-(1,1)$ hand designed for the first two fingers of the task in Fig. 2

Joint	Solution
Wrist (S_0)	$1k + \varepsilon(3.46i + 2.34j)$
Finger 1 (S_1)	$-0.31i - 0.24j + 0.92k + \varepsilon(4.57i - 3.13j + 0.70k)$
Finger 2 (S_2)	$0.10i + 0.23j + 0.97k + \varepsilon(4.08i - 1.07j - 0.15k)$

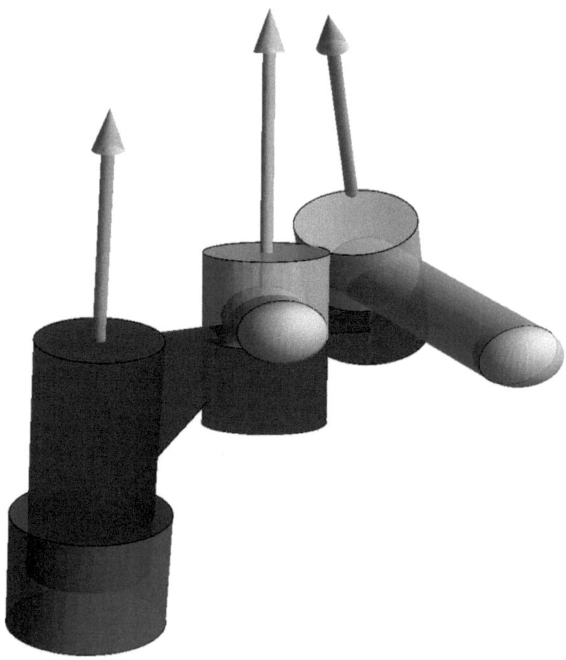

Fig. 5 The two-fingered design for the first two finger tasks in Fig. 2

$m = 2.33$ positions; if the designer imposes some constraints in the design parameters, the hand is solvable for exactly $m = 2$ positions. In particular, we need to define or constraint three of the unknown parameters.

For this example, the direction and relative angle of the wrist joint S_0 are predefined to values selected by the designer: $s_0 = (0,0,1)$ and $\Delta\theta_0 = \pi/4$. This is one of the many possibilities in which the design can be partially pre-determined; in this case, a linear system with a single solution is obtained. The example uses the positions for fingers 1 and 2 from Table 2 to yield the design of Table 4 and Fig. 5.

Similarly, two constraints can be imposed on the design parameters of the $1-(1,1,1)$ hand and one constraint on the $1-(1,1,1,1)$ hand in order to solve these for two exact positions. The solution process is similar to the one indicated for the five-fingered and one-fingered hands.

7 Conclusions

Kinematic synthesis can be used to yield multi-fingered hand designs in which the task of each finger is simultaneously defined. A big variety of hand topologies and tasks can be used to yield hand designs. In this work, we present the solution of one of the simplest family of hands, that of a wristed hand with a single revolute joint at the wrist and a number of fingers, each of them attached to the palm by a single revolute joint. It can be proved that we can design hands with two to five fingers able to perform a pick-and-place task where each finger's position can be exactly specified.

For these four hand topologies, a solution method is developed and shown to yield two hand solutions in the general case. Two examples are presented, one for the five-fingered hand and another one for the two-fingered hand.

Task-oriented hand design may lead to more optimal end effectors. The task specification can be complemented with the definition of velocities and accelerations for grasping characterization. Those constraints can be added to the design process without increasing the complexity of the system of equations.

Acknowledgments This work is supported by the National Science Foundation under Grant No. 1208385. The content is solely the author's responsibility.

References

1. Chen, I., Yang, G., Kang, I.: Numerical inverse kinematics for modular reconfigurable robots. J. Robotic Syst. **16**(4), 213–225 (1999)
2. Garcia de Jalon, J., Bayo, E.: Kinematic and Dynamic Simulation of Multibody Systems: The Real-Time challenge. Springer, New York (1994)
3. Jain, A.: Graph-theory roots of spatial operators for kinematics and dynamics. In: Proceedings of the 2010 International Conference on Robotics and Automation, pp. 2745–2750. Anchorage, Alaska (2010)
4. Lee, J.J., Tsai, L.: Structural synthesis of multi-fingered hands. ASME J. Mech. Des. **124**, 272–276 (2002)
5. Makhal, A., Perez-Gracia, A.: Solvable multi-fingered hands for exact kinematic synthesis . In: Advances in Robot Kinematics. Ljubljiana, Slovenia (2014)
6. Selig, J.M.: Geometric Fundamentals of Robotics (Monographs in Computer Science). Springer (2004)
7. Simo-Serra, E., Moreno-Noguer, F., Perez-Gracia, A.: design of non-anthropomorphic robotic hands for anthropomorphic tasks. In: ASME Design Engineering Technical Conferences. Washington DC, USA (2011)
8. Simo-Serra, E., Perez-Gracia, A.: Kinematic synthesis using tree topologies. Mech. Mach. Theory **72 C**, 94–113 (2014)
9. Simo-Serra, E., Perez-Gracia, A., Moon, H., Robson, N.: Design of multi fingered robotic hands for finite and infinitesimal tasks using kinematic synthesis. In: Advances in Robot Kinematics. Innsbruck, Austria (2012)
10. Stramigioli, S.: Modeling and IPC control of interactive mechanical systems—a coordinate-free approach, vol. LNCIS 266. Springer (2001)

11. Tischler, C., Samuel, A., Hunt, K.: Kinematic chains for robot hands—1. orderly number synthesis. Mech. Mach. Theory **30**(8), 1193–1215 (1995)
12. Tsai, L., Roth, B.: A note on the design of revolute-revolute cranks. Mech. Mach. Theory **8**, 23–31 (1973)
13. Tsai, L.W.: Mechanism Design: Enumeration of Kinematic Structures According to Function. CRC Press, Boca Raton (2001)

A Modular Shape-Adaptive Mechanism for Robust Robotic Grasping

Carl A. Nelson

Abstract Underactuated and adaptive robotic hands are known for their ability to achieve multiple contact points on arbitrarily shaped objects with relatively simple actuation, improving grasp stability. However, in some cases the sequence in which these contact points are initiated does not promote robust capture of the grasped object. This paper presents the design of a new type of underactuated grasper based on an asymmetric pantograph structure. The new grasper achieves robust enveloping capture of arbitrarily shaped objects (including non-convex shapes) while maintaining balanced forces at multiple contact points. The design is easily adjustable for differently sized objects.

Keywords Underactuated grasper · Robotic grasping · Grasp stability · Pantograph mechanism

1 Introduction

In robotic grasping, one of the most important issues is that of stability of the grasp. Pinch-type grasps rely heavily on friction at the contact points, since they can involve as few as two contact points [1], and without a significant frictional effect this would not lead to stability under general loading conditions (see, for example, [2], p. 402). Enveloping grasps can involve more contact points, but one of the challenges for arbitrarily shaped objects is maintaining some uniformity in the magnitudes of the contact forces to decrease sensitivity of the grasp to disturbances (and thereby increase robustness) [1]. In contrast to dexterous hands, which are mechanically very complex [3, 4], underactuated graspers are favored for their relative mechanical simplicity and their adaptability (without the need for feedback control) [5], but one drawback of typical underactuated grasper designs has to do

C.A. Nelson (✉)
Department of Mechanical and Materials Engineering,
University of Nebraska-Lincoln, Lincoln, NE, USA
e-mail: cnelson5@unl.edu

S. Bai and M. Ceccarelli (eds.), *Recent Advances in Mechanism Design for Robotics*, Mechanisms and Machine Science 33,
DOI 10.1007/978-3-319-18126-4_3

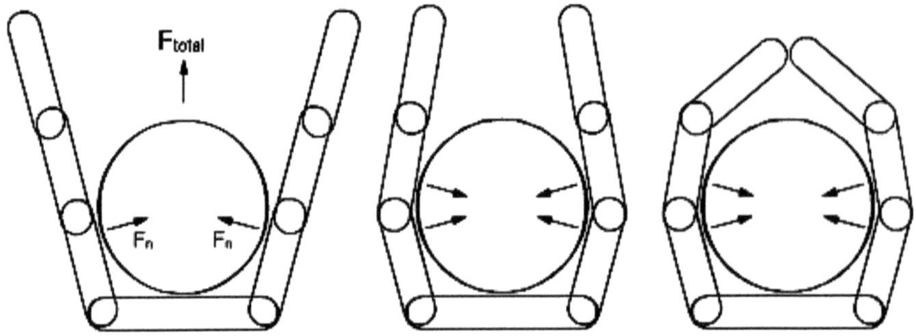

Fig. 1 Typical closure sequence for underactuated grasper: initial normal forces tend to push object out of grasp

with the sequence in which contact is made. As illustrated in Fig. 1, initial contact can push the object away from the grasper before the later contacts, which would stabilize the grasp, are made. Even the most recently developed underactuated graspers, whether developed for prostheses [6] or for industrial applications [7], can exhibit this characteristic. A grasper combining the benefits of underactuation (balanced multi-point contact) with a more robust approach to enveloping would be desirable to resolve this issue. The ability to easily reconfigure the grasper for different sizes and types of objects is also of interest.

The remainder of the paper is organized as follows. First, a design strategy for robust grasping of arbitrarily shaped objects is described, and a mechanism following that strategy is put forward. Analysis is then presented to elucidate the design parameters and their effects on the grasping behavior. Prototype fabrication and testing are then described, and conclusions are presented.

2 Grasper Design

Based on the limitations illustrated in Fig. 1, the functional design requirements are for the grasper to achieve a multi-point contact on the grasped object, and to at least partially envelope the object prior to initial contact (to avoid pushing the object out of the grasp). To do this robustly requires a mechanism which extends and encircles the grasped object and then closes inward on the object, or which extends/encircles and closes inward simultaneously. The design presented here was inspired by the extension behavior of the well-known pantograph mechanism [2, 8]. In this case, however, instead of extending in a rectilinear fashion, the intent is to extend along a curved path which eventually encloses the grasped object. As shown in [8], moving the pivot point of the scissor-like pantograph cell away from the links' midpoint produces an arc-shaped extension (Fig. 2). This provides the desired enveloping behavior, and adjusting the distance of the pivot point from the center of each link can be used to tailor the grasper's enveloping radius to the object(s) being grasped.

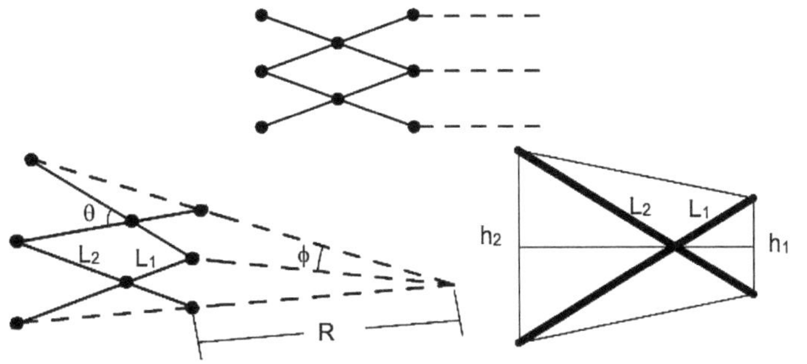

Fig. 2 Pantograph (*top*); modified pantograph and associated geometry (*left*); right-triangle construction geometry for modified pantograph cell (*right*)

Satisfaction of the second design requirement can be pursued by noting that as the modified pantograph extends along its arc, the line segment joining opposing pin joints of each scissor-like cell decreases in length, and remains oriented radially inward (see Fig. 3). By allowing a spring-loaded contacting element (hereafter referred to as a finger) to slide along this line segment, extension of the modified pantograph can also drive the inward motion of the finger. Arrangement of fingers in each of the modified pantograph cells produces a complete grasper which simultaneously envelopes and collapses to produce multi-point contact.

To ensure that contact forces remain well balanced for different sizes and shapes of grasped objects, the fingers are adjustably spring loaded, as shown in Fig. 4. During initial extension of the pantograph, only the return spring (with a relatively small spring constant) is active, keeping the fingers retracted. At some point in the extension, a stop is engaged and contacts (activates) the second, stiffer spring, whose function is to deploy the finger. Adjustment for different object sizes is made by changing the position of the stop on the finger, thus changing both the amount of enveloping which occurs prior to finger deployment and the overall finger extension (related to the diameter of the grasped object). Because the engagement of each finger deployment spring is based only on the position of the stop and not on the contacts of the other fingers (as in many other underactuated grasper designs), the contact forces generally remain balanced, and can be further tailored by adjusting the stops on the fingers non-uniformly if desired.

Fig. 3 Sliding finger placement between two pantograph cells

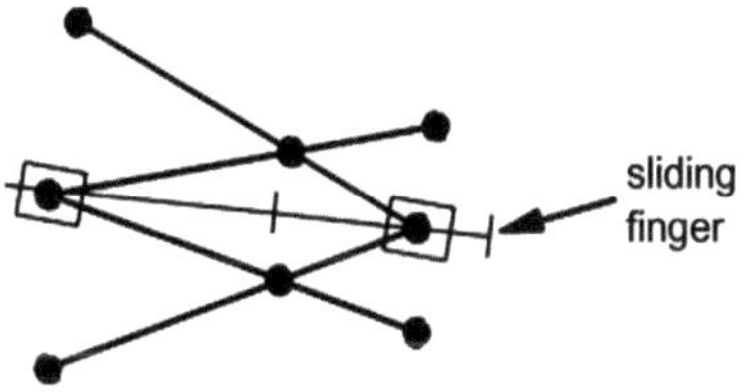

Fig. 4 Spring placement on sliding finger; extension spring k_1 is less stiff, and compression spring k_2 is more stiff

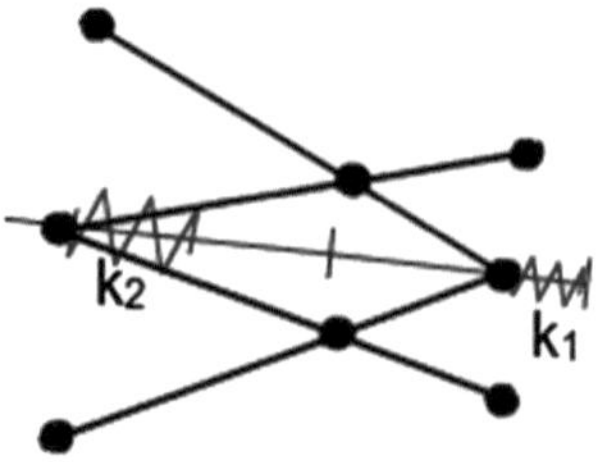

3 Analysis

As shown in the trapezoid-shaped construction of Fig. 2, and using the notation indicated therein, the modified pantograph results in similar triangles such that

$$\sin(\theta/2) = h_1/2L_1 = h_2/2L_2 \tag{1}$$

It can also be seen that the radius of curvature of the enclosed (grasped) space is given by

$$r_1 = h_1/[2\sin(\phi/2)] = L_1\sin(\theta/2)/\sin(\phi/2) \tag{2}$$

Still using the construction of Fig. 2, the relation between the pantograph input angle θ and the single-cell inclination angle ϕ is

$$\tan(\phi/2) = (\Delta L/L)\tan(\theta/2) \tag{3}$$

where ΔL is defined as $L_2 - L_1$, and L is defined as $L_1 + L_2$. Therefore, with an n-cell symmetric arrangement (two identical gripper halves each with n modified pantograph cells), a closure limit is reached when the total horizontal travel of the distal cell is equal to half the initial spacing of the two gripper sides, or

$$d = 2h_1[\cos(\phi) + \cos(2\phi) + \cdots + \cos(n\phi)] \tag{4}$$

where d is the initial separation between gripper halves (or the palm width).

Although identities exist to evaluate $\cos(\tan^{-1}(\phi))$ and so forth for combining Eqs. (3) and (4), the sum in Eq. (4) makes a closed-form expression in θ cumbersome; therefore, it seems more appropriate to employ a computational approach to solve for the limiting value of θ_{max} for a given grasper configuration. This would influence the parameter values used in setting up a grasper of this type for handling objects in different size ranges.

Finally, as the input to the pantograph mechanism is more likely to be a slider displacement than an angular motion in θ, we can note that the ideal fully collapsed cell has width L and that the slider displacement relation, measured from this fully collapsed position, is that of a slider-crank [2]:

$$x_{input} \approx L_1(1-\cos\theta) + L_1^2\sin^2\theta/2L_2 \quad (5)$$

Equation (5) likewise is well suited to a computational rather than a closed-form solution. Therefore, a typical procedure might be to use Eqs. (2)–(3) together to find a desirable combination of L_1 and L_2 which produces a value of r_1 which is well matched for the object being grasped, and then select the initial spacing d according to Eq. (4). Equation (5) can be used if precise control of the closure is required.

Now, as concerning contact forces F_c (axially along each finger), with the return spring (k_1) and the finger deployment spring (k_2) chosen such that $k_2 \gg k_1$,

$$F_c = k_2[x_f + L(1-\cos\theta) - y] - k_1 x_f \quad (6)$$

$$y = L - k_{2free} - x_s \quad (7)$$

where x_f is the finger displacement measured from the inner pin joint, y is the amount of finger displacement which occurs prior to contact with the finger deployment spring (with free length k_{2free}), and x_s is the stop position measured from the inner pin joint. The actual values of F_c and x_f will vary for different grasped objects depending on the object shape and the equilibrium of forces achieved, and the constraint that $x_f \leq x_s$ (see Fig. 4).

4 Prototype and Testing

A proof-of-concept prototype grasper was fabricated in acrylic using a laser cutter and is shown in Fig. 5. The pantograph components include multiple pivot locations (L_1/L_2 = {0.4, 0.65, 1, 1.55, 2.5}) so that the angular dependency described in Eq. (3) can be adjusted. Elastic bands serve as the finger return springs, and each wire finger includes an adjustable stop as previously mentioned.

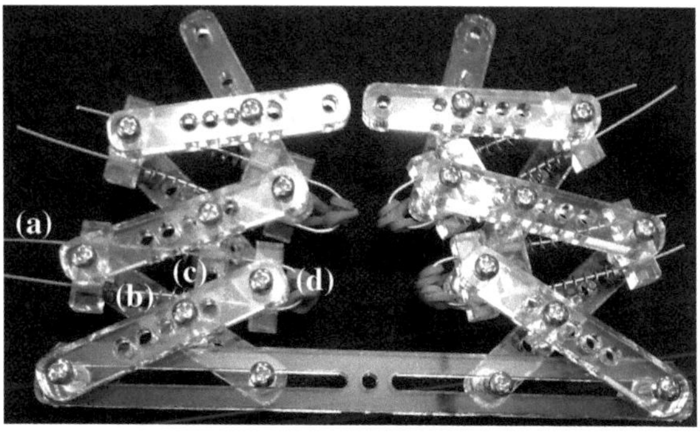

Fig. 5 Prototype grasper: **a** finger, **b** finger deployment spring, **c** stop, **d** return spring

To test the grasper, several objects were used, each with a different shape and size, as shown in Fig. 6. The grasp for each object was customized using different pivot point settings (different enveloping curvatures) as appropriate to achieve the grasp. In each case, the grasp can be considered successful within the limitations of materials and workmanship of the prototype grasper (the grasper functioned as intended and as described above). The possible grasp configurations are summarized in Table 1, with the cases from Fig. 6 indicated. (Some of these are not necessarily practical, such as those with all $L_1/L_2 > 1$, since the grasper arms would splay outward rather than curve inward to envelope the grasped object.) In general, the number of possible grasper configurations is m^n, where m is the number of pivot adjustments and n is the number of pantograph cells.

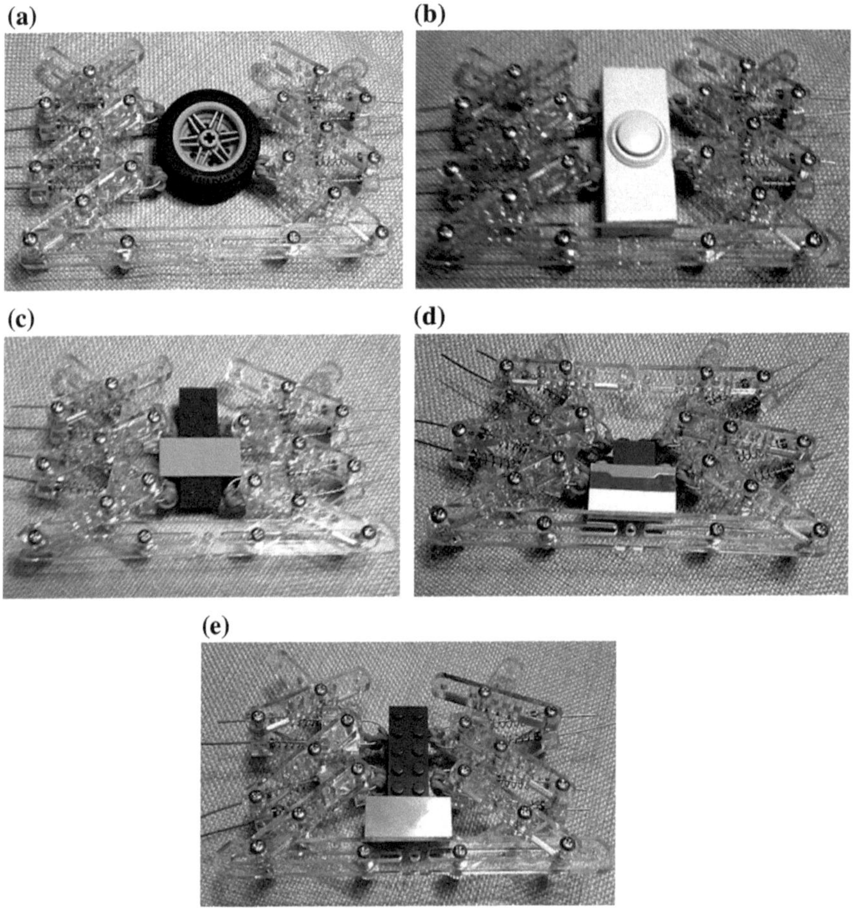

Fig. 6 Sample grasps with the prototype; pantograph cells with L_1/L_2: **a**, **b** 1, 1, 1; **c** 0.65, 1, 1; **d** 0.65, 0.65, 1, **e** 1.55, 0.65, 0.65

Table 1 Possible grasper configurations for three length ratios in each of three pantograph cells

Configuration	L_1/L_2 (cell 1)	L_1/L_2 (cell 2)	L_1/L_2 (cell 3)	Indicated in Fig. 6
1	Low	Low	Low	–
2	Low	Low	Medium	(d)
3	Low	Low	High	–
4	Low	Medium	Low	–
5	Low	Medium	Medium	(c)
6	Low	Medium	High	–
7	Low	High	Low	–
8	Low	High	Medium	–
9	Low	High	High	–
10	Medium	Low	Low	–
11	Medium	Low	Medium	–
12	Medium	Low	High	–
13	Medium	Medium	Low	–
14	Medium	Medium	Medium	(a, b)
15	Medium	Medium	High	–
16	Medium	High	Low	–
17	Medium	High	Medium	–
18	Medium	High	High	–
19	High	Low	Low	(e)
20	High	Low	Medium	–
21	High	Low	High	–
22	High	Medium	Low	–
23	High	Medium	Medium	–
24	High	Medium	High	–
25	High	High	Low	–
26	High	High	Medium	–
27	High	High	High	–

5 Conclusions

A novel adaptive grasper was presented which provides stable, multi-contact grasping based on a modified pantograph mechanism. It achieves stability by partially enveloping the grasped object prior to radially collapsing inward to engage the contact points. Experimental testing demonstrated the desired functionality, and analysis provides additional insights into design parameters for practical implementation.

Acknowledgments The author acknowledges support from the UNL College of Engineering.

References

1. Xiong, C., Ding, H., Xiong, Y.: Fundamentals of Robotic Grasping and Fixturing. World Scientific, Singapore (2007)
2. Mabie, H.H., Reinholtz, C.F.: Mechanisms and Dynamics of Machinery, 4th edn. Wiley, New York (1984)
3. Jacobsen, S.C., Wood, J.E., Knutti, D.F., Biggers, K.B.: The UTAH/M.I.T. dextrous hand: work in progress. Int. J. Robot. Res. **3**(4), 21–50 (1984)
4. Butterfaß, J., Grebenstein, M., Liu, H., Hirzinger, G.: DLR-hand II: next generation of a dextrous robot hand. In: Proceedings of the 2001 IEEE International Conference on Robotics & Automation, May 21–26, pp. 109–114, Seoul, Korea (2001)
5. Birglen, L., Laliberté, T., Gosselin, C.: Underactuated Robotic Hands. Springer, New York (2008)
6. Dechev, N., Cleghorn, W.L., Naumann, S.: Multiple finger, passive adaptive grasp prosthetic hand. Mech. Mach. Theory **36**, 1157–1173 (2001)
7. Odhner, L.U., Jentoft, L.P., Claffee, M.R., Corson, N., Tenzer, Y., Ma, R.R., Buehler, M., Kohout, M., Howe, R.D., Dollar, A.M.: A Compliant, underactuated hand for robust manipulation. Int. J. Robot. Res. **33**(5), 736–752 (2014)
8. You, Z., Chen, Y.: Motion Structures: Deployable Structural Assemblies of Mechanisms. Spon Press, New York (2012)

Mass Equivalent Dyads

V. van der Wijk

Abstract In this paper it is shown how a general 2-DoF dyad can be designed mass equivalent to a general (1-DoF) link element. This is useful in the synthesis of balanced mechanisms, for instance to increase or reduce the number of DoFs of a balanced mechanism maintaining its balance. Also it can be used as a simple approach for synthesis of complex balanced mechanisms. For finding the parameters for mass equivalence, a mass equivalent model with real and virtual equivalent masses is used. First the characteristics of this model are explained, then the properties of a mass equivalent dyad are shown. Subsequently with two methods the parameters of a mass equivalent dyad are derived and application examples are illustrated and discussed.

Keywords Mass equivalent modeling · Shaking force balance · Static balance · Mass equivalent dyad · Pantograph linkage

1 Introduction

Figure 1a shows a mechanism link with two joints A_i and A_{i+1} and a center of mass (CoM) in S_i, which is a point defined by parameters e_i and f_i as illustrated. In [2] a new mass equivalent model of this link was proposed and validated, which is shown in Fig. 1b. Here an equivalent mass m_i^a is located in A_i, an equivalent mass m_i^b is located in A_{i+1}, and an equivalent mass m_i^c is located in both J_{i1} and J_{i2}. J_{i1} is located at a distance $s_{i1} = \sqrt{e_i^2 + f_i^2}$ from S_i normal to the line A_iS_i, as illustrated, and J_{i2} is located at a distance $s_{i2} = \sqrt{(l_i - e_i)^2 + f_i^2}$ from S_i normal to the line $A_{i+1}S_i$, as illustrated. As in Fig. 1a S_i is the CoM of the four equivalent masses.

V. van der Wijk (✉)
Faculty of Electrical Engineering, Mathematics and Computer Science, Robotics and Mechatronics Group, University of Twente, Enschede, The Netherlands
e-mail: v.vanderwijk@kineticart.nl

S. Bai and M. Ceccarelli (eds.), *Recent Advances in Mechanism Design for Robotics*, Mechanisms and Machine Science 33,
DOI 10.1007/978-3-319-18126-4_4

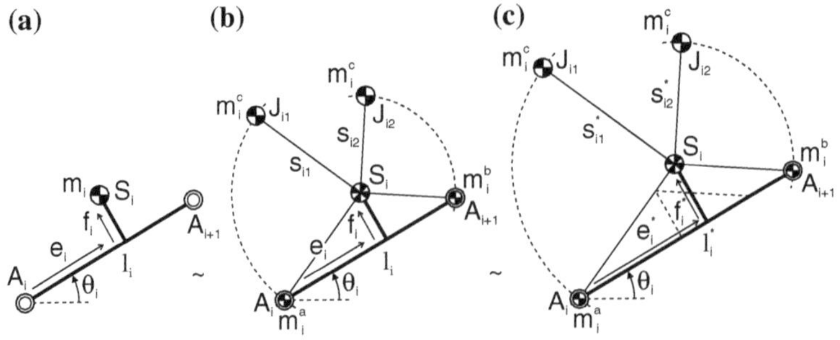

Fig. 1 **a** Link with two joints A_i and A_{i+1} and a general CoM S_i; **b** Mass equivalent model of the link with a real equivalent mass m_i^a in A_i, a real equivalent mass m_i^b in A_{i+1}, and a virtual equivalent mass m_i^c in both J_{i1} and J_{i2}; **c** The values of m_i^a, m_i^b, and m_i^c are independent of the scale of the link

Special of this mass model is that it allows the link CoM to be off the line through the joints, i.e. f_i can be nonzero, which is not possible with other mass equivalent modeling methods [1, 6].

In [2, 3] it was shown how with this model the motion of the masses of a general closed kinematic chain can be analyzed as an open chain where one of the elements is considered with equivalent masses. With this method it was possible to synthesize general inherently force balanced closed-chain linkages. The aim of this paper is to show that this mass equivalent model can represent not only the mass of a general single element, but also the mass of multi-DoF linkages. For all motion these multi-DoF linkages then have to be mass equivalent to the model in Fig. 1b. This is shown and explained by the synthesis of a 2-DoF dyad linkage. The application of mass equivalent linkages was shown in a preliminary study presented in [4] where mass equivalent linkages were applied to add DoFs to a balanced mechanism, maintaining its balance. Also at the end of this paper application examples are shown and discussed. For a realistic interpretation, all illustrations in this paper are drawn to scale.

2 Mass Equivalent Dyad

The conditions for which the model in Fig. 1b is mass equivalent with the link in Fig. 1a can be written as $m_i^a e_i = m_i^b(l_i - e_i)$ and $m_i^c l_i = (m_i^a + m_i^b)f_i$, which define the CoM of the model in S_i, and $m_i^a + m_i^b = m_i$ which is the total mass of the model equal to the mass of the link. Since mass $m_i^c = m_i f_i / l_i$ is not part of the 'real' mass of the link, this mass is referred to as a virtual equivalent mass while masses $m_i^a = m_i(1 - e_i/l_i)$ and $m_i^b = m_i e_i / l_i$ are referred to as the real equivalent masses.

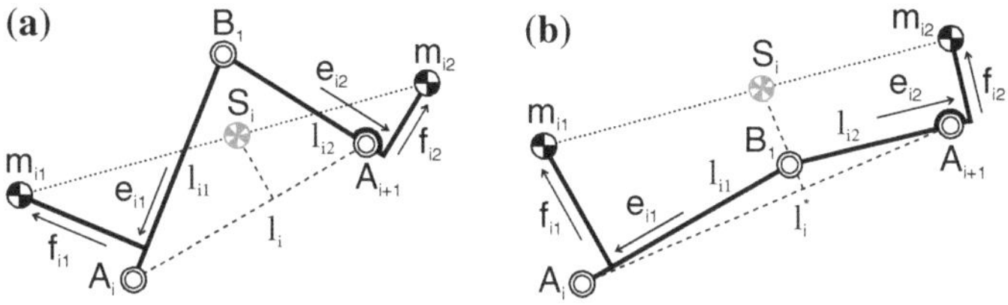

Fig. 2 Dyad that is mass equivalent with the link in Fig. 1a, shown in two poses and drawn to scale with parameters $\lambda_1 = 0.5991$, $\lambda_2 = 0.2658$, $l_{i2}/l_{i1} = 0.6978$, and $m_{i1}/m_{i2} = 0.8750$

In Fig. 1c it is shown that when the model is scaled, the equivalent masses remain unchanged. The values of the equivalent masses are independent of the size of the model. They can be described by $m_i^a = m_i(1 - \lambda_1)$, $m_i^b = m_i\lambda_1$, and $m_i^c = m_i\lambda_2$ with the constants $\lambda_1 = e_i/l_i$ and $\lambda_2 = f_i/l_i$.

While the length of the rigid link in Fig. 1a is fixed, a changing length as l_i^* in Fig. 1c can be obtained for instance by the distance between the extremities of a dyad. Figure 2 shows a dyad with link lengths l_{i1} and l_{i2} that is mass equivalent to the link in Fig. 1a. The link masses m_{i1} and m_{i2} are located such that their common CoM is exactly in S_i of the model in Fig. 1c for any of its poses and $m_{i1} + m_{i2} = m_i$. Practically this could mean that in a balanced linkage a general single link can be substituted with a mass equivalent dyad without affecting the balance. Alternatively it could mean that the mass motion of the dyad can be represented by the mass motion of a single link.

3 Method of Rotations About the Principal Joint (RAPJ)

The calculations of the design parameters of the dyad for mass equivalence can be quickly derived with the *method of rotations about the principal joints* (RAPJ) [2]. The dyad has one principal joint which is B_1. Figure 3a shows the mass equivalent model of the dyad for this method where real equivalent masses m_i^a and m_i^b are located in A_i and A_{i+1}, respectively, and virtual equivalent mass m_i^c is located both at l_{i1} and l_{i2} distance from B_1, normal to their respective links as illustrated. The complete procedure to derive the parameters of the mass equivalent dyad in Fig. 2 from the link in Fig. 1a is to first make the mass equivalent model of the link as in Fig. 1b, subsequently to make the mass equivalent model of the dyad as in Fig. 3a and from this to derive the mass parameters of the dyad in Fig. 3b.

First, the total real mass of the mass equivalent model in Fig. 3a has to be equal to the mass of the dyad in Fig. 3b, which is written as $m_i^a + m_i^b = m_{i1} + m_{i2}$. The other conditions can be derived from the linear momentum equations of each DoF individually, where the linear momentum of the mass equivalent model must equal the linear momentum of the mass equivalent dyad. Figure 4a shows the motion of DoF 1 where link B_1A_{i+1} is fixed and link A_iB_1 rotates about B_1. Figure 4b shows

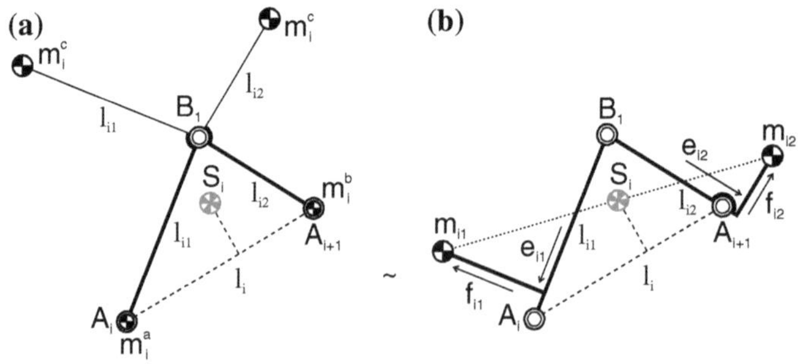

Fig. 3 a The mass equivalent model for analysis with the RAPJ-method of the b mass equivalent dyad

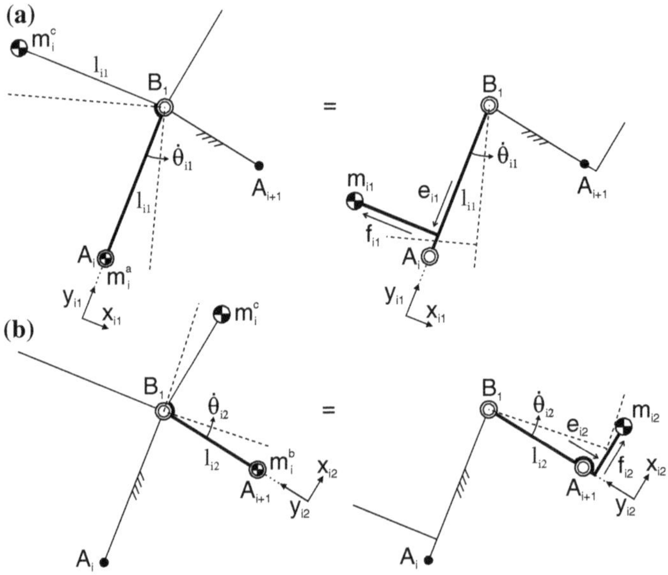

Fig. 4 For the individual motion of a DoF 1 and b DoF 2, the linear momentum of the mass equivalent model equals the linear momentum of the dyad

the motion of DoF 2 where link B_1A_i is fixed and link B_1A_{i+1} rotates about B_1. The linear momentum L_{i1} and L_{i2} of these individual motions can be written with respect to their relative reference frames $x_{i1}y_{i1}$ and $x_{i2}y_{i2}$, which are aligned with lines A_iB_1 and $A_{i+1}B_1$, respectively, as

$$\frac{\overline{L}_{i1}}{\dot{\theta}_{i1}} = \begin{bmatrix} m_i^a l_{i1} \\ -m_i^c l_{i1} \end{bmatrix} = \begin{bmatrix} m_{i1} e_{i1} \\ -m_{i1} f_{i1} \end{bmatrix} \qquad \frac{\overline{L}_{i2}}{\dot{\theta}_{i2}} = \begin{bmatrix} m_i^b l_{i2} \\ m_i^c l_{i2} \end{bmatrix} = \begin{bmatrix} m_{i2} e_{i2} \\ m_{i2} f_{i2} \end{bmatrix} \tag{1}$$

From these equations the conditions for mass equivalence follow as

$$e_{i1} = \frac{m_i^a l_{i1}}{m_{i1}}, \quad f_{i1} = \frac{m_i^c l_{i1}}{m_{i1}}, \quad e_{i2} = \frac{m_i^b l_{i2}}{m_{i2}}, \quad f_{i2} = \frac{m_i^c l_{i2}}{m_{i2}} \tag{2}$$

Herewith the conditions for which the dyad in Fig. 2 is mass equivalent to the link in Fig. 1a are obtained.

4 Method of Rotations About Principal Points (RAPP)

This section aims at clarifying the modeling approach in Fig. 3a. Therefore the conditions of the mass equivalent dyad are derived with the more fundamental method of *rotations about principal points* (RAPP) [2]. As shown for the two poses of the dyad in Fig. 5, the motion of masses m_{i1} and m_{i2} can be traced graphically with respect to their common CoM in S_i with a pantograph or 2-DoF principal vector linkage. Then links A_iB_1 and B_1A_{i+1} are the principal elements of the principal vector linkage and P_1 and P_2 are the principal points in each principal element. Within their element, P_1 and P_2 are defined with principal dimensions a_1 and a_2 from B_1 along the line from B_1 through m_{i1} and m_{i2}, respectively. With the parallelogram with sides a_1 and a_2, S_i is traced for all motion for the conditions $m_{i1}p_1 = m_{i2}a_1$ and $m_{i2}p_2 = m_{i1}a_2$ [2]. Because of similarity, the mass parameters of each principal element are related as

$$\frac{p_1 + a_1}{a_1} = \frac{e_{i1}}{b_1} = \frac{f_{i1}}{c_1}, \qquad \frac{p_2 + a_2}{a_2} = \frac{e_{i2}}{b_2} = \frac{f_{i2}}{c_2} \tag{3}$$

where parameters b_i and c_i define the principal points P_1 and P_2 relative to lines B_1A_i and B_1A_{i+1}, respectively, as illustrated in Fig. 5. The mass parameters of each element then can be derived to depend on b_i and c_i as

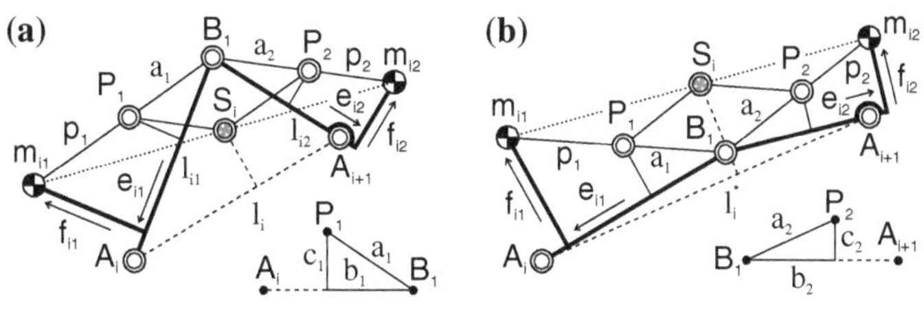

Fig. 5 Mass equivalent dyad for analysis with rotations about the principal points (RAPP-method) drawn for two poses. The dyad is part of a pantograph which traces S_i for all motion

$$
\begin{aligned}
e_{i1} &= (\frac{p_1}{a_1}+1)b_1 = (\frac{m_{i2}}{m_{i1}}+1)b_1, \qquad & f_{i1} &= (\frac{p_1}{a_1}+1)c_1 = (\frac{m_{i2}}{m_{i1}}+1)c_1 \\
e_{i2} &= (\frac{p_2}{a_2}+1)b_2 = (\frac{m_{i1}}{m_{i2}}+1)b_2, \qquad & f_{i2} &= (\frac{p_2}{a_2}+1)c_2 = (\frac{m_{i1}}{m_{i2}}+1)c_2
\end{aligned} \tag{4}
$$

To calculate b_i and c_i, Fig. 6a shows the mass equivalent model of the dyad for the RAPP-method. As in the model in Fig. 3a the real equivalent masses m_i^a and m_i^b are in joints A_i and A_{i+1}, respectively. The virtual equivalent masses m_i^c here are projected twice about each principal point P_1 and P_2 as illustrated. This projection is explained in [2], where it is also shown how in the limits $a_1 \to 0$ and $a_2 \to 0$ the model in Fig. 3a is obtained which is also applicable if a_1 and a_2 have general values.

The model in Fig. 6a is analyzed with the linear momentum of the rotations about the principal points P_1 and P_2 of the two relative motions of the dyad. Figure 6b shows the relative motion of DoF 1 where link A_iB_1 rotates about P_1 and link B_1A_{i+1} solely translates. This means that if the pantograph is regarded a real linkage, this is motion with link P_1S_i fixed. Since a virtual equivalent mass has the property that its mass acts as a real mass for rotational motion while it has zero mass for translational motion of its element [2], for the relative motion of DoF 1 the virtual equivalent masses in link A_iB_1 act as a real mass while the virtual equivalent masses in link B_1A_{i+1} are zero. Figure 7a shows the relative motion of DoF 2 where link S_iP_2 is regarded fixed such that link B_1A_{i+1} rotates about P_2 and link A_iB_1 solely translates. For this motion the virtual equivalent masses in link B_1A_{i+1} act as a real mass while the virtual equivalent masses in link A_iB_1 are zero.

The linear momentum of the motions in Figs. 6b and 7a can be represented with the reduced mass models in Fig. 7b, c, respectively. For rotational motion about P_1 of the model in Fig. 7b the linear momentum is equal to the linear momentum of the motion in Fig. 6b. For rotational motion about P_2 of the model in Fig. 7c the linear

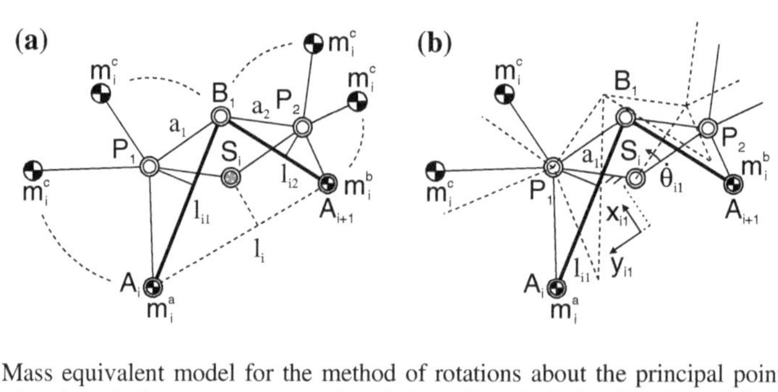

Fig. 6 **a** Mass equivalent model for the method of rotations about the principal points (RAPP) where the virtual equivalent masses are projected about the principal points P_1 and P_2. **b** Relative motion of DoF 1 where link A_iB_1 rotates about P_1 and link B_1A_{i+1} solely translates

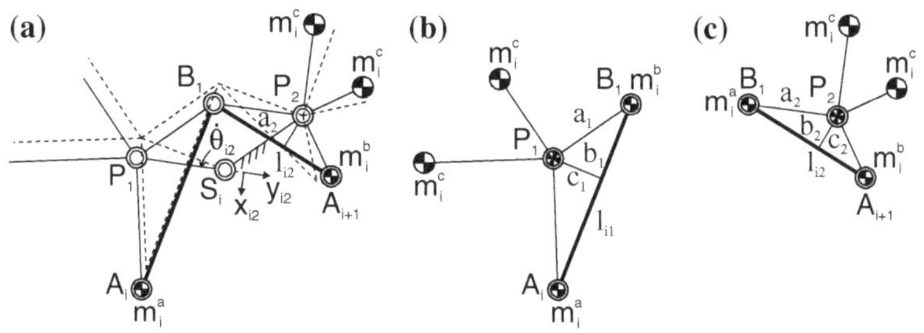

Fig. 7 **a** Relative motion of DoF 2 where link B_1A_{i+1} rotates about P_2 and link A_iB_1 solely translates. **b** Reduced mass model of DoF 1 where P_1 is the common CoM; **c** Reduced mass model of DoF 2 where P_2 is the common CoM

momentum is equal to the linear momentum of the motion in Fig. 7a. Therefore these models are also named Equivalent Linear Momentum Systems (ELMS) [2]. As a result, the principal points P_1 and P_2 are the common CoMs of the reduced mass models. Therefore the parameters in each reduced mass model are related by

$$\begin{aligned} m_i^a(l_{i1} - b_1) &= m_i^b b_1, \quad m_i^c l_{i1} = (m_i^a + m_i^b)c_1 \\ m_i^a b_2 &= m_i^b(l_{i2} - b_2), \quad m_i^c l_{i2} = (m_i^a + m_i^b)c_2 \end{aligned} \tag{5}$$

From these equations b_1, c_1, b_2, and c_2 are derived as

$$\begin{aligned} b_1 &= \frac{m_i^a l_{i1}}{m_i^a + m_i^b} = (1 - \lambda_1)l_{i1}, \quad c_1 = \frac{m_i^c l_{i1}}{m_i^a + m_i^b} = \lambda_2 l_{i1} \\ b_2 &= \frac{m_i^b l_{i2}}{m_i^a + m_i^b} = \lambda_1 l_{i2}, \quad c_2 = \frac{m_i^c l_{i2}}{m_i^a + m_i^b} = \lambda_2 l_{i2} \end{aligned} \tag{6}$$

Subsequently with (4) parameters e_{i1}, f_{i1}, e_{i2}, and f_{i2} are obtained and the principal dimensions a_1 and a_2 are calculated as

$$a_1 = \sqrt{b_1^2 + c_1^2} = l_{i1}\sqrt{(1 - \lambda_1)^2 + \lambda_2^2}, \quad a_2 = \sqrt{b_2^2 + c_2^2} = l_{i2}\sqrt{\lambda_1^2 + \lambda_2^2} \tag{7}$$

with which p_1 and p_2 are found from $m_{i1}p_1 = m_{i2}a_1$ and $m_{i2}p_2 = m_{i1}a_2$ or from (3) as

$$p_1 = \frac{m_{i2}}{m_{i1}}a_1, \quad p_2 = \frac{m_{i1}}{m_{i2}}a_2 \tag{8}$$

5 Discussion

From the two methods in Sects. 3 and 4 the RAPJ-method shows to gain results relatively quickly while the RAPP-method gives more fundamental insight. For instance the principal points in the reduced mass models in Fig. 7b, c are related with the so called 'barycenters' in [5]. Where the barycenters are found by a trick, the principal points have physical meaning as being the joints of a pantograph with principal dimensions, as shown in Fig. 5. Therefore this method may be valuable also in general dynamical analysis. It is interesting to note that the reduced mass models are of similar composition as the mass equivalent models in Fig. 1b, c.

Regarding the pantograph in Fig. 5, since the triangle $A_iA_{i+1}S_i$ remains similar for all motion, joints A_i and A_{i+1} are found as a set of similarity points of the pantograph in Fig. 5 and therefore they are also a set of similarity points of the mass equivalent dyad in Fig. 2.

In a preliminary study of this work in [4] it was shown how also the lines P_1S_i and P_2S_i of the pantograph in Fig. 5 can be carried out as real links with each an arbitrarily located CoM which is included in the common CoM in S_i. It was shown how such a mass equivalent pantograph can substitute a coupler link of a general force balanced 4R four-bar linkage to obtain a force balanced 5R five-bar linkage, increasing the number of DoFs with one. Also the other three links of a general force balanced 4R four-bar linkage can be substituted similarly, increasing the number of DoFs to six in total.

Figure 8 shows the result of a general force balanced 4R four-bar linkage $A_0A_1A_2A_3$ of which the coupler link A_1A_2 is substituted with the mass equivalent

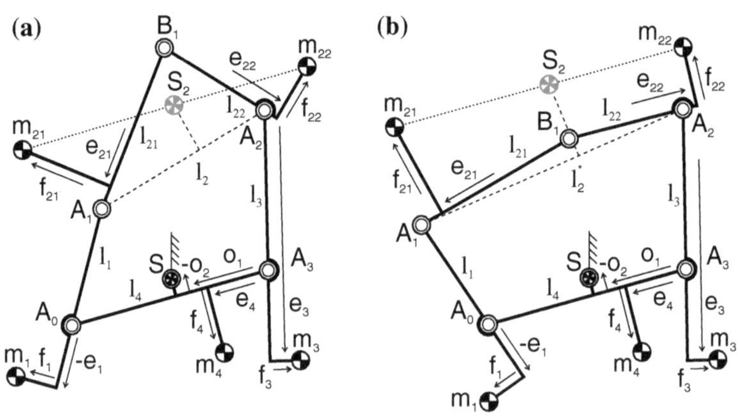

Fig. 8 Force balanced 5R five-bar linkage obtained from a general force balanced 4R four-bar linkage $A_0A_1A_2A_3$ with coupler link A_1A_2 substituted with the mass equivalent dyad of Fig. 2, illustrated for two poses

dyad in Fig. 2. Since the distance l_2 is no longer fixed, one DoF is gained, obtaining a force balanced 5R five-bar linkage of which the common CoM is in S for all motion. The mass parameters of the other links, i.e. the CoMs of m_1, m_3, and m_4, and the location of the common CoM in S are not affected by this substitution while force balance is maintained.

Another way of applying mass equivalent dyads is shown in Fig. 9. Figure 9a shows an immovable structure of a 4R four-bar linkage $A_0A_1A_2A_3$ with an addition link A_4A_5 mounted in between. All elements have a general CoM and the common CoM is in S as if it was a movable linkage. Figure 9b shows how this structure with one overconstraint (0 DoF) is transformed into a movable force balanced 3-RRR manipulator linkage with 3 DoF motion. Therefore links A_0A_1, A_2A_3, and A_4A_5 are each substituted with a mass equivalent dyad. Also here the mass parameters of the other links, i.e. the CoMs of m_2 and m_4, and the location of the common CoM in S are not affected by this substitution while force balance is maintained.

Figure 10 shows an alternative view on the force balanced 3-RRR manipulator linkage in Fig. 9. The configuration can also be regarded as that linkage $A_1A_2B_2$ is a mass equivalent dyad with the CoM in S_2 and link 3 is unchanged as in Fig. 9a. The result is equal and other ways of regarding the configuration are also possible. Another observation is that for an element with more than two joints only two joints have to be selected and used for the mass equivalent modeling. This means that the closed loop into which the model is applied needs to be selected. Link $A_1A_2A_5$ in Fig. 10, for instance, has three joints, while for the mass equivalent dyad $A_1A_2B_2$ only joints A_1 and A_2 in the closed loop $A_0B_1A_1A_2B_2A_3$ are involved.

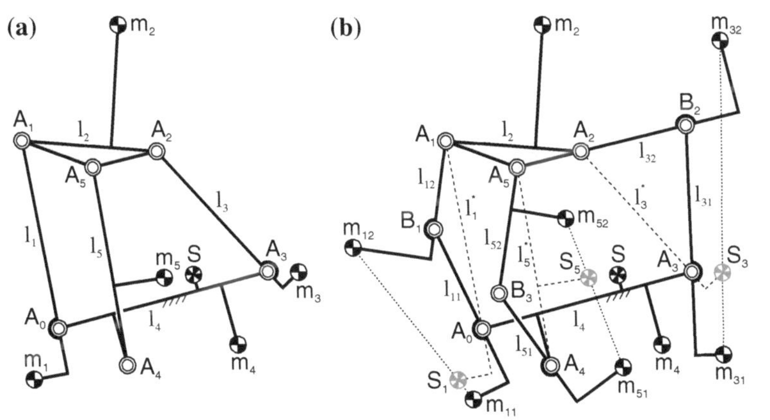

Fig. 9 **a** From a force balanced 4R four-bar linkage with additional link A_4A_5 a **b** force balanced 3-RRR manipulator linkage can be obtained by substituting links A_0A_1, A_2A_3, and A_4A_5 each with a mass equivalent dyad

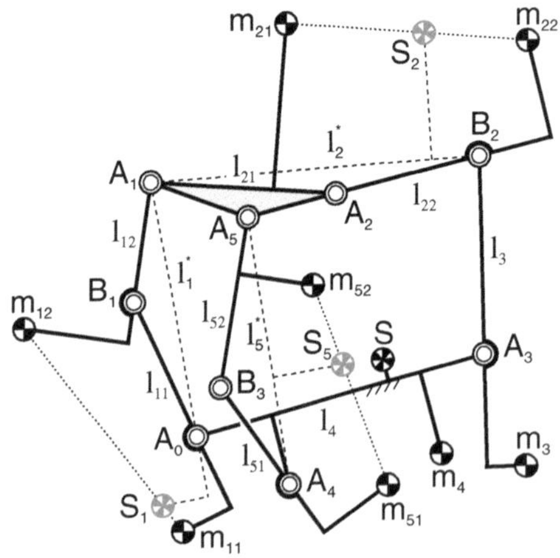

Fig. 10 The force balanced 3-RRR manipulator linkage in Fig. 9b can also be regarded with linkage $A_1A_2B_2$ as a mass equivalent dyad instead of link 3

6 Conclusion

In this paper it was shown how a general 2-DoF dyad can be designed mass equivalent to a general (1-DoF) link element. For finding the parameters for mass equivalence, a mass equivalent model with real and virtual equivalent masses was used. First the characteristics of this model were explained and subsequently the properties of a mass equivalent dyad were shown. Both with the RAPJ-method and the RAPP-method the parameters of a mass equivalent dyad were derived. With the RAPJ-method the solutions were found quickly while the RAPP-method showed fundamental insight. As an application example it was shown how a force balanced 5R five-bar linkage can be obtained from a force balanced 4R four-bar linkage by substituting the coupler link with a mass equivalent dyad. It was also shown how a force balanced 3-RRR parallel manipulator can be found from a rigid 0-DoF structure by substituting three links each with a mass equivalent dyad.

References

1. Chaudhary, H., Saha, S.K.: Balancing of shaking forces and shaking moments for planar mechanisms using the equimomental systems. Mech. Mach. Theory **43**, 310–334 (2008)
2. Van der Wijk, V.: Methodology for analysis and synthesis of inherently force and moment-balanced mechanisms - theory and applications (dissertation). University of Twente (free download: http://dx.doi.org/10.3990/1.9789036536301) (2014)
3. Van der Wijk, V.: Closed-chain principal vector linkages. In: Flores, P., Viadero, F. (eds.), New Trends in Mechanism and Machine Science, vol. 24, pp. 829–837. Springer, London (2015)

4. Van der Wijk, V., Herder, J.L.: On the addition of degrees of freedom to force-balanced linkages. Proceedings of the 19th CISM-IFToMM Symposium on Robot Design, Dynamics, and Control, pp. 2012–025 (2012)
5. Wittenburg, J.: Dynamics of Systems of Rigid Bodies. B.G. Teubner Stuttgart (1977)
6. Wu, Y., Gosselin, C.M.: On the dynamic balancing of multi-dof parallel mechanisms with multiple legs. Mech. Des. **129**, 234–238 (2007)

A Light Weight Arm Designed with Modular Joints

Honggen Fang, Lijie Guo and Shaoping Bai

Abstract The conventional industrial manipulator has some drawbacks such as low payload-weight ratio, bulky structure and high power consumption, which limit their applications in such areas a space, anti-terrorism, service and medical robots. To overcome these shortcomings, a novel lightweight arm was developed based on modular joints, modular connection and light shaft structures. This paper discusses the general requirements for lightweight robots, upon which the new robot was designed. Both mechanics and electronics designs are presented. The development work of a prototype is described. Preliminary tests were conducted to evaluate the performance of the light weight arm. The results demonstrate the good performances of the prototype and validate the feasibility of the new robot system.

Keywords Lightweight robotic arm · Modular joint · Modular construction

1 Introduction

The first industrial manipulator is dated back to 1962. Since then, robots have been increasingly deployed in flexible manufacturing systems and factory automations. In the new century, robot applications have been expanded from automotive industry to mining, building, defense and military, medical and domestic services. On the other hand, some shortcomings of industrial robots become obvious: low

H. Fang (✉) · L. Guo
Shanghai Aerospace Equipments Manufacturer, Shanghai, China
e-mail: fangdanwei@163.com

L. Guo
e-mail: guolijie149@163.com

S. Bai
Department of Mechanical and Manufacturing Engineering,
Aalborg University, Aalborg, Denmark
e-mail: shb@m-tech.aau.dk

S. Bai and M. Ceccarelli (eds.), *Recent Advances in Mechanism Design for Robotics*, Mechanisms and Machine Science 33,
DOI 10.1007/978-3-319-18126-4_5

payload-weight ratio, bulky structure and high power consumption. To address these problems, research efforts have dedicated to the development of lightweight robots. Some notable examples of lightweight robots include the LWR and MIRO developed by German Aerospace Center [1–3], the RPL and PBA lightweight manipulators developed by SHUNK company [4], the light weight manipulator designed for mars exploration by JPL [5], the ROBONAUT for space working developed by NASA [6, 7], the lightweight manipulator deployed in anti-terrorism developed by Foster-Miller and iRobot company [8]. Other examples of lightweight robots include the Barrett WAM™ Arm [9], the UR industrial robot, the da Vinci Surgical manipulator [10], the Lightweight Anthropomorphic Arm [11], the KUKA LWA-4D [12], etc.; These lightweight robots have common features, namely, compact structure, high payload-weight ratio, lightweight and more control modes.

In this paper, a light weight arm designed and developed based on modular joint and modular construction is introduced. Modular joints are designed to reduce the robot mass and enhance the payload-weight ratio. It helps also improve the mechanical bandwidth. The development work is described, with preliminary results of robot testing included.

2 Analysis of Design Requirements

Industrial robots are designed with rigid links for high-accuracy movement in a known environment. Normally no additional sensors (e.g. force/torque sensor, additional position sensor) are needed. Occasionally, the link elasticity is considered for improvement of the accuracy of industrial robots [13].

In contrast, lightweight robots, for which a high payload-to-weight ratio (e.g. 1:1 with the LWR III) is desirable, are built with relatively flexible links and joints. The elasticity is not negligible, which is a major factor contributed to the pose error of the tool center point (TCP). Due to the compact structures, the structure stiffness of a lightweight robot is inherently low. Approaches to improve the stiffness have to be considered.

To get an overview of the lightweight robots, some available lightweight robots are compared, as listed in Table 1 Most of these robots have six or seven joints which enables the capability of collision avoidance. Their payload-to-weight ratios are far better than those of standard industrial robots. Furthermore, extra torque and position sensors able to sensing the arm deflection are applied to detect possible collisions with other objects.

Form the comparison, it is seen that following features are desirable for lightweight robots:

(I) a reduced inertia.
(II) sensing of position, force/torque, for impedance control on Cartesian and joint spaces.
(III) back-drivability of the drive train.

Table 1 Comparison of typical robotic systems

Robot	LAA	WAM	LWA-4D	PA10-7C	UR-5	LWR-III
Kinematics	Serial	Serial	Serial	Serial	Serial	Serial
DOF	5	7	7	7	6	7
Maximum reach	1.32 m	1 m	1.1 m	1 m	0.85 m	0.9 m
Payload	4 kg	4 kg	10 kg	10 kg	5 kg	14 kg
Total weight	10 kg	25 kg	18 kg	35 kg	18.4 kg	14 kg
Position control	Yes	Yes	Yes	Yes	Yes	Yes
Joint torque control	No	Yes	No	No	Yes	Yes

(IV) sensing external loads for collision detection.
(V) accounting for elasticity influence.
(VI) easy and convenient construction, installation and setup.

3 Design of Modular Joints

The joints of lightweight robots are normally designed by two approaches. One approach is to design with highly integrated components. The most representative mechatronic integrated type is the DLR LWR modular joint which integrates electronics and mechanics components, including sensor, electric braking unit, harmonic drive, optimized design motor, position limit sensor, communication and control units. Moreover, composite materials are used for the robot links to further reduce the arm weight. The design of the entire system (controllers, power supply) is optimized for weight reduction in order to enable the mobile application of the systems.

The other approach is to allocate motors in the base and transmit their motions to joints by tendon-like mechanisms, like the design of Barrett WAM™ arm. This type of design leads to highly dexterous, naturally backdrivable and compliant actuations. The arm has a faster response time than conventional manipulator. However, as its drive units are away from joints, this type of robot has a large footprint with lower interchangeability.

In the proposed design of robots, we aimed to design a robot with modular joints which has higher reliability, better interchangeability, more cost-effective and greater payload-weight ratio. A designed modular joint is shown in Fig. 1. The joint carries the features listed in Sect. 1. The motor adopts high power-density permanent magnet synchronous motors with hollow shaft (its specifications shown in Table 2). Within the big hollow shaft, the cables are routing. The safety brake, which stops the arm for instance at power drop-out, is custom designed for guaranteeing the joint compact structure. Stress and strain torque sensors are used to measure the real driving torque at the output shafts. Different from the most lightweight arms with all circuit board integrated into one modular joint, some

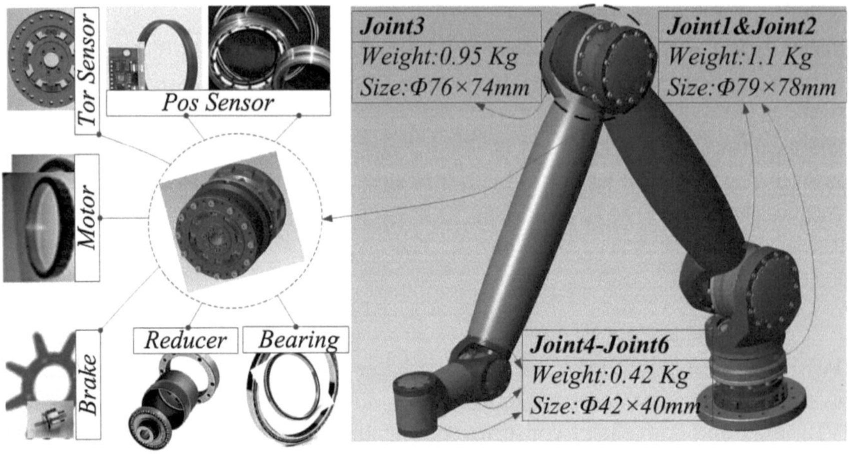

Fig. 1 An overview of a modular joint and the robot arm

Table 2 Three types of motors used in the lightweight robot arm

Type	Rated torque (Nm)	Rated voltage (V)	Maximum rotation speed (RPM)	Rated current (A)	Motor size diameter × height (mm)	Weight (g)
M-I	0.5	48	4500	5	50 × 16	140
M-II	0.25	48	4500	4.2	50 × 8	100
M-III	0.045	24	8000	3	25 × 3.5	30

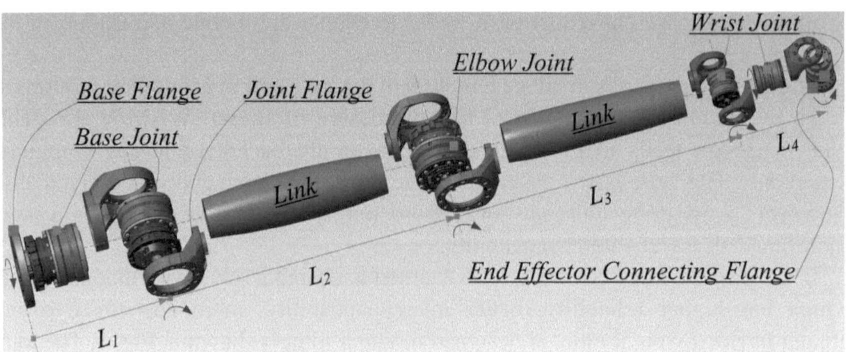

Fig. 2 Exploded view of arm

electric boards are placed inside links, which helps to reduce the modular joint volume and making the mechatronic system easy to maintain.

With the designed modular joints, a six-DOF lightweight arm is further designed, as shown in Fig. 2. Some major specifications of the arm are displayed in Table 3.

Table 3 Arm specifications

DOF	Maximum reach (mm)	Maximum width (mm)	Designed payload (kg)	Total weight (kg)
6	880	92	4	7

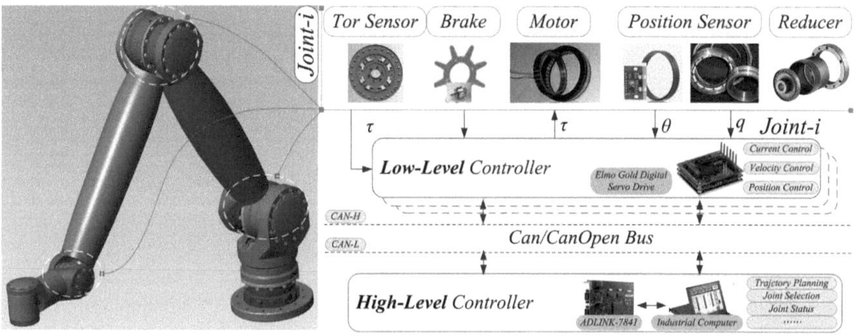

Fig. 3 Hardware system architecture

4 Robot Development

The robot control system was developed, consisting of low-level and high-level controllers as shown in Fig. 3. Elmo Gold Digital Servo Drive is adopted in the low-level control, which has a compact structure and better performances than other controllers. The controller supports multiple control modes including position control, current control, advance position control, filter design, velocity control and parameter tuning.

The high-level controller is an industrial computer based on ADLINK 7841 CAN board. All servo drives are connected to the CANOPEN bus serially.

The user interface software system was developed using visual C++. Figure 4 displays two modules, namely, single joint control module, multi-joint control module. Other function modules include CAN bus communication settings, dynamic sensor data tracking and view, teaching module, trajectory planning, parameter settings and error detecting, etc., have also been built. The users can easily operate the manipulator though the software.

5 Robot Testing

The designed robot was fabricated and constructed. Figure 5 shows the prototype of the robot, in which the wrist is not included. Several testings of the robot have been conducted. At first, payload capability was tested. Then, position controls were carried out with the robot, done at both single and multiple joint levels. At single

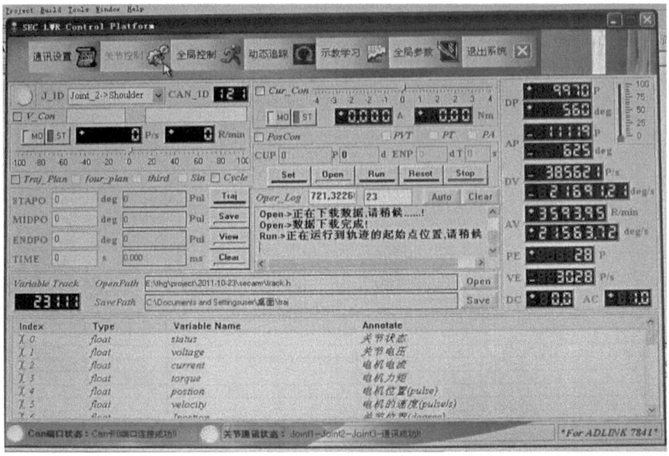

Fig. 4 Main UI for robot control

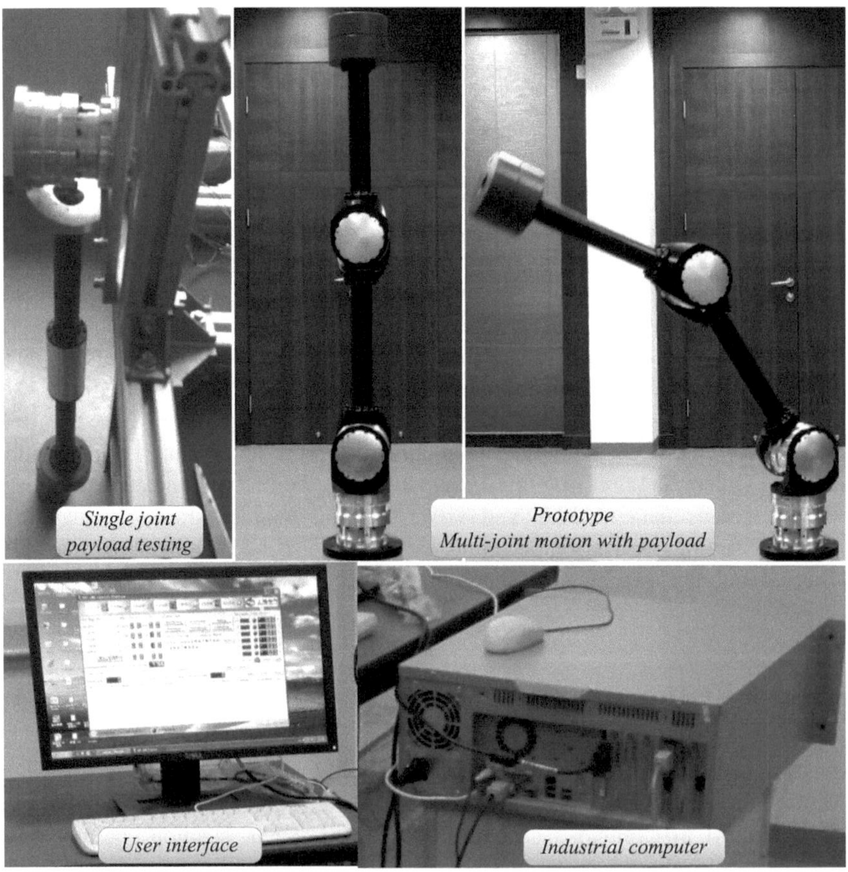

Fig. 5 Robot prototype and experiment platform

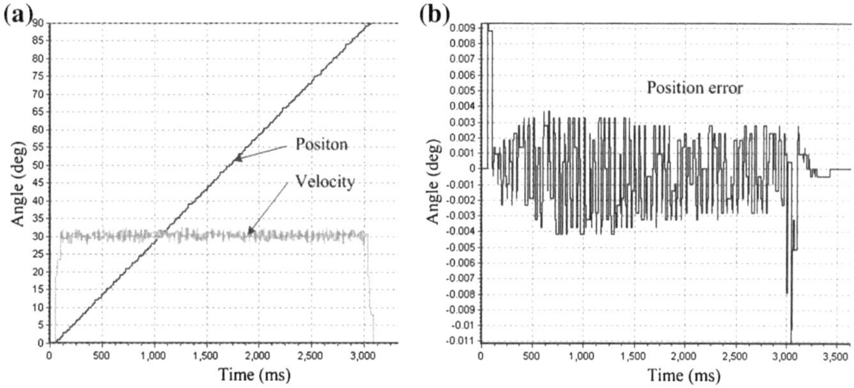

Fig. 6 Shoulder joint motion control. **a** Position and velocity. **b** Position error

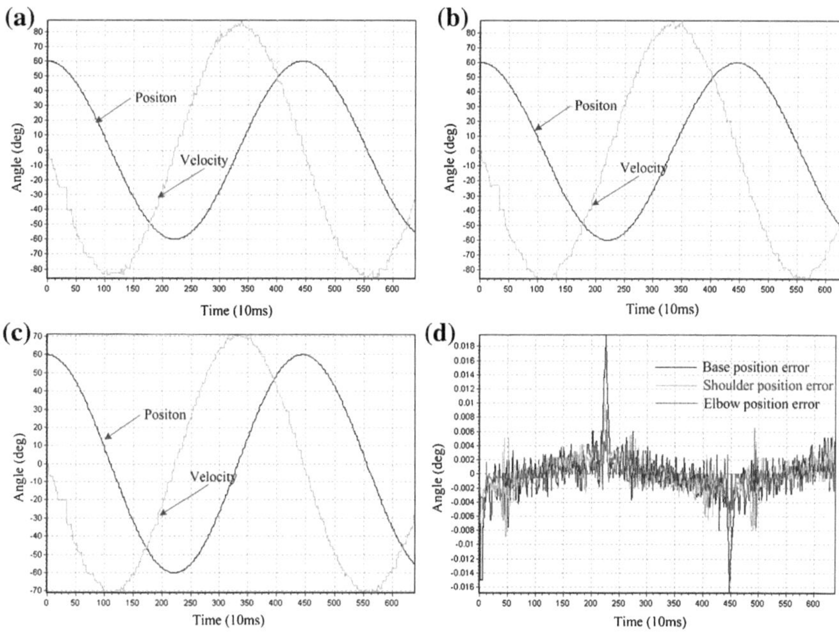

Fig. 7 Multi joints motion control. **a** Base joint position and velocity. **b** Shoulder joint position and velocity. **c** Elbow joint position and velocity. **d** Position error

joint motion control test, a trajectory command for 90° rotation in 3 s is sent to the shoulder joint. The results are shown in Fig. 6. During the period of acceleration and deceleration, the dynamic error is within 0.01°, while in the steady-state, the error is within 0.004° at the stage of uniform motion (Fig. 7).

6 Conclusions

In this work, a modular lightweight arm was developed. Modular joints were developed, in which all mechanics and electronics components are integrated for improving the payload-weight ratio and structure compactness. With the modular joints, a six-DOF lightweight robot was designed. A prototype was built, with which testings were conducted. The preliminary results show that a good performance of the robot is achieved. Future development with more degrees of freedom and joint/Cartesian space compliance control towards practical applications are considered.

Acknowledgments The first author would like to acknowledge the support of the Chinese CSC.

References

1. Hirzinger, G., Albu-Schaffer, A., Haahnle, M., Schaefer, I., Sporer, N.: On a new generation of torque controlled light-weight robots. In: IEEE International Conference on Robotics and Automation, pp. 3356–3363 (2001)
2. Abadia, D.: Comparative analysis development of control systems for the DLR light weight robot. Master's thesis, DLR, University of Zaragossa, (2000)
3. Hagn, U., Konietschke, R., Tobergte, A., Nickl, M.: DLR MiroSurge: a versatile system for research in endoscopic telesurgery. Int J CARS **5**, 183–193 (2010)
4. http://www.schunk-modular-robotics.com/
5. Stieber, M.E.: Vision-based sensing and control for space robotics applications. IEEE Trans. Instrum. Meas. **48**(4), 807–812 (1999)
6. Bluethmann, W., Ambrose, R., Diftler, M., Askew, S., Huber, E., Goza, M.: Robonaut: a robot designed to work with humans in space. Auton. Robot. **14**, 179–197 (2003)
7. Diftler, M.A., Mehling, J.S., Abdallah, M.E., Radford et al, N.A.: Robonaut 2-the first humanoid robot in space. IEEE International Conference on Robotics and Automation, Shanghai, China, May 9–13, 2011
8. Mavroidis. Development of a New Arm For The Foster-Miller TALON Robot. Department of Mechanical, Industrial and Manufacturing Engineering, Northeastern University (2005)
9. Townsend, W.T., Guertin, J.A.: Teleoperator slave—WAM design methodology. Ind. Robot. Int. J. **26**(3), 167–177 (1999)
10. http://www.universal-robots.com/
11. Zhou, L., Bai, S.: A new approach to design of a lightweight anthropomorphic arm for service applications. ASME J. Mech. Robot. **7**(3), 031001 (2015)
12. http://mobile.schunk-microsite.com/en/produkte/produkte/dextrous-lightweight-arm-lwa-4d.html
13. Weiss, M.: Robot control, industrial robot and method for obtaining an absolutely precise model. European Patent Application EP1 980374A2

Dynamic Balancing Conditions of Planar Parallel Manipulators

Mario Acevedo and José María Reyes

Abstract Force and moment balancing (dynamic balancing) of rigid body linkages with constant mass links is a traditional but still very active area of research in machine dynamics and robotics. The shaking force and the shaking moment caused by all moving links can be reduced in different ways but all having a common difficulty named to derive the so-called balancing conditions, that in general can be cumbersome. In this article a novel method to find the dynamic balancing conditions based on the use of Natural Coordinates is introduced. The method is direct, efficient, and easy to automate through the application of a computer algebra system. It can be used to obtain the shaking force and the shaking moment balancing conditions for planar and spatial mechanisms.

Keyword Dynamic balancing · Parallel manipulators · Natural Coordinates

1 Introduction

Force and moment balancing (dynamic balancing) of rigid body linkages with constant mass links is a traditional but still very active area of research in mechanical engineering [1–6], and its benefits are well known [7, 8]. On the other hand parallel manipulators are excellent candidates for advanced robotic applications by virtue of their low moving inertia, high stiffness, high dexterity, compact size, and high power-to-weight ratio [9]. However, they exert forces and moments on their base while moving, causing fatigue, vibration, noise and disturbances in their supporting structure and even on the mobile end-effector platform [10]. To improve these dynamic characteristics dynamic balancing is needed allowing their better use. An overview of dynamic balancing techniques and methods can be found in [11–13].

M. Acevedo (✉) · J.M. Reyes
Faculty of Engineering, Universidad Panamericana, Mexico City, Mexico
e-mail: mario.acevedo@up.edu.mx

S. Bai and M. Ceccarelli (eds.), *Recent Advances in Mechanism Design for Robotics*, Mechanisms and Machine Science 33,
DOI 10.1007/978-3-319-18126-4_6

But dynamic balancing of linkages has some difficulties, named to derive the so-called balancing conditions [14] and to obtain the mechanism configurations that can be balanced. So in this work a novel and general method to find the dynamic balancing conditions based on the use of Natural Coordinates [15] is introduced. The method is direct due to the fact that with the Natural Coordinates the whole system can be represented by a set of points, avoiding the explicit use of angles, facilitating the generation of the expressions of the linear momentum and the angular momentum required to obtain the shaking force and the shaking moment balancing conditions for the linkages in the plane and in space, although at this time is presented for planar parallel manipulators. In this way these balancing conditions are expressed entirely in terms of the location of the center of mass, the mass and the moment of inertia. In addition the method can be easily automated, mainly if a computer algebra system is available.

Once the balancing conditions are found, these can be used to identify the mechanism configurations that can be balanced, and to the effective design of reactionless manipulators either using counter-inertias [16] or counter-rotary counterweights [6, 17].

2 Balancing Conditions Using Natural Coordinates

For the effective design of reactionless mechanisms it is necessary first to obtain the dynamic balancing conditions. In this section a new method based on Natural Coordinates is presented. The method is straightforward and can be easily automated, mainly it has the advantage of being suitable for the application of a computer algebra system.

For a mechanism to be dynamically balanced, the global centre of mass of all the moving bodies must remain stationary (zero linear momentum) and the angular momentum must remain constant (zero) with respect to a fixed point. So a mechanism must be force-balanced in order to be dynamically balanced, see [18] for further explanation.

A mechanism is force balanced if its linear momentum, $\mathbf{l}$, is conserved. This condition in general can be expressed as:

$$\mathbf{l} = \sum_{k=1}^{n} \mathbf{l}_k = cnt. \tag{1}$$

where n is the number of total moving elements in the linkage, $\mathbf{l}_k = m_k \mathbf{v}_k$, while $\mathbf{v}_k$ and m_k are the velocity of the center of mass and the mass of the k_{th} moving body, respectively.

On the other hand a mechanism is moment balanced if its angular momentum, $\mathbf{h}$, is conserved. This condition is expressed as:

$$\mathbf{h} = \sum_{k=1}^{n} \mathbf{r}_k \times (\mathbf{l}_k) + \mathbf{h}_k = cnt. \quad (2)$$

where $\mathbf{r}_k$ is the position vector of the center of mass, $\mathbf{h}_k = \mathbf{I}_k \omega_k$, while $\mathbf{I}_k$ and ω_k are the inertia tensor with respect to the center of mass and the angular velocity of the k_{th} body respectively.

So it is necessary to find equivalent expressions to Eqs. (1) and (2) in Natural Coordinates, then finding the complete set of force and moment balancing conditions.

2.1 Linear Momentum Using Natural Coordinates

In this section the Eqs. (1) and (2) are expressed in Natural Coordinates.

When dealing with mechanical systems in the plain, Natural Coordinates introduce a set of points to define a body, the basic points, see [15] for a detailed explanation. So a body can be modeled in Natural Coordinates with a pair of points, i and j, as seen in Fig. 1. In this figure it can be noted an inertial fixed reference frame **XY**, and a local reference frame **xy** attached to the moving body at the basic

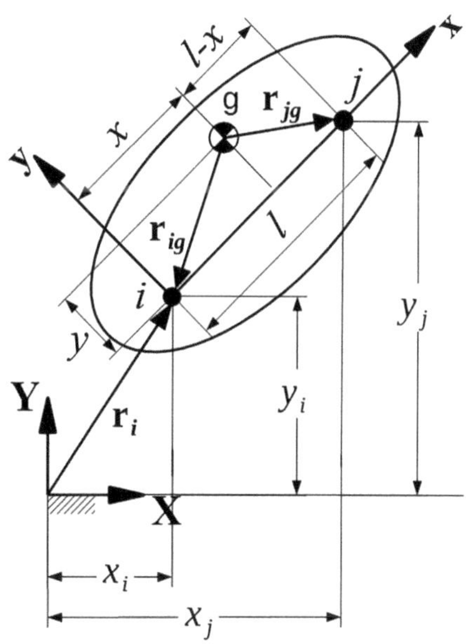

Fig. 1 A general model of a body using Natural Coordinates. Two basic points, i and j, with a local moving reference frame attached to the body at point i, the origin

point i. It is also noted that the second basic point j has its position in local coordinates at $(l, 0)$, and that the center of mass of the body, point g, has local coordinates (x, y).

The pair of basic points introduce a vector of four coordinates represented by $\mathbf{q}$, the positions, and its velocities $\dot{\mathbf{q}}$ as:

$$\mathbf{q} = [x_i \quad y_i \quad x_j \quad y_j]^T; \quad \dot{\mathbf{q}} = [\dot{x}_i \quad \dot{y}_i \quad \dot{x}_j \quad \dot{y}_j]^T \tag{3}$$

Using this pair of points and considering that the body has a total mass m, and a inertia tensor I with respect to the origin of the local reference frame (point i), the constant mass matrix of a body can be expressed as (see [15]):

$$\mathbf{M} = \begin{bmatrix} m - \frac{2mx}{l} + \frac{I}{l^2} & 0 & \frac{mx}{l} - \frac{I}{l^2} & -\frac{my}{l} \\ 0 & m - \frac{2mx}{l} + \frac{I}{l^2} & \frac{my}{l} & \frac{mx}{l} - \frac{I}{l^2} \\ \frac{mx}{l} - \frac{I}{l^2} & \frac{my}{l} & \frac{I}{l^2} & 0 \\ -\frac{my}{l} & \frac{mx}{l} - \frac{I}{l^2} & 0 & \frac{I}{l^2} \end{bmatrix} \tag{4}$$

The complete method to obtain this mass matrix is described in detail in [15].

So it is possible to calculate a set of the linear momentum vectors associated to the basic points in the body as:

$$\mathbf{M}\dot{\mathbf{q}} = \begin{bmatrix} \mathbf{l}_i \\ \mathbf{l}_j \end{bmatrix} \tag{5}$$

where $\mathbf{l}_i$ and $\mathbf{l}_j$ are the linear momentum associated to the points i and j respectively. They can be expressed as:

$$\mathbf{l}_i = \begin{bmatrix} \left(\frac{I}{l^2} - \frac{2mx}{l} + m\right)\dot{x}_i + \left(\frac{mx}{l} - \frac{I}{l^2}\right)\dot{x}_j - \frac{my}{l}\dot{y}_j \\ \left(\frac{I}{l^2} - \frac{2mx}{l} + m\right)\dot{y}_i + \frac{my}{l}\dot{x}_j + \left(\frac{mx}{l} - \frac{I}{l^2}\right)\dot{y}_j \end{bmatrix} \tag{6}$$

$$\mathbf{l}_j = \begin{bmatrix} \left(\frac{mx}{l} - \frac{I}{l^2}\right)\dot{x}_i + \frac{my}{l}\dot{y}_i + \left(\frac{I}{l^2}\right)\dot{x}_j \\ -\frac{my}{l}\dot{x}_i + \left(\frac{mx}{l} - \frac{I}{l^2}\right)\dot{y}_i + \left(\frac{I}{l^2}\right)\dot{y}_j \end{bmatrix} \tag{7}$$

Then the total linear momentum of the body, $\mathbf{l}_b$, is:

$$\mathbf{l}_b = \mathbf{l}_i + \mathbf{l}_j = \begin{bmatrix} \left(m - \frac{mx}{l}\right)\dot{x}_i + \left(\frac{my}{l}\right)\dot{y}_i + \left(\frac{mx}{l}\right)\dot{x}_j - \left(\frac{my}{l}\right)\dot{y}_j \\ \left(\frac{my}{l}\right)\dot{x}_i + \left(m - \frac{mx}{l}\right)\dot{y}_i + \left(\frac{my}{l}\right)\dot{x}_j - \left(\frac{mx}{l}\right)\dot{y}_j \end{bmatrix} \tag{8}$$

Then it is possible to calculate the total linear momentum of the mechanism as:

$$\mathbf{l} = \sum_{b=1}^{n} \mathbf{l}_b \tag{9}$$

where $\mathbf{l}_b$ is the linear momentum of a body and can be calculated using Eq. (8).

2.2 Angular Momentum Using Natural Coordinates

The angular momentum of the body, represented by a pair of masses at the basic points i and j, with respect to the general (inertial) reference frame can be calculated as:

$$\mathbf{h}_b = \mathbf{r}_i \times \mathbf{l}_i + \mathbf{r}_j \times \mathbf{l}_j \tag{10}$$

where $\mathbf{r}_i$ and $\mathbf{r}_j$, are the position vector of points i and j with respect to the general (inertial) reference frame.

So the angular momentum of the whole system can be easily calculated by

$$\mathbf{h} = \sum_{b=1}^{n} \mathbf{h}_b \tag{11}$$

where $\mathbf{h}_b$ is the angular momentum of a body that can be calculated using Eq. (10).

3 Balancing Conditions of a 2RRR Planar Parallel Manipulator (Five-Bar Mechanism)

As an application of the previous described method to the planar parallel manipulators consider a general 2RRR manipulator as the one shown in Fig. 2, that is modeled in Natural Coordinates. In this case body one is defined with points A and B, body two is defined with points B and C, body three is defined with points D and C, and body four is defined with points E and D. The origin of the local reference frames are also indicated, being at A, B, D and E respectively.

The total linear momentum of this mechanism can be obtained by calculating the linear momentum of each body and then applying the Eq. (9) to get:

$$\mathbf{l} = \begin{bmatrix} a\dot{x}_B + b\dot{y}_B + c\dot{x}_C + d\dot{y}_C + e\dot{x}_D + f\dot{y}_D \\ \\ a\dot{y}_B + b\dot{x}_B + c\dot{y}_C + d\dot{x}_C + e\dot{y}_D + f\dot{x}_D \end{bmatrix} \tag{12}$$

Fig. 2 A model of the 2RRR planar parallel manipulator using Natural Coordinates

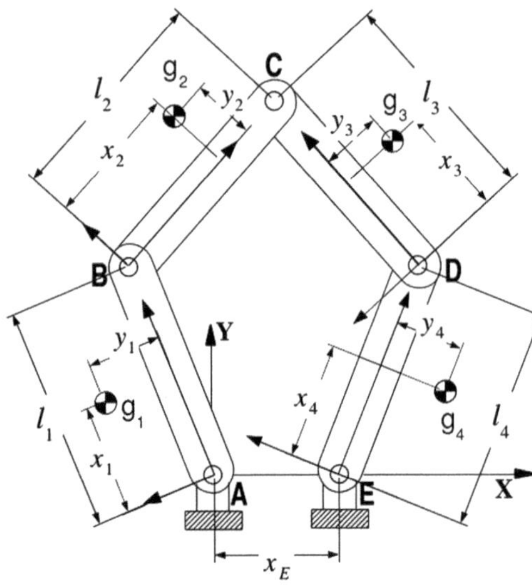

where:

$$a = \left(-\frac{m_2 x_2}{l_2} + \frac{m_1 x_1}{l_1} + m_2\right);\ b = \left(\frac{m_2 y_2}{l_2} - \frac{m_1 y_1}{l_1}\right);\ c = \left(\frac{m_3 x_3}{l_3} + \frac{m_2 x_2}{l_2}\right)$$
$$d = \left(-\frac{m_3 y_3}{l_3} - \frac{m_2 y_2}{l_2}\right);\ e = \left(\frac{m_4 x_4}{l_4} - \frac{m_3 x_3}{l_3} + m_3\right);\ f = \left(\frac{m_4 y_4}{l_4} - \frac{m_3 y_3}{l_3}\right) \tag{13}$$

All these expressions can be calculated with little effort using a computer algebra system. Equations (13) equated to zero are the force balancing conditions of the manipulator.

On the other hand the angular momentum of the mechanisms can be obtained using Eq. (11). In this case the angular momentum of the system is:

$$
\begin{aligned}
\mathbf{h} = & \left[\frac{I_1}{l_1^2} - \frac{m_2 x_2}{l_2} + m_2\right](\mathbf{r}_B \times \mathbf{v}_B) - \left(\frac{m_2 x_2}{l_2}\right)(\mathbf{r}_C \cdot \mathbf{v}_B) \\
& + \left[\frac{I_2}{l_2^2} - \frac{m_2 x_2}{l_2}\right](\mathbf{r}_B \times \mathbf{v}_B - \mathbf{r}_C \times \mathbf{v}_B - \mathbf{r}_B \times \mathbf{v}_C) + \left[\frac{I_3}{l_3^2} + \frac{I_2}{l_2^2}\right](\mathbf{r}_C \times \mathbf{v}_C) \\
& + \left[\frac{I_3}{l_3^2} - \frac{m_3 x_3}{l_3}\right](\mathbf{r}_C \times \mathbf{v}_D - \mathbf{r}_D \times \mathbf{v}_C - \mathbf{r}_D \times \mathbf{v}_D) + \left[\frac{I_4}{l_4^2} + m_3\right](\mathbf{r}_D \times \mathbf{v}_D) \\
& - \left[\frac{I_4}{l_4^2}\right](\mathbf{r}_D \times \mathbf{v}_D) + \left(\frac{m_3 y_3}{l_3}\right)(\mathbf{r}_B \cdot \mathbf{v}_C + \mathbf{r}_D \cdot \mathbf{v}_C - \mathbf{r}_C \cdot \mathbf{v}_D) \\
& + \left(\frac{h_4 m_4}{l_4}\right)(x_4 \dot{y}_4 + x_4 \dot{x}_4)
\end{aligned} \tag{14}
$$

Substituting the force balancing conditions in the previous equation it can be obtained the angular momentum of the force balanced 2RRR planar parallel manipulator as:

$$
\begin{aligned}
\mathbf{h} = & \left[\frac{I_1}{l_1^2} + m_2\right](\mathbf{r}_B \times \mathbf{v}_B) \\
& + \left[\frac{I_2}{l_2^2}\right](\mathbf{r}_B \times \mathbf{v}_B - \mathbf{r}_C \times \mathbf{v}_B - \mathbf{r}_B \times \mathbf{v}_C + \mathbf{r}_C \times \mathbf{v}_C) \\
& + \left[\frac{I_3}{l_3^2}\right](\mathbf{r}_C \times \mathbf{v}_C - \mathbf{r}_D \times \mathbf{v}_C - \mathbf{r}_C \times \mathbf{v}_D + \mathbf{r}_D \times \mathbf{v}_D) \\
& + \left[\frac{I_4}{l_4^2} + m_3\right](\mathbf{r}_D \times \mathbf{v}_D - \mathbf{r}_E \times \mathbf{v}_D)
\end{aligned} \tag{15}
$$

where I_i, $i = 1 \cdots 4$, is the moment of inertia of body i with respect to an axis passing through the origin of its local reference frame (not though the center of mass).

The angular momentum, Eq. (15), can be reorganized simplifying the corresponding cross products as:

$$
\begin{aligned}
\mathbf{h} = & \left[\frac{I_1}{l_1^2} + m_2\right](\mathbf{r}_B \times \mathbf{v}_B) + \left[\frac{I_2}{l_2^2}\right](\mathbf{r}_{CB} \times \mathbf{v}_{CB}) \\
& + \left[\frac{I_3}{l_3^2}\right](\mathbf{r}_{CD} \times \mathbf{v}_{CD}) + \left[\frac{I_4}{l_4^2} + m_3\right](\mathbf{r}_{DE} \times \mathbf{v}_D)
\end{aligned} \tag{16}
$$

where $\mathbf{r}_{ij}$ and $\mathbf{v}_{ij}$, are the position and velocity vectors of point i with respect to point j, respectively. This equation can in turn be reduced applying the triple product property of vectors, this is:

$$\mathbf{r}_B \times \mathbf{v}_B = \mathbf{r}_B \times (\omega_1 \times \mathbf{r}_B) = \omega_1(\mathbf{r}_B \cdot \mathbf{r}_B) - \mathbf{r}_B(\mathbf{r}_B \cdot \omega_1) = \omega_1 l_1^2 \mathbf{k}$$

$$\mathbf{r}_{CB} \times \mathbf{v}_{CB} = \mathbf{r}_{CB} \times (\omega_2 \times \mathbf{r}_{CB}) = \omega_2(\mathbf{r}_{CB} \cdot \mathbf{r}_{CB}) - \mathbf{r}_{CB}(\mathbf{r}_{CB} \cdot \omega_2) = \omega_2 l_2^2 \mathbf{k}$$

$$\mathbf{r}_{CD} \times \mathbf{v}_{CD} = \mathbf{r}_{CD} \times (\omega_3 \times \mathbf{r}_{CD}) = \omega_3(\mathbf{r}_{CD} \cdot \mathbf{r}_{CD}) - \mathbf{r}_{CD}(\mathbf{r}_{CD} \cdot \omega_3) = \omega_3 l_3^2 \mathbf{k}$$

$$\mathbf{r}_{DE} \times \mathbf{v}_D = \mathbf{r}_{DE} \times (\omega_4 \times \mathbf{r}_{DE}) = \omega_4(\mathbf{r}_{DE} \cdot \mathbf{r}_{DE}) - \mathbf{r}_{DE}(\mathbf{r}_{DE} \cdot \omega_4) = \omega_4 l_4^2 \mathbf{k}$$

then the final form of the angular momentum of the force balanced 2RRR planar parallel manipulator is:

$$\mathbf{h} = \left(I_1 + m_2 l_1^2\right)\omega_1 + (I_2)\omega_2 + (I_3)\omega_3 + \left(I_4 + m_3 l_4^2\right)\omega_4 \tag{17}$$

Equation (17) in turn can be taken to define the moment balancing condition of this mechanism as:

$$\begin{aligned}\mathbf{h} = {} & \left(I_1 + m_2 l_1^2\right)\omega_1 - k_1\omega_1\left(I_1'\right) \\ & + (I_2)\omega_2 - k_2\omega_2\left(I_2'\right) \\ & + (I_3)\omega_3 - k_3\omega_3\left(I_3'\right) \\ & + \left(I_4 + m_3 l_4^2\right)\omega_4 - k_4\omega_4\left(I_4'\right) = 0\end{aligned} \tag{18}$$

where I_i', $i = 1 \cdots 4$ are the four counter-inertias, see [19], or the four counter-rotary counterweights, see [6], rotating in opposite direction with respect to their associated body, and by a factor of k_i, $i = 1 \cdots 4$. In general a greater k gives a lesser thickness.

These additional rotary elements must be added to the system in order to get the constant (zero) angular momentum. A possible design solution of a force balanced manipulator with four counter-inertias can be seen in Fig. 3. This proposal has been designed using a general purpose system for the dynamics of multi-body systems. A similar result has been proposed in [19].

The proposed solution at Fig. 3 can be difficult to produce in practice, a substantial designing process has to be done. In addition this solution may be impractical but on the contrary it is very useful to show the application of the method and to understand all terms in the equations. An equivalent more efficient solution can be found using counter-rotary counterweights, see [6].

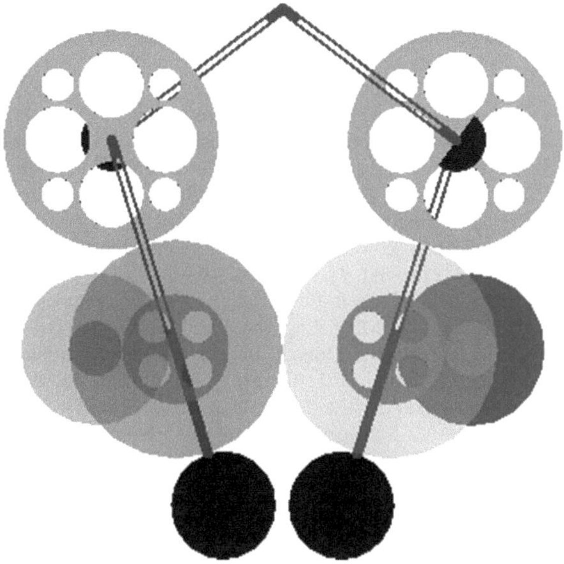

Fig. 3 A prototype of the dynamic balanced parallel manipulator

4 Conclusions

A novel efficient and easy to automate method to find the dynamic balancing conditions of planar parallel manipulators, based on the use of Natural Coordinates, has been introduced. The method is direct due to the fact that with the Natural Coordinates the whole system can be represented by a set of points, avoiding the explicit use angles (trigonometric expressions), facilitating the generation of the linear momentum and the angular momentum required to obtain the shaking force and the shaking moment balancing conditions. In this way these conditions are expressed entirely in terms of the location of the center of mass, the mass and the moment of inertia. Its application to the dynamic balancing of a 2RRR planar parallel manipulator is shown, proving that it can be obtained the same results as the obtained with other methods described elsewhere.

References

1. Briot, S., Arakelian, V.: Complete shaking force and shaking moment balancing of in-line four-bar linkages by adding a class-two RRR or RRP Assur group. Mech. Mach. Theory **57**, 13–26 (2012)
2. Erkaya, S.: Investigation of balancing problem for a planar mechanism using genetic algorithm. J. Mech. Sci. Technol. **27**(7), 2153–2160 (2013)
3. van der Wijk, V., Krut, S., Pierrot, F., Herder, J.L.: Design and experimental evaluation of a dynamically balanced redundant planar 4-RRR parallel manipulator. Intech Int. J. Adv. Robot. Syst. **23**, (2013)

4. Wang, K., Luo, M., Mei, T., Zhao, J., Cao, Y.: Dynamics analysis of a three-DOF planar serial-parallel mechanism for active dynamic balancing with respect to a given trajectory. Int. J. Robot. Res. **23**, 744–759 (2013)
5. Acevedo, M.: Active dynamic balancing of the redundantly actuated 2RR planar parallel manipulator. In: The 3rd Joint International Conference on Multibody System Dynamics and The 7th Asian Conference on Multibody Dynamics, Busan, Korea, June 30–July 3 (2014)
6. Acevedo, M.: Design of reactionless mechanisms with counter-rotary counter-masses. In: Zhang, D., Wei, B. (eds.) Dynamic Balancing of Mechanisms and Synthesizing of Parallel Robots. Springer, Berlin (in press 2015)
7. Berkoft, R.S., Lowen, G.G.: A new method for completely force balancing simple linkages. ASME J. Eng. Ind. **91**(1), 21–26 (1969)
8. Lowen, G.G., Berkoft, R.S.: Theory of shaking moment optimization of force-balanced four-bar linkages. ASME J. Eng. Ind. **93**(1), 53–60 (1971)
9. Merlet, J.P.: Parallel Robots, 2nd edn. Springer, Dordrecht (2006)
10. Wu, Y., Gosselin, C.M.: On the dynamic balancing of multi-dof parallel mechanisms with multiple legs. ASME J. Mech. Des. **129**(2), 234–238 (2006)
11. Arakelian, V., Dahan, M., and Smith, M.: A historical review of the evolution of the theory on balancing of mechanisms. In: Ceccarelli, M. (ed.) Proceedings HMM2000, International Symposium on History of Machines and Mechanisms—Cassino, pp 291–300. Springer, Berlin, Heidelberg, New York (2000)
12. Arakelian, V.G., Smith, M.R.: Shaking force and shaking moment balancing of mechanisms: a historical review with new examples. ASME J. Mech. Des. **127**, 334–339 (2005)
13. Arakelian, V.G., Smith, M.R.: Erratum: shaking force and shaking moment balancing of mechanisms: a historical review with new examples. ASME J. Mech. Des. **127**, 1035 (2005)
14. Dresig, H., Rockhausen, L., and Naake, S.: Balancing conditions for planar mechanisms, flexible mechanisms, dynamics and analysis. In: 22nd ASME Mechanism Conference, New York, DE, vol. 127, pp. 67–73 (1992)
15. García de Jalón, J., and Bayo, E.: Kinematic and dynamic simulation of multibody systems: the real-time challenge. Springer, Berlin Heidelberg, New York (1994)
16. Arakelian, V.G., Smith, M.R.: Design of planar 3-DOF 3-RRR reactionless parallel manipulators. Mechatronics **18**, 601–606 (2008)
17. van der Wijk, V., Herder, J.L.: Synthesis of dynamically balanced mechanisms by using counter-rotary countermass balanced double pendula. ASME J. Mech. Des. 131: 111003-1–111003-8 (2009)
18. Gosselin, C.: Gravity compensation, static balancing and dynamic balancing of parallel mechanisms. In: Wang, L., Xi, J. (eds.) Smart Devices and Machines for Advanced Manufacturing, pp. 27–48. Springer, London (2008)
19. Van Khang, N., Van Phong Dien, N., Van Son, P.: Balancing conditions of planar mechanisms with multi-degree of freedom. Vietnam J. Mech. **27–4**, 204–212 (2005)

Design and Characterization of a New 5-DOF Arc Welding Robot

T. Li, F.Y. Guo, M.Z. Luo, M. Ceccarelli, X. Liu, S.X. Chen and L. Fu

Abstract In this paper, a new 5-DOF arc welding robot is presented from aspects of mechanical design, control system and handheld welding teaching (HWT) method. Structure of this robot is inspired by selective compliance assembly robot arm (SCARA). With the aim to achieve HWT, magnet-clutches and encoders are adopted into the transmission chains which have large reduction ratio. In addition, a gravity balance mechanism is developed to reduce the load of operators. The processing of HWT operation is presented with optimal space arc interpolation algorithm. Two main design characteristics of this robot are low cost and easy operation features as compared to conventional 6-DOF industrial robots. Finally, experiments are discussed as from testing the welding performance and reported results validate the efficiency of the proposed solution.

T. Li (✉) · F.Y. Guo · M.Z. Luo · X. Liu · S.X. Chen · L. Fu
Hefei Institute of Physical Science Chinese Academy of Sciences,
Institute of Advanced Manufacturing Technology, Hefei, China
e-mail: roboylee@163.com

F.Y. Guo
e-mail: gfy2011@mail.ustc.edu.cn

M.Z. Luo
e-mail: lmz@iim.ac.cn

X. Liu
e-mail: liu2186@126.com

S.X. Chen
e-mail: chensx499796981@126.com

L. Fu
e-mail: longfu@iamt.ac.cn

F.Y. Guo
University of Science and Technology of China, Hefei, China

M. Ceccarelli
University of Cassino and South Latium, Cassino, Italy
e-mail: ceccarelli@unicas.it

S. Bai and M. Ceccarelli (eds.), *Recent Advances in Mechanism Design for Robotics*, Mechanisms and Machine Science 33,
DOI 10.1007/978-3-319-18126-4_7

Keywords Industrial robot · Welding robot · Robot design · Robot operation · Mechanism design

1 Introduction

As one kind of automatic welding devices, industrial robotic arms are widely used in nowadays industry and play an important role because of their high repeatability accuracy and stability of operations. Four main international robot suppliers ABB, KUKA, FANUC and YASKAWA occupy about 50 % of the market share. It is well known that their robotic arms are well used in high-end of welding industry [1], like auto industry.

Most of the industrial robotic arms in practical production are still on-line teaching and playback type [2, 3], which cannot meet efficiency requirement of various products. In order to make the robotic welding process easier and more effective, researchers proposed three main ways to solve such problems. One is using robot vision based on image processing, which recognizes the welding trajectory by vision sensors to guide the robot for arc welding [4]. The second one is adopting off-line programming based on computing graphics simulation in a virtual CAD-based environment [5]. Another one is using walk-through programming techniques with a wrist force/torque sensor [6]. These methods increase the cost or have higher requirements for the operators. This is a rather common and crucial aspect in a variety of different robotic tasks, especially for the small-medium enterprises, where mid-low size production does not allow such a costly and time consuming set-up. In other words, these robotic arms are not welcomed in medium and small enterprises, which representative the mid-low end of the welding markets, due to two main reasons. One is that these robotic arms are not easy to use, because they ask for operators with professional skills, like programming skill. The other one is their high cost. Generally, medium and small enterprises do not wish to hire operators with professional skills and buy those expensive robotic arms. Undoubtedly, this hinders the process of industrial automation [7]. But now the mid-low end of the welding markets has a more and more strong demands in automatic welding devices, due to the increasing of labor cost and few people want to be engaged in welding work, which is dirty, boring and harmful.

In this paper, a novel teaching function based on a physical interaction between human operator and the robot is proposed, which represents a step forward in making the programming phase fairly simpler, intuitive and faster, and even in promoting an increased autonomy and cognitive ability of the robotic system. Following are the main contributions of this paper: the proposed HWT (handheld welding teaching) programming function is fast and simple; by employing this equipment, the teaching process can be more efficient; under the premise of meet the requirements of mid-low end industrial application, cost saving has been considered seriously in the proposed process of designing mechanical structure and developing robot control system.

2 Mechanical Design

Based on a survey of local enterprises in Changzhou, which work on welding fabrication as mid-low end of welding market in China, we find that it is not necessary for them to use conventional 6-DOF robotic arm because nearly 90 % of their welding work products have simple welding seams and open space conditions. Usually the welding trajectories are line segments, circular arc or combines. Thus we carry out a conceptual design of a new welding robot as a SCARA inspired structure with 5 DOFs. A kinematic scheme containing DOF configuration of this structure is illustrated in Fig. 1 with each DOF driven by one stepper motor. Stepper motors are chosen because of their lower cost as compared with servo motors. In the first two DOFs, electromagnetic clutches are assembled between reducers and gear pairs. They are used to isolate reducer's resistance in order to make the operator able to drag the welding torch easily. In the other 3 DOFs, no clutches are used because the resistance is small and acceptable when dragging the welding torch. Encoders are used with the aim to record the displacement of each joint during the HWT process. Detailed explanation will be given in later paragraphs.

After carefully evaluating the working condition of the chosen representative enterprises, design specifications of the new welding robot are specified and they are listed in Table 1. Maximum speed of the 1st, 2nd, and 5th revolute joints is 60°/s, maximum speed of the 4th joint is 120°/s, and maximum speed of the translational joint 3 is 150 mm/s. Motion range of the 1st to 5th joints are 360°, 180°, 480 mm, ±180° and ±45°, respectively. Positioning accuracy and repeat positioning accuracy are 0.5 and 0.05 mm, respectively.

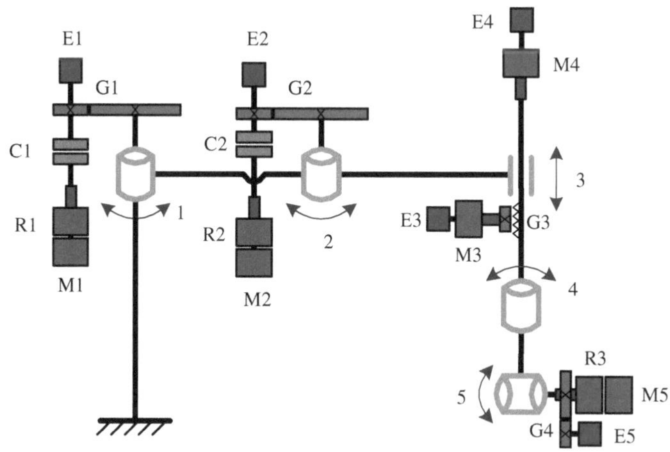

Fig. 1 DOF configuration of the proposed new arc welding robot (*M1–M2* servo motors; *M3–M5* stepper motors; *C1–C2* electromagnetic clutches; *R1–R3* reducers; *G1*, *G2* and *G4* gear pairs; *G3* rack-and-pinion; *E1–E5* encoders; *1*, *2*, *4* and *5* revolute joints; *3* translational joint)

Table 1 Design specifications of the welding robot (#: DOF)

1-# max speed	*2-# max speed*	*3-# max speed*	*4-# max speed*	*5-# max speed*
60°/s	60°/s	150 mm/s	120°/s	60°/s
1-# motion range	*2-# motion range*	*3-# motion range*	*4-# motion range*	*5-# motion range*
360°	180°	480 mm	±180°	±45°
Maximum turning radius	*Positioning accuracy*	*Repeat positioning accuracy*	*Position measurement*	*Weight*
1200 mm	0.5 mm	0.05 mm	Incremental encoder	270 kg

After the conceptual design and specifying the design parameters, mechanical design of this robot is worked out. The outward appearance of the whole welding robot system is shown in Fig. 2. The robot system mainly consists of two bases, a big arm, a small arm, a vertical arm, a gravity balance mechanism, a wire feeder, a welding torch, and a teach pendent. In addition, it also has a control cabinet and a welding power source, which are not shown in the figure.

Assuming that a workpiece is already fixed on a jog and fixture, a general operation procedure of the HWT can be described as: firstly, operator switch on the

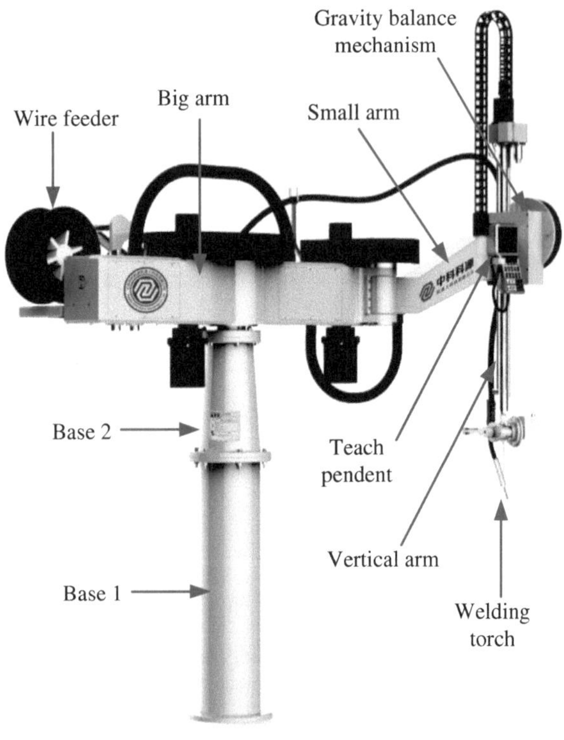

Fig. 2 The designed welding robot system (control cabinet and welding power are not shown)

power supply for the robot body, the wire feeder and the welding power source; the two electromagnetic clutches are switched off in order to disconnect the 1st and 2nd transmission chains; then the operator drag the welding torch with a proper gesture to a feature point of the seam and use the teach pendent to record the position of this point (position are recorded by encoders in the 5 robot joints); and so on other feature points are recorded in the same way; after that, the control system choose a proper interpolation algorithm to generate the trajectory of the seam; and then the electromagnetic clutches are switched on to connect the transmission chains; finally, the operator use the teach pendent to start the automatic welding program and the robot start to weld the workpiece.

In particular, joints 1, 2 and 3 are used for locating the welding torch in 3D space, and joints 4 and 5 are used for adjusting the gesture of the welding torch. Electromagnetic clutches and gravity balance mechanism are developed especially for HWT process. In particular, electromagnetic clutches are used for eliminating the 1st and 2nd reducer's resistance during HWT stage, whereas the gravity balance mechanism is used for balancing the gravity of the vertical arm. Because the vertical arm is about 20 kg, without the mechanism, the operator cannot lift the vertical arm moving up and down to execute HWT. With the help of them, operator can easily drag the welding torch along with the robot body to those determined feature points on a seam under welding.

Figure 3 gives a section view of the electromagnetic clutch assembling. The electromagnetic clutch is composed of a driving part and a flying wheel. The driving part is fixed to the reducer shaft, whereas the flying wheel is fixed to the flange of the output shaft. A thrust bearing is used for keeping the working gap within a specified range. In addition, a double-gear group is used in order to decrease the backlash of the gear pairs.

The gravity balance mechanism at the end of the SCARA arm is composed of a gear box and a turbination spring box. Its detailed structure is shown in Fig. 4. In Fig. 4a, the gear box is fixed to the end of the small arm. Inside the box, there are a small gear and a big gear which are fixed on a motor driving shaft and a turbination spring shaft, respectively. Motor 3 is installed on its frame at the left side of the

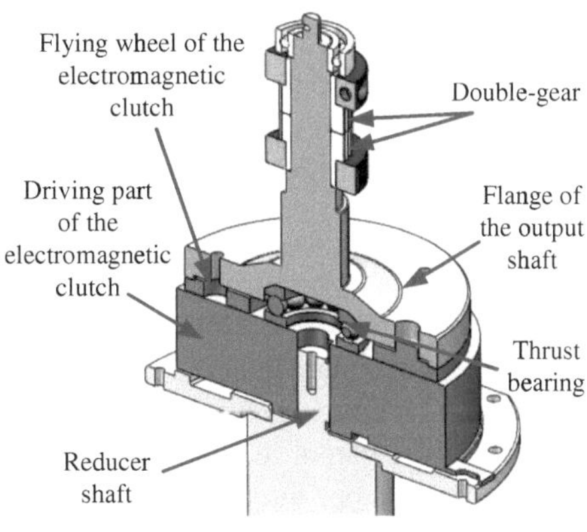

Fig. 3 A section view of the electromagnetic clutch and the double-gear assembling

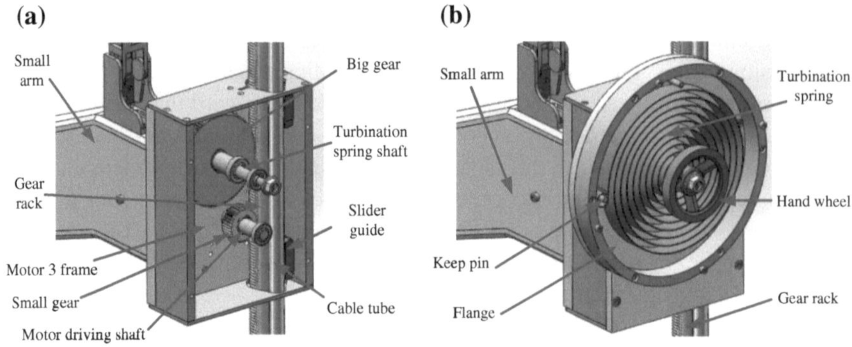

Fig. 4 Views of the structure of the gravity balance mechanism: **a** the gear box; **b** the turbination spring box

driving shaft. The two gears form two rack-and-pinion pairs. A slider guide is fixed to the back side of the gear rack. In addition, a cable tube is also fixed to the gear rack. They form the vertical arm, which can move up and down along two sliders assembled on the right board of the gear box. In Fig. 4b, the inner end of the turbination spring is fixed to the spring shaft, and the outer end is fixed to a keep pin, which is fixed to a flange. Besides, a hand wheel, which is used for moving the vertical arm, is fixed to the right end of the spring shaft.

Working principle of the gravity balance mechanism can be described as follows: The motor drives the small gear to move the vertical arm through the small rack-and-pinion pair, then the vertical arm drives the big gear through the big rack-and-pinion pair, and simultaneously makes the turbination spring to rotate. It is worth to note that first the turbination spring needs an proper initial tightening when assembled into the mechanism in order to provide a suitable reaction force to balance the gravity of the vertical arm; and use a small gear and a big gear are used for making that the big gear rotate less than the small gear with the aim to provide relatively constant reaction force of the spring when the vertical arm moves within its motion range. Benefits of the gravity balance mechanisms can be identified in two aspects: First as mentioned above, it can reduce the load on human's hand in the process of HWT, thus can let the operator move the torch easily and flexibly to a desired point. Then it can prevent possible collisions of the vertical arm against the part under welding, when there is a sudden power-off.

The relationship between rotation angle φ and torque T can be expressed as

$$\varphi = \frac{12K_1 Tl}{Ebh^3} \tag{1}$$

where E is the Young material elastic modulus, h is the thickness of the spring, b is the width of the spring, T is the spring torque and l is the spring length. Because of Eq. (1) for the output force as function of the rotation angel of the spring, rotation angle φ and output torque T has a nearly linear characteristic. In the design, the length of the coil spring is ten times the length of the needed length, and the spring balance position is set at the center of the vertical stroke of the spring motion. The mechanical design of the proposed new solution is based on the mechanisms for the electromagnetic clutch in Fig. 3 and gravity balance system in Fig. 4.

3 Handheld Welding Teaching (HWT) Method

The control system of the proposed welding robot contains teach pendent, control equipment, stepper motors and absolute encoders. The control equipment works as based on the optimal arc interpolation algorithm and offline motion generation. In general, traditional online teaching function is time consuming and inefficient. In order to operate the proposed HWT efficiently, a teaching storage is built as based on workpiece. The HWT procedure is illustrated in Fig. 5. In HWT process, the only thing that the operator needs to do is to record positions of the feature points. It costs about 2 s for the operator to record each feature point. After the points are recorded, the robot will program automatically. In addition, the welding quality is usually better than manual welding in the mid-low end of the welding market. Thus the performance of the HWT method is fairly satisfactory.

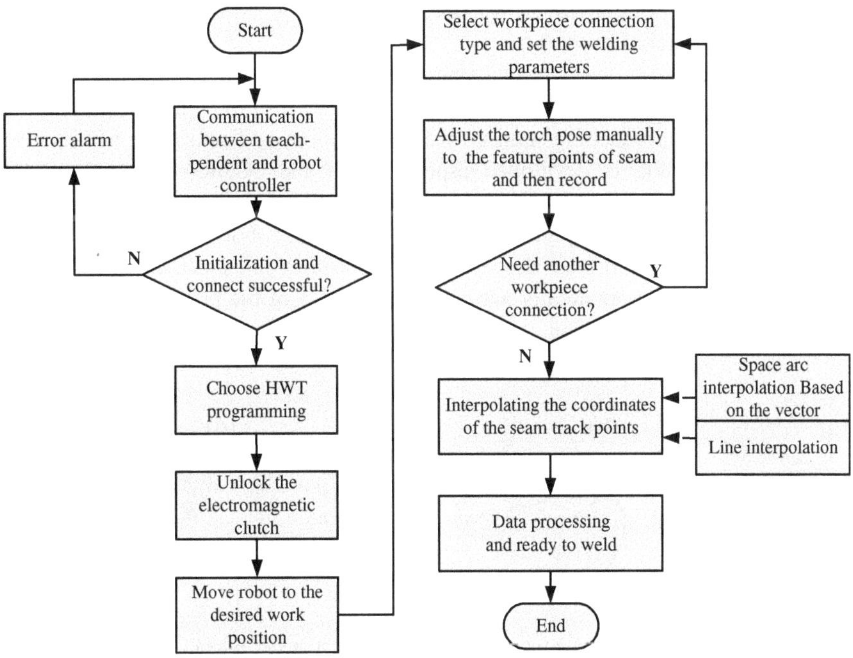

Fig. 5 A flowchart of the HWT procedure for the welding robot in Fig. 2

Short explanation of steps in the flowchart can be outlined as follows:

First to start the system: Teach-pendent, robot controller and motor driver are the core device, after start the robot, it will initialize the device and check their communication. If there's any error, the robot will alarm automatically and display the error (if it has) on the teach-box.

Then to do HWT: After initialization, the operator can choose HWT function and disconnect the clutch, move torch to the workpiece by hand. Choose workpiece type, welding parameters, set the interpolation mode, adjust the torch pose by hand, and then record the key points of seam and normal track. If there's other workpiece, the operator can repeat the HWT operation.

Then to Weld: After record the entire welding trajectory and interpolating the coordinates of the trajectory track points in the 3D reconstruction is done, welding is ready to begin.

Mapping the design features to machining features is the basis of robot navigation. Each group of designed feature mapping of the workpieces is defined in the form of D = {C, P, R}, where C is the setting of connection type, whose examples are shown in Fig. 6; P is defined in the form of P = (seam(x, y, z), normal(x, y, z), M), where seam(x, y, z) is the sequential setting of the seams and normal(x, y, z) contains the start point, end point and also the other points without welding; x, y and z are the points' position in the Cartesian coordinate; M = {Vx, V, I, L} denotes the requirements of welding method, Vx is the motion velocity, V is the welding voltage, I is the welding current and L is the welding number. R denotes the requirements of welding method, R = {w, t, r}, where w is the seam width, t is the depth, r is the material property.

Figure 7 shows an example of using the path generation via HWT. After the HWT computation, the robot system has all the information for performing sequentially arm approaching motion, welding motion, and returning arm motion. The HWT procedure is developed especially for the SCARA inspired 5-DOF welding robot, which is totally different from traditional 6-DOF industrial robots. Thus it is not possible to update a part of the existing robot program when using HWT method. As mentioned above, the operator does not need to program. The robot can program automatically and it cost about 1/5 of the programming time of the traditional industrial robot.

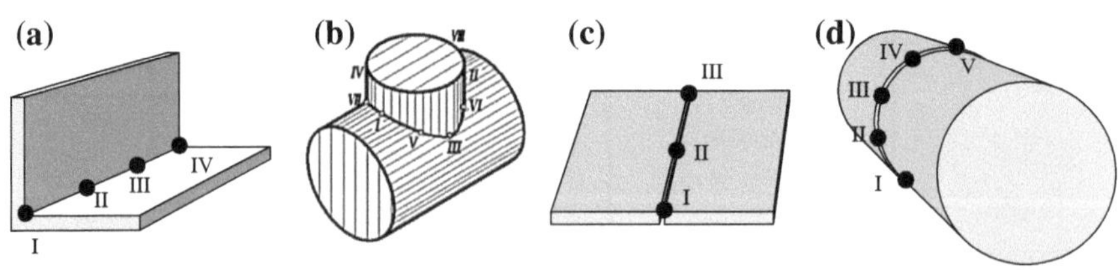

Fig. 6 Different connection types for arc welding as: **a** right angle; **b** tubular; **c** plate; **d** arc

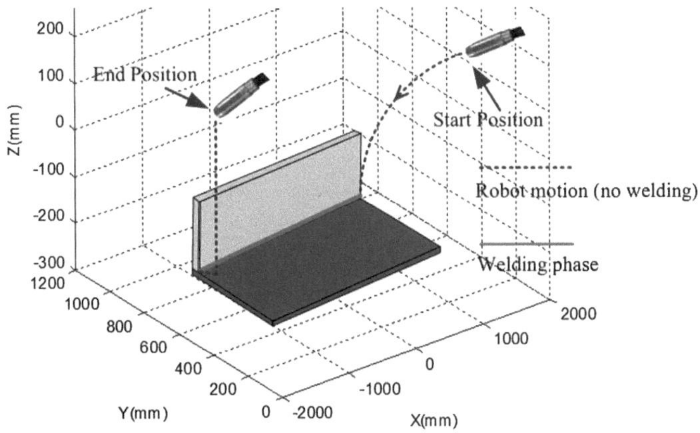

Fig. 7 An example of the trajectory of torch when welding the right angle connection as computed by HWT

4 Experiments

A scheme of the experimental layout is shown in Fig. 8 with 4 parts: shielding gas source, wire feeder, digital welding power source and welding robots system. With the experiment platform, welding of different type of typical workpieces have been carried out. Figure 9 shows an example of using HWT procedure and the experimental results are reported in photos of Fig. 10 for tests, which indicate a very satisfactory capability of the new proposed welding robot system.

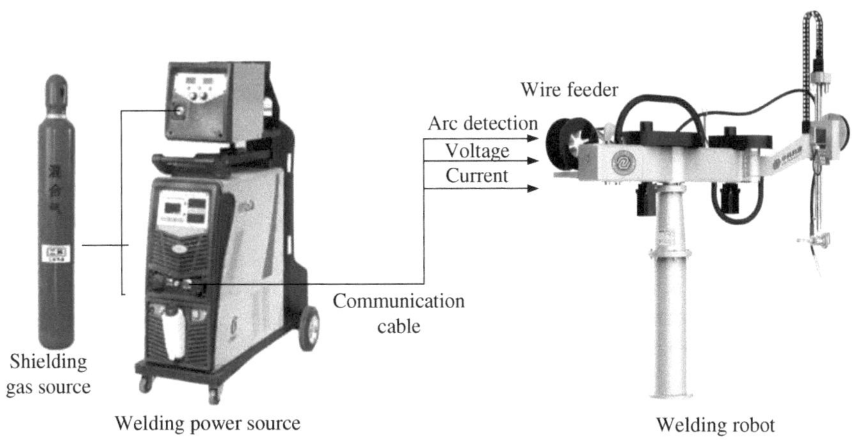

Fig. 8 An experimental layout

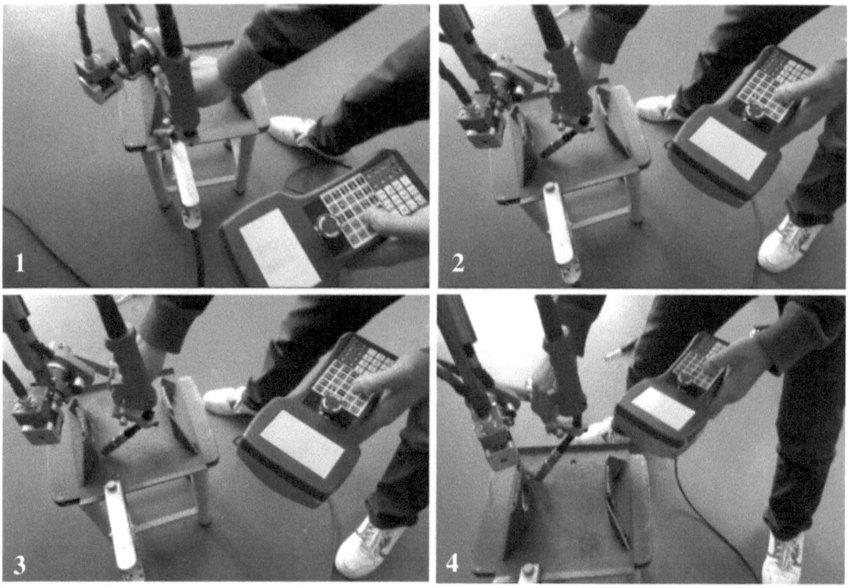

Fig. 9 An experimental test using HWT procedure in a welding case

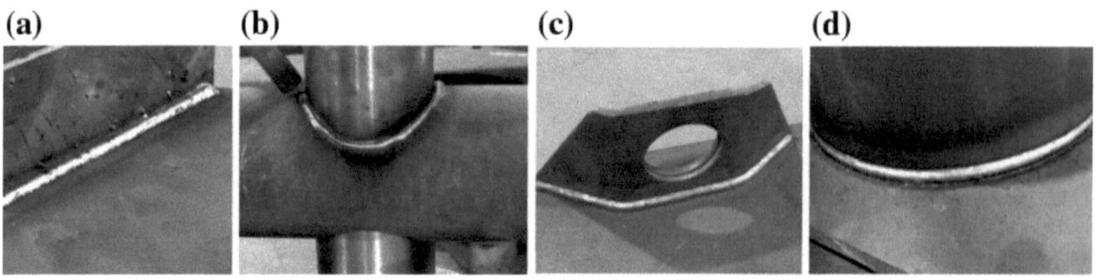

Fig. 10 Test results of welding by using HWT procedure for the cases in Fig. 9

5 Conclusions

A new welding robot system is presented in this paper. Main low-cost easy-operation features are implemented in a SCARA structure design for the robot body and developing of a specific HWT procedure. As a novel teaching function, HWT is based on a physical interaction between human operator and robot, and it represents a step forward in making the programming phase fairly simpler, intuitive and faster as suitable for industrial operator. The welding trajectory can be very clearly demonstrated because of the feature mapping from workpiece to robot. The application of this robot can effectively improve the competitiveness of small and medium-sized enterprises in welding industry.

Acknowledgments The presented results are obtained within a project (No. BY2014048) under the support of Jiangsu Prospective Joint Research Project.

References

1. Bolmsjo, G., Olsson, M., Cederberg, P.: Robotic arc welding trends and developments for higher autonomy. Ind. Robot Intern. J. **29**(2), 98–104 (2002)
2. Bascetta, L., Ferretti, G., Magnani, G., Rocco, P.: Walk-through programming for robotic manipulators based on admittance control. Robotica **31**, 1143–1153 (2013)
3. Chen, S.B., Qiu, T., Lin, T., Wu, L., Tian, J.S., Lv, W.X.: Intelligent technologies for robotic welding, Robotic Welding. Lect. Notes Control Inf. Sci. **299**, 123–143 (2004)
4. Kim, J.Y.: CAD-based automated robot programming in adhesive spray systems for shoe outsoles and uppers. J. Robot. Syst. **21**(11), 625–634 (2004)
5. Fridenfalk, M., Bolmsjo, G.: Design and validation of a universal 6D seam tracking system in robotic welding based on laser scanning. Ind. Robot Intern. J. **30**(5), 437–448 (2003)
6. Mitsi, S., Bouzakis, K.D., Mansour, G., Sagris, D., Maliaris, G.: Off-line programming of an industrial robot for manufacturing. Int. J. Adv. Manuf. Technol. **26**(3), 262–267 (2005)
7. Deb, K., Pratap, A., Agarwal, S., Meyarivan, T, A fast and elitist multiobjective genetic algorithm: NSGA-II. IEEE Trans. Evolut. Comput. **6**(2), 182–189 (2002)

Optimization of Grinding Parameters for Twist Drill in Biglide Parallel Machine

M.I. Kim and Ping Zou

Abstract In view of the Biglide parallel machine, this paper presented the theoretical and experimental basis for grinding the designated geometrical parameters of twist drill based on optimizing of the grinding parameters. The kinematics structure of the Biglide parallel machine was firstly introduced for grinding of the twist drill, and then the mathematical model of the twist drill flank was derived. And the relations between the geometric parameters of the drill point and grinding parameters of the grinding machine were analyzed, through which the optimal grinding parameters were obtained by using genetic algorithm. It was verified that the optimized grinding parameters improved the grinding precision,depending on the customers' demands.

Keywords Biglide parallel machine · Twist drill · Grinding of drill point · Grinding parameters

1 Introduction

Within the recent years, the parallel mechanisms have attracted many researchers' and machine manufacturers' attractions as machine tools because of their conceptual potentials in high motion dynamics and accuracy combined with high structural rigidity [1, 2].

Due to its high precision and versatility for various machining, PKM became increasingly attractive for grinding of twist drill-complex geometry of cutting tool, using the grinding system based on the Stewart platform [3]. But because of its

M.I. Kim · P. Zou (✉)
School of Mechanical Engineering & Automation,
Northeastern University, Boston, People's Republic of China
e-mail: pzou@mail.neu.edu.cn

M.I. Kim
e-mail: mingsun819@163.com

S. Bai and M. Ceccarelli (eds.), *Recent Advances in Mechanism Design for Robotics*, Mechanisms and Machine Science 33,
DOI 10.1007/978-3-319-18126-4_8

disadvantages such as small workspace and complex foreword kinematics, many researchers paid many attentions to PKM with limited degrees of freedom [4]. Using the fluted drill for grinding work piece in many cases, its grinding movement is relatively simple, its process does not need a big workspace and its grinding force is relatively small, therefore, the Biglide parallel machine can be reasonable to grind twist drill in high precision by only using the grinding parameters [5, 6].

The geometry of twist drill is more complex than the other cutting tool, so many modeling methods and numeric approximation approaches have been reported [7, 8]. However, it is still remained as a difficult problem to grind out the required geometry of twist drill along with the demand of users for improving its cutting performance [5, 9, 10].

In this paper, the grinding system of twist drill is constructed using parallel kinematics machine and the modeling of twist drill flank is made based on grinding parameters. For the practical grinding of twist drill, those key geometric parameters are analyzed and optimized based on the grinding parameters.

Optimizing the grinding parameters based on genetic algorithm, the theoretical basis is obtained for grinding of the high-precision twist drill flank according to the user's needs of the twist drill geometry.

2 Kinematics of Biglide Parallel Machine

The structure of the Biglide parallel machine, as shown Fig. 1, has three translations, which is comprised of a certain length of biglides, coupling with the moving platform and the slide units by revolute joint. The work table moves up and down by the unit of worm and worm wheel. The linear movement of each slider is provided with the ball screw and actuated by a stepping motor.

The velocity of the drill point can be decomposed into v_x and v_z, according to the relationship as shown in Fig. 2; they are given as:

$$v_x = \frac{v}{2},\ v_z = \frac{v(L_1 - vt - 2r_0)}{2\sqrt{4L_2^2 - (L_1 - vt - 2r_0)^2}} \tag{1}$$

The motion of the drill point in z-axis generates acceleration, however, the value of the acceleration is very small (about 0.008 mm/s^2) and the grinding time is very short (2 s), thus its motion can be linearized.

In order to guarantee the whole surface of the drill flank to be ground, the angle between the drill axis and the moving direction of the drill point in grinding should not be larger than the angle between the grinding wheel axis and the drill axis.

$$\tan^{-1}\left(\frac{v_x}{v_z}\right) \leq \theta \tag{2}$$

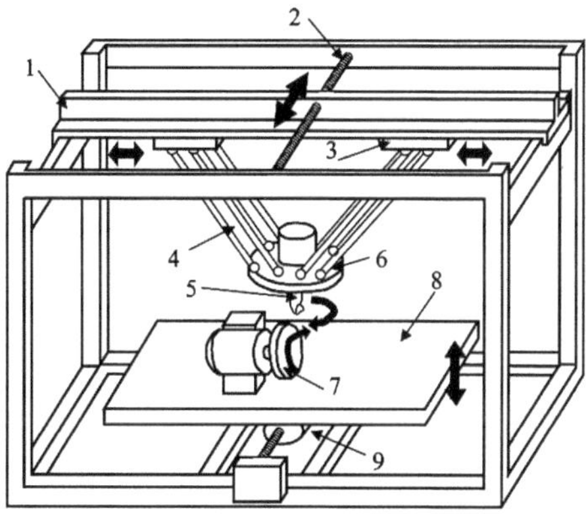

Fig. 1 Structure chart of Biglide parallel machine: *1* sliding carriage, *2* ball screw, *3* slider, *4* rod, *5* drill, *6* moving platform, *7* grinding wheel, *8* work table, *9* worm gearing

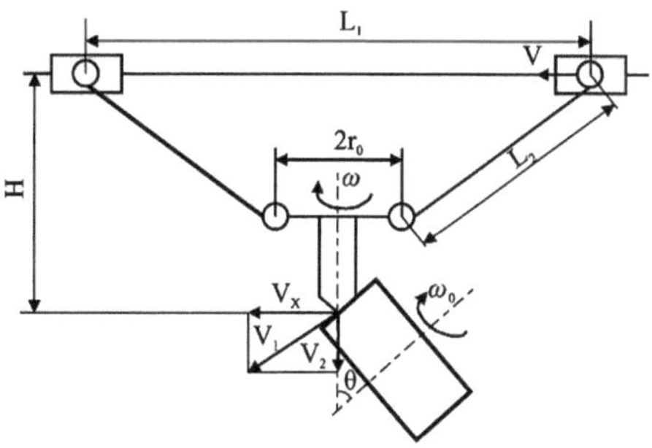

Fig. 2 Schematic movement Biglide parallel machine

3 Modeling of Twist Drill Flank

The drill point geometry is uniquely configured with the drill flank and of the flute. The flute feature being designed by the manufacturer, the drill flank one becomes the principal factor for developing new drill point geometry, as well as improving the grinding accuracy and efficiency for the drill point concerned by its user and manufacturer.

As shown in Fig. 3, the moving coordinate system $S_2(x_2, y_2, z_2)$ performs a screw motion with respect to global coordinate system $S(x, y, z)$ and together with a translation along negative X-direction. The angle of rotation and axial displacements in motion are described with φ, $k_z\varphi$ and $k_x\varphi$, respectively.

Here k_x and k_z are given by:

$$k_z = \frac{v_z}{\omega},\; k_x = \frac{v_x}{\omega} \tag{3}$$

When angle of drill revolution is φ, an arbitrary point on the intersection curve of the drill flank and the grinding wheel is expressed as $P(x_p, y_p, z_p)$ in coordinate

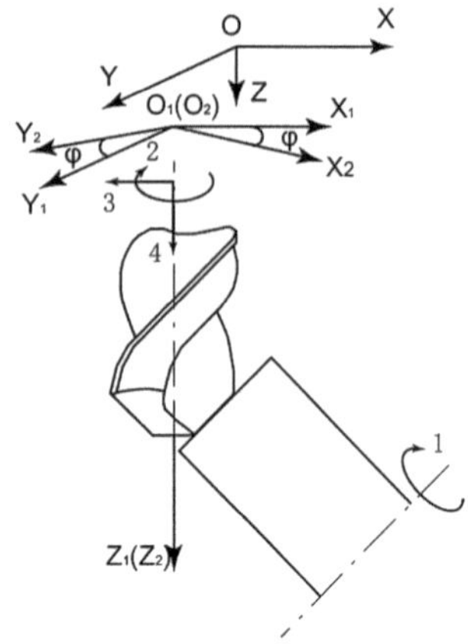

Fig. 3 Coordinate system

system S_2. Then, in coordinate system S, the ground flank surface can be represented as following.

$$\begin{bmatrix} x \\ y \\ z \\ 1 \end{bmatrix} = \begin{bmatrix} \cos\varphi & -\sin\varphi & 0 & -k_x\varphi \\ \sin\varphi & \cos\varphi & 0 & 0 \\ 0 & 0 & 1 & k_z\varphi \\ 0 & 0 & 0 & 1 \end{bmatrix} \begin{bmatrix} x_p \\ y_p \\ z_p \\ 1 \end{bmatrix} \quad (4)$$

$$\begin{cases} x = x_p \cos\varphi - y_p \sin\varphi - k_x\varphi \\ y = x_p \sin\varphi + y_p \cos\varphi \\ z = z_p \cot\varphi + k_z\varphi \end{cases} \quad (5)$$

4 Parametric Model of Twist Drill

In this section, three important geometric parameters for cutting force, cutting temperature and durability are represented in terms of grinding ones such as relief angle, semi-point angle and chisel angle, one by one.

As shown Fig. 4, the relief angle in an arbitrary point A on cutting edge is defined as the angle between the drill flank surface and the drill point end plane, measured on the section tangent to cylinder with radius r_a crossing the point A.

Generally, the relief angle means the one of the outside corner C.

Setting the coordinate system $O_a(x_a, y_a, z_a)$ with x_a axis passing the point A, the relief angle of its location is expressed as following.

$$\tan\alpha_{fa} = \left[\frac{dz_a}{dy_a}\right]_a \quad (6)$$

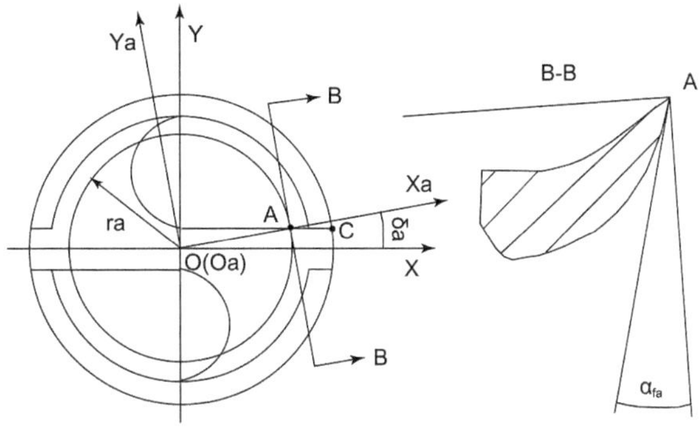

Fig. 4 Schematic figure of relief angle

The relationship equation of the coordinate system $O_a(x_a, y_a, z_a)$ and $O(x, y, z)$ is expressed as

$$\begin{bmatrix} x_a \\ y_a \\ z_a \end{bmatrix} = \begin{bmatrix} \cos\delta_a & \sin\delta_a & 0 \\ -\sin\delta_a & \cos\delta_a & 0 \\ 0 & 0 & 1 \end{bmatrix} \begin{bmatrix} x \\ y \\ z \end{bmatrix} \tag{7}$$

According to Eqs. (5) and (7), the relief angle of the drill point is given as

$$\tan\alpha_{fc} = \frac{\frac{v}{2}\left(\frac{L_1}{2} - r_0\right)}{\left(\omega r_c \cos\delta_c - \frac{v}{2}\sin\delta_c\right)\sqrt{L_2^2 - \left(\frac{L_1}{2} - r_0\right)^2}} \tag{8}$$

The semi-point angle is defined as the one on the outer corner point C, is expressed as follow.

$$\tan\rho = \left[\frac{dz}{dx}\right]_C \tag{9}$$

According to Eq. (5), the semi-point angle of the drill point is given as

$$\rho = \theta \tag{10}$$

According to Fig. 5, the chisel edge angle ψ can be expressed as follows.

$$\tan\psi = \left[\frac{dy}{dx}\right]_O \tag{11}$$

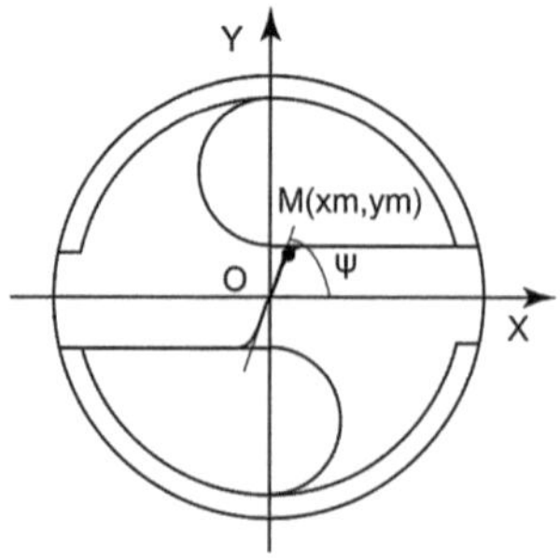

Fig. 5 Schematic figure of chisel edge angle

Considering the point M forms the profile of the chisel edge and the projected curve of the one to X axis is very small, the chisel edge angle can be simply expressed as the bevel on the point M instead of the one on the point O.

$$\tan\psi = \frac{2\omega r_m}{v} \tag{12}$$

5 Optimizing of Grinding Parameters

Compared with the other traditional optimization methods, it is known that genetic algorithm has the obvious advantages in solving nonlinear and multi-objective optimization problems.

The objective function and constraint are discussed below:

Objective function:

The objective is to maximize the fitness function as given by Eq. (13).

$$F(L_1, v, \omega) = \frac{1}{D} \tag{13}$$

$$D = \eta_\alpha |\alpha_{fc} - \alpha_0| + \eta_\psi |\psi - \psi_0| \tag{14}$$

$$\eta_\alpha + \eta_\psi = 1 \tag{15}$$

α_{fc} and ψ are get in Eqs. (8) and (9), respectively.

α_0 and ψ_0 are specified by users.

Constraint:

(a) Structural constraint
400 mm < L_1 < 1100 mm, 0 < v < 20 mm/s, 0 < ω < 15 rad/s

(b) Grinding constraint

$$\frac{\sqrt{4L_2^2 - (L_1 - vt - 2r_0)^2}}{(L_1 - vt - 2r_0)} \leq \tan\theta \tag{16}$$

6 Application of the Proposed Method

For the comparison of grinding results before and after optimization, two drill blanks are selected with the same material and dimension.

Drill parameters: radius 3 mm, material k20 (tungsten carbide), helix angle 30°, initial semi-point angle 59°, initial chisel edge angle 50°, $\delta_c = 1.66°$.

Grinding wheel parameters: radius 125 mm, width 16 mm (cylindrical grinder), power 180 w, normal revolutions 2800 r/min.

Machine parameters: $L_2 = 500$ mm, $\theta = 60°$, $r_0 = 65$ mm

Grinding parameters before optimization: $L_1 = 830$ mm, $v = 2.5$ mm/s, $\omega = 1.57$ rad/s

Grinding parameters after optimization: $L_1 = 965$ mm, $v = 2.86$ mm/s, $\omega = 2.52$ rad/s

For optimization of the grinding parameters, the following parameters are used.

Computational parameters: population size 100, maximum evolution 200, selection probability 0.5, crossover probability 0.8, mutation probability 0.05.

Target values: relief angle $\alpha_0 = 16°$, chisel edge angle $\psi_0 = 53°$, semi-point angle $\rho = 60°$,

Weight coefficient: $\eta_\alpha = 0.8$, $\eta_\psi = 0.2$

After grinding of drill point on the Biglide parallel machine as shown Fig. 1, the relief angle is measured with dial indicator installed on it.

Its measurement principle is shown in Fig. 6, r means the distance from the drill axis to contacting point of the probe and the drill flank, δ is the rotational angle

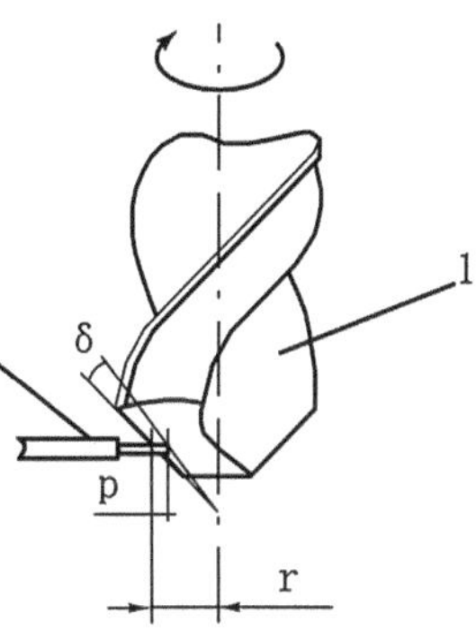

Fig. 6 Cylindrical surface measurement schematic. *1* Drill, *2* probe of dial indicator

of drill point, and p means the changed reading value of dial indicator while the drill point rotates up to δ.

$$\tan \alpha_{fc} = \frac{p}{r\delta} \tag{17}$$

The measurement is repeated ten times for each ground drill and their average value is get as measurement value. The measurement of chisel edge angel is conducted by processing image of ground drill.

The measurement values are below for drill ground before/after optimizing.

Before optimizing: α_{mfc} = 18.5° (error 15.6 %), ψ_m = 46.5° (error 12.3 %)

After optimizing: α_{mfc} = 16.3° (error 1.9 %), ψ_m = 53.5° (error 1 %)

The original drill, ground drill before/after optimizing are shown in Fig. 7, it is relatively clear that the optimized grinding parameters improve the grinding performance for twist drill on Biglide parallel machine.

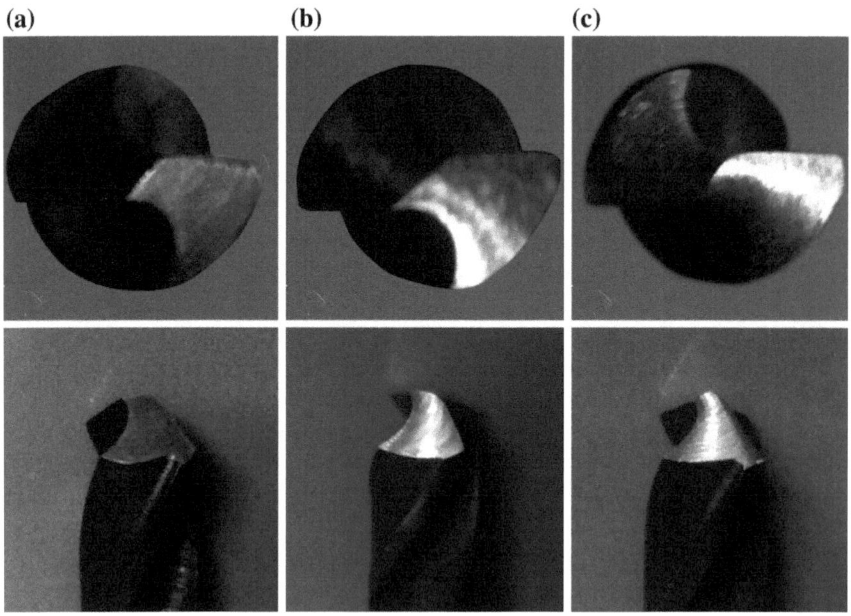

Fig. 7 Compartment of drill points. **a** Original. **b** Ground before optimizing. **c** Ground after optimizing

7 Conclusions

(1) The grinding of twist drill flank is easily achieved by using Biglide parallel machine, meeting the requirements of the grinding precision and the users.

The only two movements-translational motion of the slider and rotational motion of the grinding wheel can form the needed ones for grinding of the twist drill in the Biglide parallel machine.

(2) It is verified optimized grinding parameters based on generic algorithm help the users with improving the grinding precision of the twist drill.

This work can be further extended to the analysis and design of the reasonable geometry of the twist drill under the different cutting conditions.

References

1. Tlusty, J., Ziegert, J., Ridgeway, S.: Fundamental comparison of the use of serial and parallel kinematics for machine tools. Ann. CIRP **48**(1), 351–356 (1999)
2. Weck, M., Staimer, D.: Parallel kinematic machine tools-current state and future potentials. CIRP Ann.-Manuf. Technol. 51(2), 671–683 (2002)
3. Chen, W.Y., Chen, D.C.: Kinematic equations for a virtual axis drill point grinder. Chin. J. Mech. Eng. **34**(3), 46–50 (1998)
4. Li, B., Hu, X., Wang, H.: Analysis and simulation for a parallel drill point grinder part 1-kinematics, workspace and singularity analysis. Int. J. Adv. Manuf. Technol. **31**(9–10), 915–925 (2007)
5. Zou, P., Yang, X., Ai, M.: Mingze ai.: study on twist drill grinding with a Biglide parallel grinder. Adv. Mater. Res. **97–101**(3), 2119–2122 (2010)
6. Olarra, A., Allen, J.M., Axinte, D.A.: Experimental evaluation of a special purpose miniature machine tool with parallel kinematics architecture-Free leg hexapod. Precis. Eng. **2**, 1–16 (2014)
7. Hsieh, J.-F.: Mathematical model for twist drill. Int. J. Mach. Tools Manuf. **45**(7–8), 967–977 (2005)
8. Yan, L., Jiang, F.: A practical optimization design of helical geometry drill point and its grinding process. Int. J. Adv. Manuf. Technol. **64**(9–12), 1387–1394 (2013)
9. Hsieh, J.-F.: Mathematical modeling of complex helical dill point. J. Manuf. Sci. Eng. **131**(6), 0610061–06100611 (2009)
10. Karabay, S.: Analysis of drill dynamometer with octagonal ring type transducers for monitoring of cutting forces in drilling and allied process. Mater. Des. **28**(2), 673–685 (2007)

Kinematics and Singularity Analysis of a 3-RRR Planar Hybrid Mechanism

Ruiqin Li, Shuiqin He and Yaohong Zhao

Abstract This paper focuses on the kinematics of a 3-RRR planar hybrid mechanism, which consists of serial and parallel kinematic chains. The configuration design is analyzed. The forward position and the inverse kinematic problems are formulated. The kinematic problem is solved by applying Chaos Particle Swarm Optimization. Moreover, three kinds of singularities are analyzed. The results play an important role in practical application for the 3-RRR hybrid mechanism.

Keywords 3-RRR hybrid mechanism · Forward positional analysis · Inverse positional analysis · Singularity

1 Introduction

A mechanism can be designed with series, parallel or hybrid configuration. In hybrid configuration, a mechanism combines both series and parallel kinematic chains. Hybrid mechanisms thus have a potentiality to take advantages of series mechanism and parallel mechanism in structure and performance.

Intensive research of hybrid mechanisms has been conducted and reported in literature. Coppola et al. [1] studied a 6-DOF (Degree Of Freedom for short) hybrid mechanism with analysis of the workspace, singularity and stiffness performances. Alexandre et al. [2] developed a method for the analysis of the hybrid mechanism. Altuzarra et al. [3] proposed a novel hybrid mechanism for a five-axis machine and analyzed its kinematics and dynamics. Guo et al. [4] introduced a new hybrid mechanism with reduced DOF and nested dynamic connection. The DOF of the hybrid mechanism was calculated by using screw theory and equivalent DOF approach. Positive displacement solution and workspace were solved by using

R. Li (✉) · S. He · Y. Zhao
School of Mechanical and Power Engineering, North University of China, Taiyuan 030051, China
e-mail: liruiqin@nuc.edu.cn

S. Bai and M. Ceccarelli (eds.), *Recent Advances in Mechanism Design for Robotics*, Mechanisms and Machine Science 33,
DOI 10.1007/978-3-319-18126-4_9

numerical methods. Shen et al. [5] developed a 3-DOF hybrid mechanism platform. The inverse position of the output end of the platform was solved as a closed form solution by using matrix calculations.

This paper presents a kind of 3-RRR planar hybrid mechanism. Firstly, the configuration is analyzed. Then, the forward position and inverse position are solved by using a particle swarm algorithm. Finally, three kinds of singularities are analyzed.

2 Configuration of the 3-RRR Hybrid Mechanism

The traditional 3-DOF planar parallel mechanism is comprised of a moving platform, a static platform and three branched chains. Merlet [6] listed most possible configurations of planar parallel mechanisms. Noting that these configurations were limited to planar 3-DOF parallel mechanisms with three legs, there are still other types of mechanisms not listed.

Figure 1 shows the configuration of a 3-RRR planar hybrid mechanism. The hybrid mechanism is comprised of a static platform $O_1O_2O_3$, a moving platform $B_1(B_2)B_3$, and three limbs $O_1A_1B_1$, $O_2A_2B_2$, $O_3A_3B_3$. Each limb consists of two links and three revolute pairs. The limb $O_1A_1B_1$ and the limb $O_2A_2B_2$ intersect at one point $B_1(B_2)$ to form a series configuration. This series configuration and another limb $O_3A_3B_3$ form the parallel configuration. Three side links O_1A_1, O_2A_2, O_3A_3 are the driving links. The moving platform, i.e., the link $B_1(B_2)B_3$ is the output link.

3 Forward and Inverse Position Analysis

3.1 *Coordinate Systems and Constraint Equations of the 3-RRR Hybrid Mechanism*

As shown in Fig. 1, the static coordinate system O-xy is connected to the static platform $O_1O_2O_3$. The coordinate origin O is located at point O_1. The y-axis is aligned with the frame link O_1O_2.

The moving coordinate system P-$x'y'$ is connected to the moving platform $B_1(B_2)$ B_3. The coordinate origin P is located at the center of the B_1B_3. x'-axis is aligned with the moving link B_1B_3.

In Fig. 1, θ denotes the angle between the x'-axis in moving platform and x-axis in static platform. r is half of the length l_3 of the moving link B_1B_3, i.e., $r = (1/2)l_3$. l_1, l_1' and l_1'' are the lengths of the driving links O_1A_1, O_2A_2, O_3A_3. l_2, l_2' and l_2'' are the lengths of the links A_1B_1, A_2B_2 and A_3B_3. φ_1, φ_1' and φ_1'' are the angular positions of the driving links O_1A_1, O_2A_2, O_3A_3. φ_2, φ_2' and φ_2'' are the angular

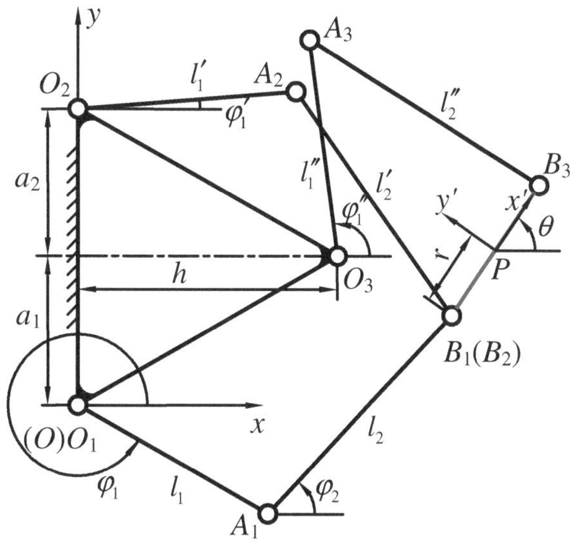

Fig. 1 The configuration and coordinate systems of the 3-RRR planar hybrid mechanism

positions of the connecting links A_1B_1, A_2B_2, A_3B_3. h is the perpendicular distance between the point O_3 and y-axis in the static coordinate system. The coordinates of the center P of the moving platform in the static coordinate system is (x_p, y_p). The coordinates of B_1, B_2 and B_3 in moving coordinate system are $B'_1(-r, 0)$, $B'_2(-r, 0)$ and $B'_3(r, 0)$. By transformation between the moving coordinate system and the static coordinate system, the coordinates of B_1, B_2 and B_3 in the static coordinate system are as follows:

$$[x_{bi} \quad y_{bi} \quad 1]^{\mathrm{T}} = A[x'_{bi} \quad y'_{bi} \quad 1]^{\mathrm{T}}, \qquad i = 1, 2, 3 \tag{1}$$

where, A is the transformation matrix.

$$A = \begin{bmatrix} \cos\theta & -\sin\theta & x_p \\ \sin\theta & \cos\theta & y_p \\ 0 & 0 & 1 \end{bmatrix} \tag{2}$$

In Fig. 1, B_1, B_2 and B_3 coordinates in the static coordinate system are calculated by Eq. (1) as follows:

$$\begin{cases} B_1(x_{b1}, y_{b1}) = B_1(-r\cos\theta + x_p, \ -r\sin\theta + y_p) \\ B_2(x_{b2}, y_{b2}) = B_2(-r\cos\theta + x_p, \ -r\sin\theta + y_p) \\ B_3(x_{b3}, y_{b3}) = B_3(r\cos\theta + x_p, \ r\sin\theta + y_p) \end{cases} \tag{3}$$

The hybrid mechanism in Fig. 1 satisfies the following constraint equations:

$$\begin{cases} \|B_1 - A_1\| = l_2 \\ \|B_2 - A_2\| = l_2' \\ \|B_3 - A_3\| = l_2'' \end{cases} \tag{4}$$

From Eq. (4), the following equations can be obtained.

$$\begin{cases} f_{(1)}^2 = (x_{b1} - l_1 \cos \varphi_1)^2 + (y_{b1} - l_1 \sin \varphi_1)^2 - l_2^2 = 0 \\ f_{(2)}^2 = \left(x_{b2} - l_1' \cos \varphi_1'\right)^2 + \left(y_{b2} - l_1' \sin \varphi_1' - a_1 - a_2\right)^2 - {l_2'}^2 = 0 \\ f_{(3)}^2 = \left(x_{b3} - l_1'' \cos \varphi_1'' - h\right)^2 + \left(y_{b3} - l_1'' \sin \varphi_1'' - a_1\right)^2 - {l_2''}^2 = 0 \end{cases} \tag{5}$$

The kinematic solutions can be found through an unconstrained optimization model,

$$\min f(x) = \sum_{i=1}^{3} f_{(i)}^2(x), \qquad i = 1, 2, 3 \tag{6}$$

3.2 Operation Steps of Chaos Particle Swarm Optimization

The unconstrained optimization problem of Eq. (6) is solved by a generic optimization method, namely, the Chaos Particle Swarm Optimization (CPSO) [7].

Compared with other optimization methods, CPSO has the advantages like run fast, simple algorithm, few setup parameters. CPSO improves the abilities of seeking the global optimum solutions and convergence speed and accuracy. The operation steps of CPSO algorithm are summarized as follows.

Step 1: Determining algorithm parameters: they include learning factor c_1 and c_2, the maximum inertia weight factor $w_{\max}$ and minimum inertia weight factor $w_{\min}$, the velocity upper limit $v_{\max}$, the error threshold of group optimal location ε, and chaos searching maximum iteration times T_0;

In order to avoid the shortcomings of the linear weight, the weight factor w in the algorithm is set up as follows.

$$w = \begin{cases} w_{\max} - \left(w_{\max} - w_{\min}\right)\left[\frac{(t-1)}{(T_0-1)}\right]^{\lambda} & t < T_0 \\ w_{\min} & t \geq T_0 \end{cases} \tag{7}$$

where, λ is a decline index.

Step 2: Randomly generating populations of particles whose number is m, and initializing the particle swarm. Getting the initial individual extreme $pbest_{id}$ and global extreme $gbest_{id}$;

Step 3: Using the velocity and position formulas of standard PSO to update iteration velocity and position of particles in the population;

Step 4: Optimizing the particle optimum position which has been searched in the above step;

Step 5: Randomly selecting a particle in the current particle swarm and replacing it with the selected chaos particle swarm optimization fitness;

Step 6: Until the required accuracy of the algorithm isn't satisfied or the maximum iteration times is reached, the algorithm stops. Otherwise, return to step 3.

3.3 Forward Position of the 3-RRR Hybrid Mechanism

The structure parameters of the 3-RRR hybrid mechanism are assigned as follows: $a_1 = 30$ mm, $a_2 = 14$ mm, $h = 0$ mm, $l_1 = l_1' = l_1'' = 30$ mm, $l_2 = l_2' = l_2'' = 40$ mm, $r = 20$ mm. The initial position is $\varphi_1 = \varphi_1' = \varphi_1'' = 0°$.

The CPSO parameters for solving the forward position are shown in Table 1. Solving Eq. (6) by using CPSO, the four optimization results are shown in Table 2. The average time for CPSO running 20 times to get the results is 2.02 s. The four groups of calculated data show four different forward positions of the 3-RRR hybrid mechanism. Four groups assembled configurations can be gained, as shown in Fig. 2.

4 Inverse Position of the 3-RRR Hybrid Mechanism

Inverse kinematics is to calculate the length of driving links l_1, l_1' and l_1'' and driving angles φ_1, φ_1' and φ_1'' in the static coordinate system under given the lengths l_3, l_2, l_2', l_2'' and the orientation angle θ of the moving platform.

Table 1 CPSO parameters

Parameter	c_1	c_2	w_{max}	w_{min}	v_{max}	ε	T_0	λ
Value	2	2	1.2	0.4	100	10^{-4}	500	0.2

Table 2 Solutions of forward position

Group no.	$\cos\theta$	$\sin\theta$	x_p/mm	y_p/mm
1	0.2800	0.9600	59.1413	41.6938
2	−0.1687	0.9875	50.8350	6.5082
3	0.4363	0.8999	0.7639	41.8956
4	−0.8464	0.5326	9.1382	6.3598

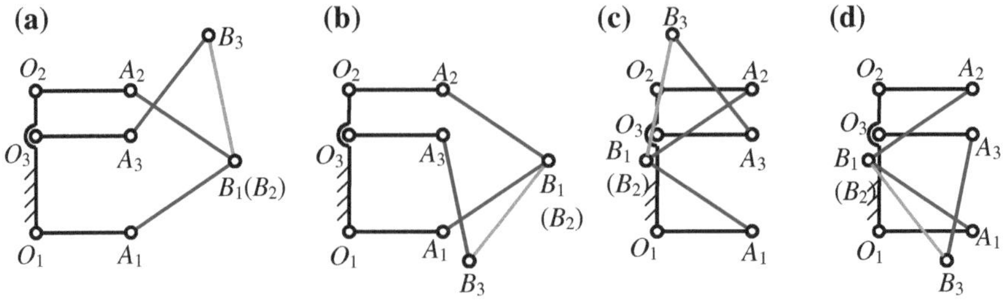

Fig. 2 Four configurations of the 3-RRR hybrid mechanism

Table 3 Solutions of inverse position

Group no.	l_1/mm	l'_1/mm	l''_1/mm	φ_1/rad	φ'_1/rad	φ''_1/rad
1	29.98	30.00	29.99	0.00	0.00	0.00
2	30.01	29.98	29.99	0.00	0.00	0.00
3	30.00	30.00	29.99	0.00	0.00	0.00
4	29.99	29.99	29.99	0.00	0.00	0.00

Based on the configuration of the hybrid mechanism in Fig. 1, substituting Eq. (3) into Eq. (5), yields

$$\begin{cases} f_{(1)}^2 = \left(-r\cos\theta + x_p - l_1\cos\varphi_1\right)^2 + \left(-r\sin\theta + y_p - l_1\sin\varphi_1\right)^2 - l_2^2 = 0 \\ f_{(2)}^2 = \left(-r\cos\theta + x_p - l'_1\cos\varphi'_1\right)^2 + \left(-r\sin\theta + y_p - l'_1\sin\varphi'_1 - a_1 - a_2\right)^2 - {l'_2}^2 = 0 \\ f_{(3)}^2 = \left(r\cos\theta + x_p - l''_1\cos\varphi''_1 - h\right)^2 + \left(r\sin\theta + y_p - l''_1\sin\varphi''_1 - a_1\right)^2 - {l''_2}^2 = 0 \end{cases} \tag{8}$$

The inverse position is mainly based on the forward position data in Table 2. Substituting the data in Table 2 into Eq. (8) to calculate the length of driving links l_1, l'_1, l''_1 and initial angles φ_1, φ'_1, φ''_1, the calculated results are shown in Table 3. Obviously, the data in Table 3 are consistent with the given link lengths l_1, l'_1, l''_1 and initial angles φ_1, φ'_1, φ''_1, in Sect. 3.3 within the error range. The above work verifies that using the chaos particle swarm algorithm to calculate forward kinematics position is feasible.

5 The Singularity of the 3-RRR Hybrid Mechanism

5.1 *Classification of Singularity Configuration*

The 3-RRR hybrid mechanism is composed of 3-DOF series and parallel closed-loop mechanism. Let $\boldsymbol{q}$ denotes the input of the driving links and $\boldsymbol{x}$ denotes the output of the moving platform. The position constraint equation of the hybrid mechanism is shown as:

$$F(\boldsymbol{x}, \boldsymbol{q}) = 0 \tag{9}$$

where, vector $\boldsymbol{q}$ and $\boldsymbol{x}$ should have the same dimension.

Differentiating Eq. (9) with respect to time, yields

$$\mathbf{J}_x\dot{\boldsymbol{x}} = \mathbf{J}_q\dot{\boldsymbol{q}} \tag{10}$$

According to the different situations about the values of $\det(\mathbf{J}_x)$ and $\det(\mathbf{J}_q)$, the singularity configurations can be divided into three types [8], namely,

(1) Forward singularity (also called Position singularity), if $\det(\mathbf{J}_x) = 0$ and $\det(\mathbf{J}_q) \neq 0$ hold simultaneously.

(2) Inverse singularity (also called Boundary singularity), if $\det(\mathbf{J}_x) \neq 0$ and $\det(\mathbf{J}_q) = 0$ hold simultaneously.

(3) Hybrid singularity (also called Configuration singularity), if $\det(\mathbf{J}_x) = 0$ and $\det(\mathbf{J}_q) = 0$ hold simultaneously.

All three types of singularities will be analyzed in this work.

5.2 Singularity of the 3-RRR Hybrid Mechanism

5.2.1 The Jacobian Matrices of the 3-RRR Hybrid Mechanism

The primary task of solving singularity is to gain the Jacobian matrices of the hybrid mechanism, i.e. $\mathbf{J}_x$ and $\mathbf{J}_q$ in Eq. (10).

Differentiating Eq. (8) with respect to time, then, Eq. (10) can be expressed as

$$\begin{bmatrix} a_{11} & a_{12} & a_{13} \\ a_{21} & a_{22} & a_{23} \\ a_{31} & a_{32} & a_{33} \end{bmatrix} \begin{bmatrix} \dot{x}_p \\ \dot{y}_p \\ \dot{\theta} \end{bmatrix} = \begin{bmatrix} J_1 & 0 & 0 \\ 0 & J_2 & 0 \\ 0 & 0 & J_3 \end{bmatrix} \begin{bmatrix} \dot{\varphi}_1 \\ \dot{\varphi}'_1 \\ \dot{\varphi}''_1 \end{bmatrix} \tag{11}$$

where, $\dot{\boldsymbol{x}} = \begin{bmatrix} \dot{x}_p & \dot{y}_p & \dot{\theta} \end{bmatrix}^T$ expresses the velocity vector of the moving platform, while $\dot{\boldsymbol{q}} = \begin{bmatrix} \dot{\varphi}_1 & \dot{\varphi}'_1 & \dot{\varphi}''_1 \end{bmatrix}^T$ is the driving velocity of the driving links.

In order to facilitate calculation, the static platform is the equilateral triangle with the side length $2a$. Hence, there exists the relationship $h = \sqrt{3}a$ in Fig. 1.

In Eq. (11), the parameters in $\mathbf{J}_x$ and $\mathbf{J}_q$ are as follows:

$$\begin{aligned}
a_{11} &= -r\cos\theta - l_1\cos\varphi_1 + x_p; \\
a_{12} &= -r\sin\theta - l_1\sin\varphi_1 + y_p; \\
a_{13} &= rx_p\sin\theta - ry_p\cos\theta + rl_1\sin(\varphi_1 - \theta); \\
a_{21} &= -r\cos\theta - l'_1\cos\varphi'_1 + x_p; \\
a_{22} &= -r\sin\theta - l'_1\sin\varphi'_1 + y_p - 2a;
\end{aligned}$$

$a_{23} = rx_p \sin\theta - ry_p \cos\theta + rl_1' \sin(\varphi_1' - \theta) + 2ar\cos\theta;$
$a_{31} = r\cos\theta - l_1'' \cos\varphi_1'' + x_p - \sqrt{3}a;$
$a_{32} = r\sin\theta - l_1'' \sin\varphi_1'' + y_p - a;$
$a_{33} = -rx_p \sin\theta + ry_p\cos\theta + 2ar\sin(\theta - \pi/6) + rl_1''\sin(\theta - \varphi_1'');$
$J_1 = rl_1\sin(\varphi_1 - \theta) - l_1x_p\sin\varphi_1 + l_1y_p\cos\varphi_1;$
$J_2 = rl_1'\sin(\varphi_1' - \theta) - l_1'x_p\sin\varphi_1' + l_1'y_p\cos\varphi_1' - 2al_1'\cos\varphi_1';$
$J_3 = rl_1''\sin(\theta - \varphi_1'') - l_1''x_p\sin\varphi_1'' + l_1''y_p\cos\varphi_1'' + 2al_1''\sin(\varphi_1'' - \pi/6).$

5.2.2 Forward Singularity

When $\det(\mathbf{J}_x) = 0$, the forward singularity can be divided into the following two situations.

(1) When the value of the third column of $\mathbf{J}_x$ is 0, i.e., $a_{13} = a_{23} = a_{33} = 0$, the deduced result is $\varphi_1 = \varphi_2 = \varphi_1' = \varphi_2' = \varphi_2'' = \theta$. However, there no exists this type of configuration.
(2) When the first two columns of the matrix $\mathbf{J}_x$ are linearly dependent, the singularity condition $\varphi_2 = \varphi_2' = \varphi_2''$ can be deduced. This shows that the links A_1B_1, A_2B_2 and A_3B_3 are collinear or parallel, as shown in Fig. 3.

5.2.3 Inverse Singularity

When $\det(\mathbf{J}_q) = 0$, there exists inverse singularity. There are three situations as follows.

(1) $J_1 = 0$, i.e.,

$$J_1 = rl_1\sin(\varphi_1 - \theta) - l_1x_p\sin\varphi_1 + l_1y_p\cos\varphi_1 = 0 \tag{12}$$

where, $x_p = l_1\cos\varphi_1 + l_2\cos\varphi_2 + r\cos\theta$, $y_p = l_1\sin\varphi_1 + l_2\sin\varphi_2 + r\sin\theta$

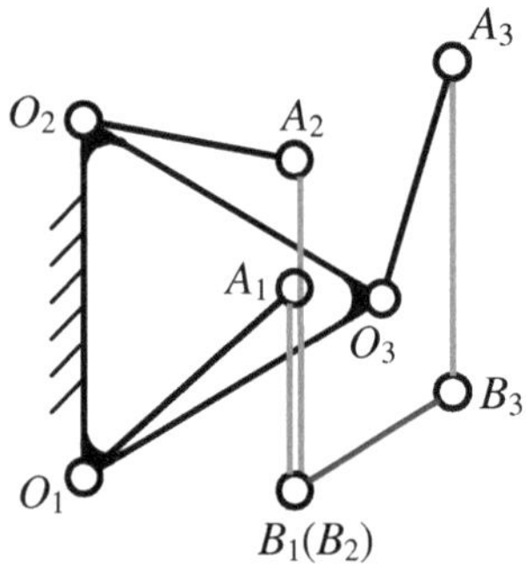

Fig. 3 Forward singularity with $\varphi_2 = \varphi_2' = \varphi_2''$

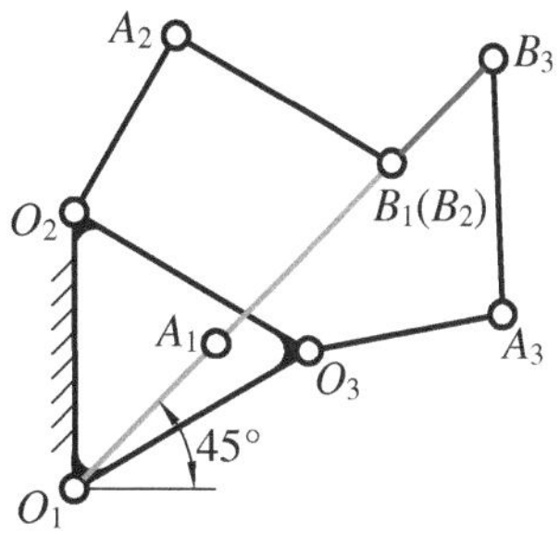

Fig. 4 Inverse singularity with $\varphi_1 = \varphi_2 = \theta = 45°$

The necessary condition of Eq. (12) is that $\varphi_1 = \theta = 45°$ or 225° and $x_p = y_p$ hold simultaneously.

It can be deduced that when $\varphi_1 = \varphi_2 = \theta = 45°$ and $\varphi_1 = \varphi_2 = \theta = 225°$, there exist inverse singularities. Figure 4 is the inverse singularity with $\varphi_1 = \varphi_2 = \theta = 45°$. It shows that the driving link O_1A_1, the connecting link A_1B_1 and the platform line B_1B_3 are collinear. For the structure parameters in Fig. 4, there is no the inverse singularity with $\varphi_1 = \varphi_2 = \theta =$ 225° because of the structural constraints.

(2) $J_2 = 0$, i.e.,

$$J_2 = rl'_1 \sin(\varphi'_1 - \theta) - l'_1 x_p \sin\varphi'_1 + l'_1 y_p \cos\varphi'_1 - 2al'_1 \cos\varphi'_1 = 0 \quad (13)$$

It is found that the singularity condition is $\varphi'_1 = \varphi'_2 = \theta$. This shows that the links O_2A_2, A_2B_2 and B_2B_3 are collinear, as shown in Fig. 5.

(3) $J_3 = 0$, i.e.,

$$J_3 = rl''_1 \sin(\theta - \varphi''_1) - l''_1 x_p \sin\varphi''_1 + l''_1 y_p \cos\varphi''_1 + 2al''_1 \sin(\varphi''_1 - \pi/6) = 0 \quad (14)$$

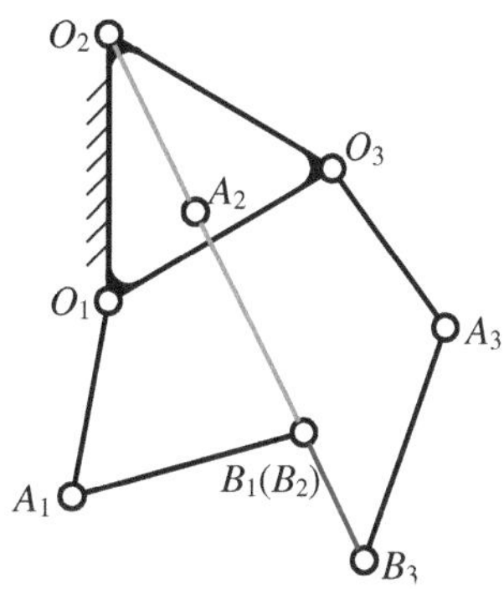

Fig. 5 Inverse singularity with $\varphi'_1 = \varphi'_2 = \theta$

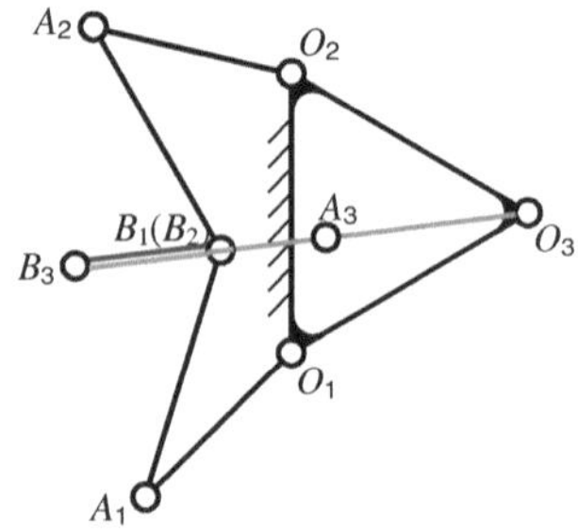

Fig. 6 Inverse singularity with $\varphi_1'' = \varphi_2'' = \theta$

It is found that the singularity position occurs when $\varphi_1'' = \varphi_2'' = \theta$. This shows the links O_3A_3, A_3 B_3 and B_1B_3 are collinear, as shown in Fig. 6.

5.2.4 Hybrid Singularity

When $\det(\mathbf{J}_x) = 0$ and $\det(\mathbf{J}_q) = 0$ hold simultaneously, the forward singularity situations and the inverse singularity situations exist at the same time. Thus, the hybrid singularity is the combination of the forward singularity and inverse singularity.

6 Conclusions

This paper analyzes the configuration of hybrid mechanism. The forward positions of the 3-RRR hybrid mechanism are calculated by using chaos particle swarm algorithm. Taking the positions of positive solutions to the mechanism constraint equations, the inverse positions of the hybrid mechanism are obtained. The accuracy of the solution forward position method is verified by inverse position data. The singularity conditions of the 3-RRR hybrid mechanism are analyzed. The configurations of forward singularity, inverse singularity and hybrid singularity are obtained, respectively. The results play an important role in practical application of the 3-RRR hybrid mechanism.

Acknowledgments The authors gratefully acknowledge the financial and facility support provided by the National Natural Science Foundation of China (Grant No. 51275486) and the Specialized Research Fund for the Doctoral Program of Higher Education (Grant No. 20111420110005).

References

1. Coppola, G., Zhang, D.: Kefu Liu. A 6-DOF Reconfigurable hybrid parallel manipulator. Robot. Comp.-Integr. Manuf. **30**, 99–106 (2014)
2. Alexandre, C., Christoph, B., Jürgen, H.: A type synthesis method for hybrid robot structures. Mech. Mach. Theory **43**(8), 984–995 (2008)

3. Altuzarra, O., Martín, Y.S., Amezua, E.: Motion pattern analysis of parallel kinematic machines: a case study. Robot. Comp.-Integr. Manuf. **25**(2), 432–440 (2009)
4. Guo, X.J., Huang, T.Y., Chang, F. Q.: Kinematics of a new hybrid mechanism with reduced DOF and nested dynamic connection. J. Mech. Eng. **49**(1): 1–6 (2013) (in Chinese)
5. Shen, Q., Wang, J.F., Shi, T.L.: Study on the reset accuracy of a 3-DOF hybrid manipulator. Mech. Sci. Technol. **31**(3), 0175–0479 (2012) (in Chinese)
6. Merlet, J.P.: Parallel Robots, pp. 27–29. Springer, Netherlands (2006)
7. Gao, Y., Xie, S.L.: Chaos particle swarm optimization algorithm. Comp. Sci. **31**(8), 13–15 (2004) (in Chinese)
8. Gosselin, C.M., Angeles, J.: Singularity analysis of closed-loop kinematic chains. IEEE Trans. Robot. Autom. **6**(3), 281–290 (1990)

A Method to Estimate the Encoder Dependent Repeatability of General Serial Manipulators

M. Brandstötter, Christoph Gruber and M. Hofbaur

Abstract A crucial parameter for the operational capability of a serial robot is its accuracy. For recurring tasks the high repeatability of the manipulator is utilized by programming the robot online. Therefore, the end effector is moved manually to all desired or necessary poses of a given task and the joint angles of the manipulator are stored at these poses. The end effector can return to these stored poses very accurately due to the high repeatability of industrial robots. But even this repeatability is subject to limits. A major influencing factor is the finite resolution of the encoders integrated within the manipulators joints. Based on the discrete encoder readings, the end effector pose can only be estimated. Despite the typically high resolution of encoders in modern industrial robots, it is worth considering this issue in detail, as one can identify interesting structural properties that are invariant with respect to the encoder resolution. In this work, we will show how to compute the upper limit of the estimation error depending on a desired pose and posture of the manipulator. Both positioning and orientation error due to this discretization effects are considered. Simulation results are illustrated by means of a general 6R serial manipulator and a path, which the end effector can follow with 16 different postures.

Keywords Serial robots · Positioning error · Orientation error · Performance criteria · Controller resolution

M. Brandstötter (✉) · C. Gruber
Department for Biomedical Informatics and Mechatronics, Institute of Automation and Control Engineering, UMIT, Hall in Tirol, Austria
e-mail: mathias.brandstoetter@umit.at

C. Gruber
e-mail: christoph.gruber@umit.at

M. Hofbaur
Institute for Robotics and Mechatronics, Joanneum Research Forschungsgesellschaft MbH, Graz, Austria
e-mail: michael.hofbaur@joanneum.at

S. Bai and M. Ceccarelli (eds.), *Recent Advances in Mechanism Design for Robotics*, Mechanisms and Machine Science 33,
DOI 10.1007/978-3-319-18126-4_10

1 Introduction

Positioning the end effector of a serial manipulator in an accurate manner is a necessary prerequisite for almost all applications. The absolute accuracy must be distinguished from the repeatability regarding the end effector positioning. Repeatability is typically preciser than absolute accuracy and is defined as how close the end effector of a robot manipulator can return to a previously met pose [12]. A high repeatability can be achieved by measuring the joint angular positions with high resolution encoders. When using the so-called online programming—a popular method to manually program industrial robots—repeatability is the main factor for end effector accuracy.

The pose of a serial robot denotes the position and orientation of the manipulator's end effector. In this paper we present a method to calculate pose errors of serial manipulators due to finite encoder resolution and confine ourselves to rigid robot structures with perfect geometry. Thereby, positioning and orientation errors means the discrepancy between a desired end effector pose and a pose based on measurements. An encoder provides discrete signals that a controller interprets incrementally to determine a given angular reading for a joint. As a consequence, it is the smallest increment of motion, i.e. the controller resolution [12], that determines the error sought for.

Besides the finite encoder resolution, there are many other sources of errors. Effects of geometrical variations on the end effector pose of a manipulator are considered in [3, 11, 13]. In addition, Jacobian matrix based probabilistic analysis are studied, amongst others, in [8–10].

Within our work we follow and extend the basic line of thoughts for resolution analysis presented by Brethé [6]. There, an actual configuration with all its uncertainties due to the controller resolution is mapped to the workspace and defines the resulting set of all end effector poses that surround the target pose as spatial resolution.

In this paper we define the maximum occurring deviation of all surrounding poses in relation to the desired pose as maximum estimation error. Therefore, we map all neighboring points in the discrete configuration space to the workspace and define a measure to quantify the maximum positioning error and orientation error. For structurally simple robots like the planar 2R manipulator one can characterize this error quantification analytically, as we have demonstrated in [5]. This work extends our framework to general 6R robots, which obviously leads to an evaluation problem of high complexity. As a consequence we evaluate the positioning and orientation errors numerically and demonstrate the proposed method to compute the encoder dependent repeatability by a path followed by the end effector of a general serial 6R manipulator. We selected this very general type of manipulator in order to present the analysis method for the most general case and, additionally, provide insight on a robot's precision with respect to multiple solutions for a predetermined end effector pose.

2 General Serial 6R Manipulators

Serial robots with general structure are characterized by a series of joint axis where consecutive axes are neither parallel to each other, nor do they intersect. Robots with general structure are currently seldomly used in robot applications. Mostly due to the complex mechanical structure that has to be built, the sophisticated calibration process for such a manipulator and computationally complex and time-consuming calculation of the inverse kinematics.

Nevertheless, such robots are used in literature to analyze and better understand general properties of serial manipulators. We follow this approach, but complement simulation studies with real-world experiments using our novel modular general serial 6R manipulator testbed [4]. The manipulator testbed consists of rotational joints together with a set of curved arm segments that are used in various configurations to link the joints and thus form a curved manipulator arm (Cuma-arm).

The forward kinematics problem of serial manipulators can be solved straightforwardly using homogeneous matrices based on the structural and geometrical characterization of the robot (e.g. by means of the Denavit-Hartenberg parameters) and the actuated joint angles q_i [1]. We concisely symbolize this well known mapping as:

$$\phi_{\text{dh}}(\mathbf{q}) = \phi_{\text{dh}}(q_1,\ q_2,\ \ldots,\ q_6) = \mathcal{P}^{\Sigma} \tag{1}$$

In (1) $\mathcal{P}^{\Sigma}$ represents a pose and is defined as the position and the orientation of the end effector in coordinate system $(O; x, y, z) \hat{=} \Sigma$. It can be described as homogeneous matrix by

$$\mathcal{P}^{\Sigma} = \begin{pmatrix} 1 & \mathbf{0}^T \\ \mathbf{t}^{\Sigma} & \mathbf{R}^{\Sigma} \end{pmatrix}, \tag{2}$$

where $\mathbf{t}^{\Sigma}$ represents the position vector and $\mathbf{R}^{\Sigma}$ the orientation matrix of $\mathcal{P}^{\Sigma}$ with respect to Σ. The rotation matrix $\mathbf{R}^{\Sigma}$ is an element of the *special orthogonal group* $\mathbf{SO}(3)$.

The computationally evaluation of the inverse kinematics for general 6R robots is more demanding, in particular when compared to the inverse kinematics of standard industrial robots. Within this paper we use the Husty-Pfurner algorithm [7] as described and analyzed in [2]. Despite its complexity, we use the following concise notation to denote the set of possible solutions for the inverse kinematics problem throughout this paper:

$$\phi_{\text{dh}}^{\text{inv}}(\mathcal{P}^{\Sigma}) = \mathbf{Q} = \begin{bmatrix} \mathbf{q}_1 \\ \vdots \\ \mathbf{q}_{16} \end{bmatrix} = \begin{bmatrix} q_{1,1} & \cdots & q_{1,6} \\ \vdots & \ddots & \vdots \\ q_{16,1} & \cdots & q_{16,6} \end{bmatrix} \tag{3}$$

The 16×6 set $\mathbf{Q}$ in (3) collects all solutions of $\mathbf{q}_i$ that define possible assignments for the six joint angles.

3 Effects of Finite Encoder Resolution

This section extends our previously published work [5] to general 6R robots and thus frames our basic concepts for numerical evaluation. We therefore start with a short recapitulation of the basic concepts and refer the interested reader to [5] for an in-depth description.

A rotary encoder is a device that converts the mechanical angular movement of a shaft into a signal to detect changes in position and can be used to determine the angular position of a joint with respect to a reference position. If a digital sensor is used a code or pulse sequence is supplied and hence, the angle measurement is discretized due to the discrete signal. The arising discretization can be modeled by

$$\kappa(\varphi) = \left\lfloor \frac{\varphi}{2\pi}\mathrm{res}_\varphi + \frac{1}{2} \right\rfloor \frac{2\pi}{\mathrm{res}_\varphi} = \left\lfloor \frac{\varphi}{\Delta\tilde{q}_\varphi} + \frac{1}{2} \right\rfloor \Delta\tilde{q}_\varphi \,, \tag{4}$$

where φ is the angular position, res_φ the number of increments (encoder ticks) per revolution, $\Delta\tilde{q}_\varphi = 2\pi/\mathrm{res}_\varphi$ denotes the minimum angle that can be distinguished, $\kappa(\varphi)$ the discretized measurement with respect to angle φ, and the symbol $\lfloor x \rfloor$ means the greatest integer less or equal than x. We will neglect other possible sources of errors affecting on positioning and orientation (e.g. dynamic influences) and only focus on the impact of this quantization error on repeatability throughout this paper. In addition, we assume that $\Delta\tilde{q}_\varphi$ is small due to high encoder resolution.

To compute estimates for the end effector poses we use measurements from joint angle encoders. Since we consider a manipulator with six revolute joints a 6-dimensional version of κ can be written as

$$\chi(q_1, \ldots, q_6) = [\tilde{q}_1 \;\; \tilde{q}_2 \ldots \tilde{q}_6]^T = [\kappa(q_1) \;\; \kappa(q_2) \ldots \kappa(q_6)]^T \,. \tag{5}$$

End effector pose estimates can be calculated from these joint angle measurements exerting the direct kinematics given in (1):

$$\tilde{\mathcal{P}}^\Sigma = \phi_{\mathrm{dh}}(\tilde{q}_1, \tilde{q}_2, \ldots, \tilde{q}_6) \tag{6}$$

Figure 1 illustrates the effect of a discrete configuration space for the simpler two-dimensional case. As can be seen from the figure, the maximum error can only occur at the edge of the region. In Fig. 2a the set of all ($8 = 3^n - 1, n = 2$) adjacent configurations of joint angles, denoted as

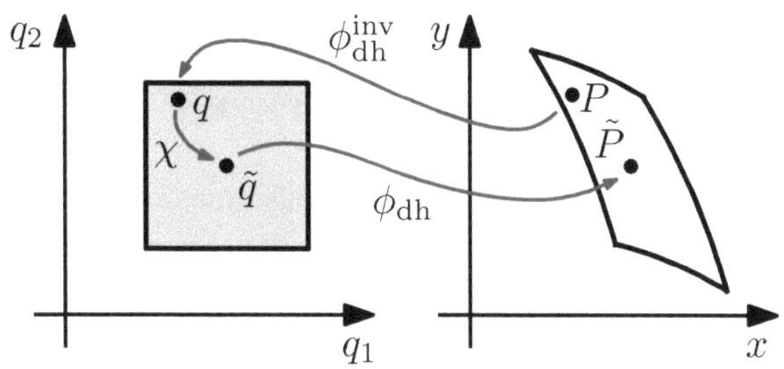

Fig. 1 A given point P is mapped by ϕ^{inv} to the configuration space with coordinates q, discretized by χ to $\tilde{q}$, and mapped back to the workspace by ϕ to the estimated point $\tilde{P}$

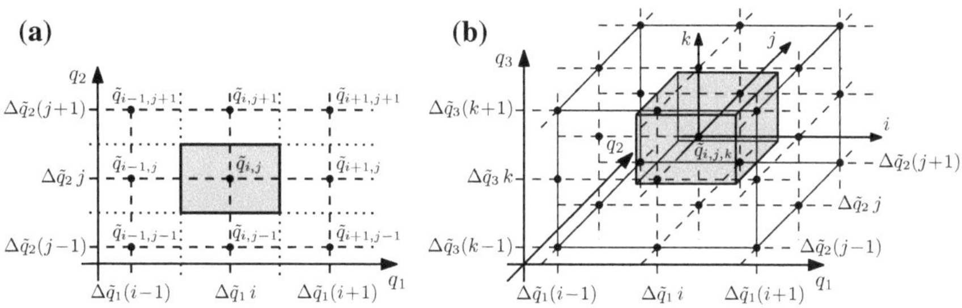

Fig. 2 Discrete configuration spaces. **a** The two-dimensional case. **b** The three-dimensional case

$$\tilde{Q}^2_{\pm 1} = \{\tilde{q}_{i-1,j-1}, \tilde{q}_{i-1,j}, \tilde{q}_{i,j-1}, \tilde{q}_{i+1,j-1}, \tilde{q}_{i+1,j+1}, \tilde{q}_{i+1,j}, \tilde{q}_{i,j+1}, \tilde{q}_{i-1,j+1}\} \tag{7}$$

of $\tilde{q}_{i,j}$, are drawn. All configurations within the marked square are estimated to $\tilde{q}_{i,j}$. The set in (7) can be easily extended for three actuators, shown in Fig. 2a, to $\tilde{Q}^3_{\pm 1}$ and for n actuators to $\tilde{Q}^n_{\pm 1}$ with $3^n - 1$ elements of adjacent configuration coordinates. Mapping $\tilde{Q}^n_{\pm 1}$ to the workspace, we get the set of all dedicated pose estimates, symbolized by

$$\tilde{\mathcal{P}}^{\Sigma}_{\pm 1} = \phi_{\mathrm{dh}}\left(\tilde{Q}^n_{\pm 1}\right). \tag{8}$$

To calculate the maximum error caused by the discretization an error metric is needed. For every pair of poses $\mathcal{P}^{\Sigma}_{A}$ and $\mathcal{P}^{\Sigma}_{B}$ a transformation matrix $\mathbf{H}^{\Sigma_A}$ can be found such that

$$\mathcal{P}^{\Sigma}_{B} = \mathcal{P}^{\Sigma}_{A} \cdot \mathbf{H}^{\Sigma_A} \tag{9}$$

and thus

$$\mathbf{H}^{\Sigma_A} = \left(\mathcal{P}_A^{\Sigma}\right)^{-1} \cdot \mathcal{P}_B^{\Sigma} = \mathbf{T}^{\Sigma_A} \cdot \mathbf{R}^{\Sigma_A}. \tag{10}$$

As depicted in (10) a general homogeneous transformation matrix $\mathbf{H}^{\Sigma_A}$ is composed by a translational part $\mathbf{T}^{\Sigma_A}$ and a rotational part $\mathbf{R}^{\Sigma_A}$. We are using the two components to define an error metric for the positioning and the orientation error.

From the translation vector $\mathbf{t}^{\Sigma_A}$ within $\mathbf{T}^{\Sigma_A}$ the Euclidean distance between $\mathcal{P}_A^{\Sigma}$ and $\mathcal{P}_B^{\Sigma}$ can be computed and used as metric for the positioning error:

$$e_{\text{pos}} = \frac{1}{2}\left|\mathbf{t}^{\Sigma_A}\right| \tag{11}$$

As we consider the transformation between two adjacent configurations in the discretized joint space we take the bisection of the Euclidean distance in task space to define the error metric in (11). As can be seen in Fig. 2a the borders of the marked region are located at $\tilde{q}_1 \pm \Delta\tilde{q}_1/2$ and $\tilde{q}_2 \pm \Delta\tilde{q}_2/2$.

As orientation error metric we use the axis-angle representation of the homogeneous transformation matrix $\mathbf{R}^{\Sigma_A}$. By this representation a rotation can be parametrized by an angle θ about an axis determined by a unit vector $\mathbf{r}$. To describe the orientation error e_{ori} the magnitude of the rotation about $\mathbf{r}$ has to be calculated and thus we obtain

$$e_{\text{ori}} = \frac{1}{2}\theta = \frac{1}{2}\operatorname{acos}\left(\frac{\operatorname{trace}\left(\mathbf{R}^{\Sigma_A}\right) - 2}{2}\right). \tag{12}$$

Based on the above considerations, the set of transformation matrices $\mathbf{H}_{\pm 1}^{\Sigma_{\tilde{\mathcal{P}}}}$

$$\mathbf{H}_{\pm 1}^{\Sigma_{\tilde{\mathcal{P}}}} = \left(\tilde{\mathcal{P}}^{\Sigma}\right)^{-1} \cdot \tilde{\mathcal{P}}_{\pm 1}^{\Sigma} = \mathbf{T}_{\pm 1}^{\Sigma_{\tilde{\mathcal{P}}}} \cdot \mathbf{R}_{\pm 1}^{\Sigma_{\tilde{\mathcal{P}}}} \tag{13}$$

between an estimated pose $\tilde{\mathcal{P}}^{\Sigma}$ and its dedicated neighbors $\tilde{\mathcal{P}}_{\pm 1}^{\Sigma}$ can be computed. Using the deviation matrices of (13), the set of position errors $\mathcal{E}_{\text{pos},\pm 1}$ and the set of orientation errors $\mathcal{E}_{\text{ori},\pm 1}$ can be found analogously to (11) and (12):

$$\mathcal{E}_{\text{pos},\pm 1} = \frac{1}{2}\left|\mathbf{t}_{\pm 1}^{\Sigma_{\tilde{\mathcal{P}}}}\right| \tag{14}$$

$$\mathcal{E}_{\text{ori},\pm 1} = \frac{1}{2}\operatorname{acos}\left(\frac{\operatorname{trace}\left(\mathbf{R}_{\pm 1}^{\Sigma_{\tilde{\mathcal{P}}}}\right) - 2}{2}\right) \tag{15}$$

The maximum positioning error $e_{\text{pos,max}}$ and maximum orientation error $e_{\text{ori,max}}$ are now given by

$$e_{\text{pos,max}} = \max\left(\mathcal{E}_{\text{pos},\pm 1}\right) \tag{16}$$

and

$$e_{\text{ori,max}} = \max\left(\mathcal{E}_{\text{ori},\pm 1}\right) \,. \tag{17}$$

4 Simulation Results

In this section we want to show how to determine the maximum pose errors of a general serial 6R manipulator due to the finite resolution of its encoders. First, we define the structure of a manipulator on which we demonstrate the findings of Sect. 3. The Denavit-Hartenberg parameter set of the used Cuma-arm configuration is listed in Table 1 and the last column indicates the offsets of each joint angle by $q_{\text{off},i}$.

Figure 3 shows a simulation of the Cuma-arm in two different poses of one specific posture. The end effector follows a linear path in task space that starts at $\mathcal{P}^{\Sigma}_{\text{start}}$ (transparent manipulator configuration) and ends at $\mathcal{P}^{\Sigma}_{\text{end}}$ (opaque configuration). The orientation of the end effector is kept constant while moving and the poses of the end effector are given by

$$\mathcal{P}^{\Sigma}_{\text{start}} = \begin{bmatrix} 1 & 0 & 0 & 0 \\ 209 & -1 & 0 & 0 \\ -187 & 0 & 1 & 0 \\ 668 & 0 & 0 & -1 \end{bmatrix}, \text{ and } \mathcal{P}^{\Sigma}_{\text{end}} = \begin{bmatrix} 1 & 0 & 0 & 0 \\ 1 & -1 & 0 & 0 \\ -473 & 0 & 1 & 0 \\ 668 & 0 & 0 & -1 \end{bmatrix}.$$

For the following results it is important to note that the end effector pose do not overlap with the last reference frame resulting from the Denavit-Hartenberg transformations. An offset transformation $\mathbf{EF}_{\text{off}}$ must be taken into account to attain the end effector pose:

$$\mathbf{EF}_{\text{off}} = \begin{bmatrix} 1 & 0 & 0 & 0 \\ -95.1 & 0 & 0 & -1 \\ 0 & 0 & 1 & 0 \\ 0 & 1 & 0 & 0 \end{bmatrix} \tag{18}$$

Table 1 DH-parameters of the used Cuma-arm

ith Link	d_i/mm	a_i/mm	α_i/deg	$q_{\text{off},i}$/deg
1	233.827	30.590	42.590	121.590
2	499.306	373.502	−67.891	76.658
3	−172.743	106.559	102.166	51.359
4	−1.646	355.206	−99.967	3.435
5	48.151	58.073	−103.621	68.288
6	16.430	0	0	126.786

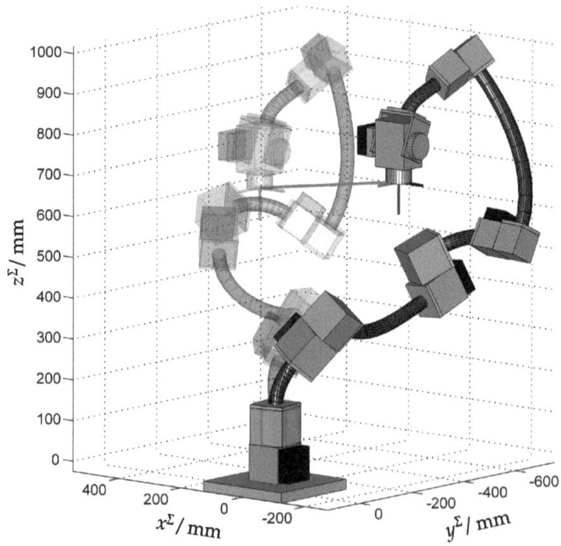

Fig. 3 The Cuma-arm moves on a straight line from $\mathcal{P}^{\Sigma}_{\text{start}}$ to $\mathcal{P}^{\Sigma}_{\text{end}}$ with posture number 7

The values in (18) correspond to dimensions of the final two-axis wrist actuator (Schunk PW70 module). This additional transformation could be also achieved by an extra line in the Denavit-Hartenberg parameter set.

The selected path for this simulation is located in a special area of the workspace for which we can find 16 individual solutions of the inverse kinematics. That is to say, the entire path can be traversed in 16 different ways. A detail planar section of the workspace is plotted in Fig. 4 and the number of solutions are indicated by different colors.

For the considered case we have chosen the minimum angular resolutions to $\Delta\tilde{q}_i = 0.01$ deg for $i = 1, \ldots, 6$. Without additional computational cost, resolutions can be selected which are different from each other. This angular resolutions allow us to determined the positioning and orientation errors for the specified path and for

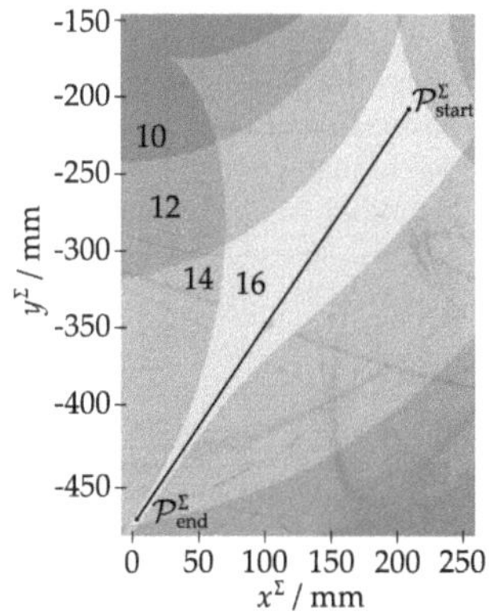

Fig. 4 A section of the workspace of the considered Cuma-arm. An area with constant height ($z^{\Sigma} = 668$ mm) and constant orientation of the end effector is shown. The colors are related to the number of solutions without an imaginary part. The inverse kinematics problem is solved with an implementation of the Husty-Pfurner algorithm [7]

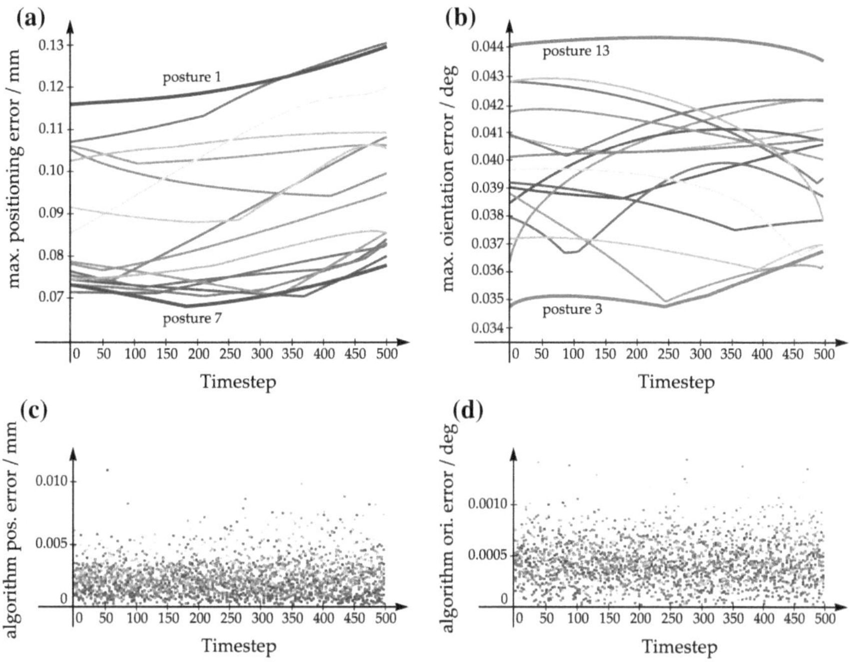

Fig. 5 Maximum errors that occur at the specified path for all postures. **a** Maximum positioning error due to the encoder resolution. **b** Maximum orientation error due to the encoder resolution. **c** Positioning error due to the limited numerical precision. **d** Orientation error due to the limited numerical precision

Table 2 Summary of data of all examined errors for the given path using the presented Cuma-arm

Posture	Sum of max. positioning errors due to the encoders	Sum of max. positioning errors due to the algorithm	Sum of max. orientation errors due to the encoders	Sum of max. orientation errors due to the algorithm
1	**18.258**	0.283	5.954	0.072
2	17.743	0.536	5.841	0.069
3	14.737	0.464	**5.343**	0.078
4	15.737	0.312	6.139	0.064
5	16.170	0.511	6.135	0.077
6	11.443	0.114	6.282	0.040
7	**10.794**	0.208	6.121	0.067
8	11.126	0.243	6.104	0.071
9	11.018	0.254	5.790	0.055
10	11.389	0.265	6.257	0.037
11	11.464	0.296	5.503	0.076
12	12.719	0.369	6.222	0.065
13	13.453	0.404	**6.689**	0.081
14	12.009	0.327	6.311	0.077
15	15.959	0.514	5.874	0.065
16	14.178	0.535	5.563	0.084
Max/min	1.692	–	1.252	–

each posture separately. Figure 5a, b illustrate the maximum position and orientation error along the traverse for all 16 different solutions. Summing up, all errors yield a measure to select a suitable posture (see Table 2). As a result, the positioning error of posture number 7 is about 1.69 times smaller than of posture number 1.

In order to describe the influence of the numerical properties of the algorithm that we use to calculate the inverse kinematics on the pose accuracy, an additional error quantity is computed.[1] Comparing each element of the set of end effector poses $\mathcal{P}^{\Sigma}_{\mathrm{alg}}$

$$\mathcal{P}^{\Sigma}_{\mathrm{alg}} = \phi_{\mathrm{dh}}\left(\phi^{\mathrm{inv}}_{\mathrm{dh}}\left(\mathcal{P}^{\Sigma}\right)\right) \tag{19}$$

with the desired end effector pose $\mathcal{P}^{\Sigma}$ as derived in Sect. 3, the error due to the inverse kinematics is indicated.[2] The calculated influence, plotted in Fig. 5c, d, suggests that the error is about one magnitude smaller than the error due to discretization.

[1]IEEE 754 double precision.

[2]The major part of the algorithm error is attributed to the inverse kinematics.

5 Conclusions

In this work we considered serial robots with encoder equipped joints and the influence of the encoders on the positioning and orientation estimation of an end effector pose. It was assumed that the resolution of the encoders is large and thus the minimum angle that can be distinguished is small. Due to the quantization effect in the configuration space and non-linear mappings into the workspace a region is formed that has to be analyzed. The maximum estimation error can be found by computing the forward kinematics of a set of configuration points which are next to a considered configuration and result from the resolution of the encoders. Defining measures for positioning and orientation errors allow to quantify the posture dependent pose error, and hence to compare the influence of the different postures. These posture dependent errors are summed up to obtain an overall assessment for a given path. The computed error curves give an upper bound of the expected encoder dependent repeatability. To demonstrate the method for general serial manipulators a path was chosen that can be followed by a Cuma-arm with 16 different solutions. The posture dependent error curves were determined and compared to each other. Hence, an assessment of a convenient posture regarding an application and its focus on positioning or orientation accuracy can be made.

References

1. Angeles, J.: Fundamentals of Robotic Mechanical Systems: Theory, Methods, and Algorithms. Mechanical Engineering Series. Springer, New York (2007)
2. Angerer, A., Hofbaur, M.: Industrial versatility of inverse kinematics algorithms for general 6R manipulators. In: IEEE International Conference on Advanced Robotics (ICAR) (2013)
3. Balli, S.S., Chand, S.: Defects in link mechanisms and solution rectification. Mech. Mach. Theory **37**(9), 851–876 (2002)
4. Brandstötter, M., Angerer, A., Hofbaur, M.: The curved manipulator (cuma-type arm): Realization of a serial manipulator with general structure in modular design. In: Proceedings of the 14th IFToMM World Congress (submitted)
5. Brandstötter, M., Gruber, C., Hofbaur, M.: On the repeatability of planar 2R manipulators with rotary encoders. In: IEEE International Conference on Advanced Robotics (ICAR 2013) (2013)
6. Brethé, J.F.: Innovative kinematics and control to improve robot spatial resolution. In: IEEE/RSJ International Conference on Intelligent Robots and Systems (IROS), pp. 3495–3500. IEEE (2010)
7. Husty, M., Pfurner, M., Schröcker, H.P.: A new and efficient algorithm for the inverse kinematics of a general serial 6 *R* manipulator. Mech. Mach. Theory **42**(1), 66–81 (2007)
8. Kotulski, Z., Szczepiński, W.: On two methods of determining the ellipses and ellipsoids of positioning accuracy of robot manipulators. Arch. Appl. Mech. **63**(1), 25–41 (1993)
9. Kumar, K.: Positioning accuracy of manipulators with encoder equipped joints. Robotics Research and Advanced Applications pp. 35–42 (1982)
10. Lee, W.J., Woo, T.C.: Tolerance volume due to joint variable errors in robots. J. Mech. Transm. Autom. Des. **111**(4), 597–604 (1988)

11. Shiakolas, P.S., Conrad, K.L., Yih, T.C.: On the accuracy, repeatability, and degree of influence of kinematics parameters for industrial robots. Int. J. Model. Simul. **22**(3), 245–254 (2002)
12. Spong, M.W., Hutchinson, S., Vidyasagar, M.: Robot modeling and control. Wiley, New York (2006)
13. Veitschegger, W., Wu, C.H.: Robot accuracy analysis based on kinematics. IEEE J. Robot. Autom. **2**(3), 171–179 (1986)

Part II
Innovative Mechanism/Robot and Their Applications

Designing and Implementing an Interactive Social Robot from Off-the-Shelf Components

Zheng-Hua Tan, Nicolai Bæk Thomsen and Xiaodong Duan

Abstract In this paper we present the design and development of the social robot called iSocioBot, which is designed to achieve long-term interaction between humans and robots in a social context. iSocioBot is 149 cm tall and the mechanical body is built on top of the TurtleBot platform and designed to make people feel comfortable in its presence. All electrical components are standard off-the-shelf commercial products making a replication possible. Furthermore, the software is based on Robot Operating Software (ROS) and is made freely available. We present our experience with the design and discuss possible improvements.

Keywords Social robot · Human-robot interaction · Multimodal sensing · Robot design

1 Introduction

Social robots have in recent years gained significant attention due to the advances in enabling technologies that make many real-world applications feasible. These applications range from social care, entertainment to education, just to name a few. According to the International Journal of Social Robotics, social robots are "robots able to interact and communicate among themselves, with humans and with the environment within the social and cultural structure in which they function" [10]. From this definition, it is obvious that the capability of interaction and communication is central for a social robot to have. Furthermore, the interaction should be

Z.-H. Tan (✉) · N.B. Thomsen · X. Duan
Department of Electronic Systems, Aalborg University, Aalborg, Denmark
e-mail: zt@es.aau.dk

N.B. Thomsen
e-mail: nit@es.aau.dk

X. Duan
e-mail: xd@es.aau.dk

S. Bai and M. Ceccarelli (eds.), *Recent Advances in Mechanism Design for Robotics*, Mechanisms and Machine Science 33,
DOI 10.1007/978-3-319-18126-4_11

natural, e.g., through speech and vision, in order to fit itself into a social environment.

Social robots are used for interacting with humans and improving humans well-being, not necessary with a goal of achieving well-defined, measurable tasks as industrial or even service robots do. With this philosophy in mind, we aim at developing robots that are suitable for various social setups with a proper height, able to move at human walking pace and rotate in place, and are a humanoid.

There are existing social robots, both commercial and proprietary. Nao[1] is one of the most widely used commercial social robots (e.g., [1, 6]). It has a number of different kinds of sensors and many degrees of freedom. On the other hand, it is small, 58 cm height, and moves slowly. PR2[2] robot is an advanced human size robot, which is rather expensive and thus in many ways not economically feasible or necessary. Furthermore, with a bulky body and strong arms, PR2 is more a service robot. Then there are proprietary robots that are developed by companies or research institutes and not for sale, which, among others, include Robovie [4], Maggie [8], and Rubi [3]. There is also the iCub [7], which is a fully open-source child-like robot, however it is designed specifically for interacting with children and very expensive and unnecessarily complex for this project. Due to the lack of existing robots able to fulfil our requirements, we turned to build our own social robot called iSocioBot using off-the-shelf components and making it open source.

The primary objective of developing iSocioBot is to make a social robot being able to establish durable relationship with humans through natural interaction.[3] This relies on developing advanced capabilities to sense and express. The sensing capability is enhanced by a fusion concept we call *reinforcement fusion*, combining sensor signals in an interactive way: e.g., when a robot detects a sound direction, it turns towards the direction to see better and moves towards it to hear better. Reinforcement fusion is analogous to reinforcement learning [5], a known term in machine learning. The implication on the robot is that it should be able to move and rotate at a fast speed, e.g., that close to humans, and have a height suitable for sensing humans while they stand or sit. To maintain long-term relationship with humans, the social robot should have a model of its user and respond socially. We propose a concept of *social behaviour entrainment* to adapt behaviours to the user's. Entrainment is an interesting phenomenon found in human-human communication, e.g., speech entrainment occurs during spoken interaction where dialogue partners tend to become similar to each other in style [2]. In expressing itself, the robot will rely on both verbal and non-verbal communications. This also highlights the importance of facial expressions and overall appearance including a comfortable height. In this paper, we present the design of iSocioBot in terms of appearance, mechanics and software, and furthermore discuss some experience gained from deploying the robot in real and uncontrolled environments.

---

[1]https://www.aldebaran.com/en/humanoid-robot/nao-robot.

[2]https://www.willowgarage.com/pages/pr2/overview.

[3]http://socialrobot.dk.

2 ISocioBot

For a robot to engage in natural and efficient interaction with persons, it needs to be able to sense what the persons are doing and be able to express itself in a similar way as they do. This requires integrating various sensors and actuators into a single system. The physical appearance and construction along with considerations regarding choice of sensors are described in this section.

2.1 Mechanical Design

The preliminary design of the robot body is seen in Fig. 1 including specification of size. We have decided on a humanoid robot with a soft appearance because it is to interact with humans for long periods in a social context, thus it cannot resemble an industrial robot. The height of 1.49 m is chosen to make users feel comfortable both while standing and sitting. In addition, it is also good for camera view on the top of the robot in terms of face detection and recognition. Figure 2 shows the actual implementation of the design. Due to the payload capacity of the base, we needed to limit the weight of the robot body, hence plastic and light wood materials were used to build the body of the robot. Acoustic clothes was used to cover the robot to make it more friendly to users, less mechanical looking and reduce the weight. The wide crack in the body of the robot below the face in Fig. 1 was originally intended for a microphone array, however after having chosen the acoustic clothes, the microphone array was placed completely hidden inside the body with satisfactory performance. This was also the case with the loudspeaker. We built the robot head using LED array and strips, such that the robot can show different types of expressions and even animations regarding the user's varying activities.

2.2 Electronic Design

iSocioBot is self-contained since all the electronic parts are inside its body and run on battery. The electronic parts are commercial off-the-shelf components making it easy to replicate the iSocioBot. The electronic system of our iSocioBot consists of computer, movement base, sensors, feedback module and power module. The details of all electronic parts can be found in Table 1.

- **Computer**: We use a DELL E6540 laptop to process the information from every sensor and control the robot. All the other electronic parts are connected to this computer through USB port.

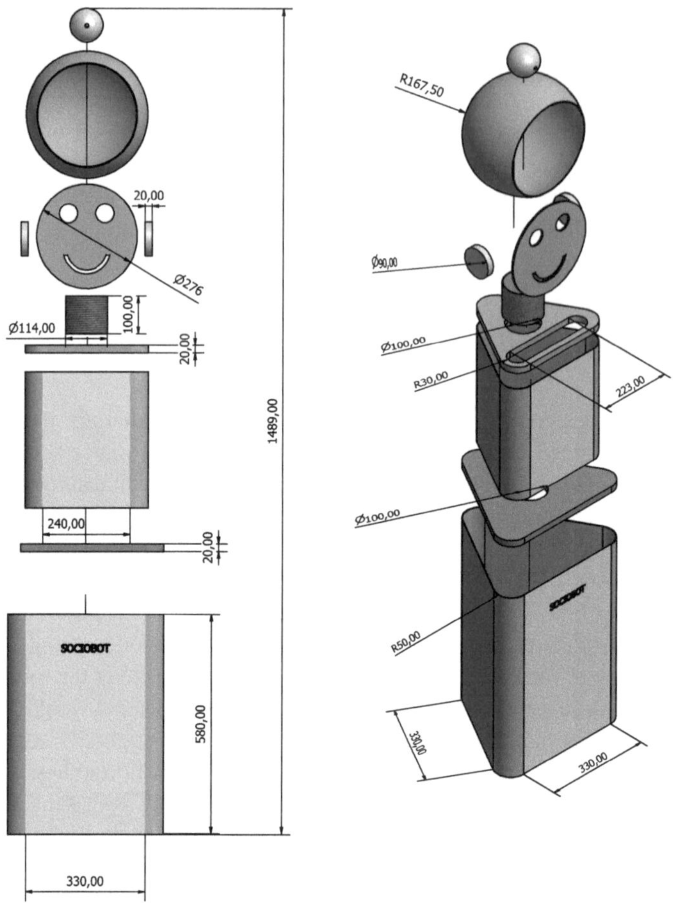

Fig. 1 The mechanical design of iSocioBot

- **Movement base**: The ability to move is achieved through a Kobuki low-cost mobile research base which is a part of the TurtleBot.[4]
- **Sensors**: Four types of sensors are used to gather information for the robot; Microsoft Kinect, Singstar microphone, camera and laser scanner. The Kinect, which is hidden by the acoustic clothes due to design consideration, is only used as microphone array. Singstar microphone is a wireless handhold microphone, which is used for close-talking speech recognition. To gather visual information, we install a High-Definition camera Logitech HD Pro Webcam C920 on the top of our iSocioBot. For localization purpose, a laser scanner, Hokuyo URG-04LX-UG01, is equipped in the middle of the robot body.

[4]http://www.turtlebot.com.

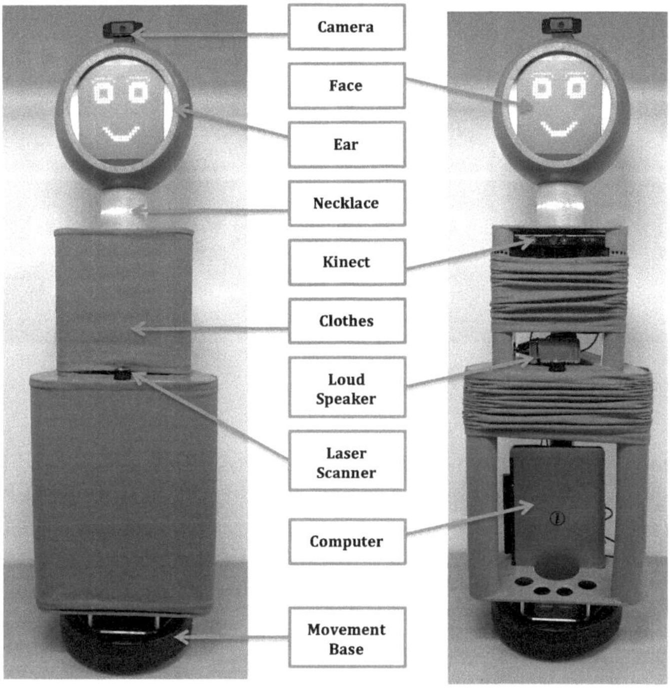

Fig. 2 The iSocioBot

Table 1 Electronic components used for iSocioBot

Part	Model	Note
Computer	Dell E6540	Intel Core2 i7-4800MQ processor, 16 GB RAM and 256 GB SSD hard-drive
Microphone array	Microsoft Kinect	
Wireless microphone	Singstar	
Camera	Logitech C920	Full HD 1080p recording
Laser scanner	Hokuyo URG-04LX-UG01	Range (5600 mm × 240°), Accuracy (±30 mm)
Loudspeaker	JABRA SOLEMATE MINI	
Head	Custom	Arduino (Mega 2560) for control, one 32 × 32 RGB matrix LED panel for face, and three Pololu RGB LED strips
Movement base	Kobuki	Maximum translational velocity: 70 cm/s, maximum rotational velocity: 180°/s, payload: 5 kg (hard floor), 4 kg (carpet)

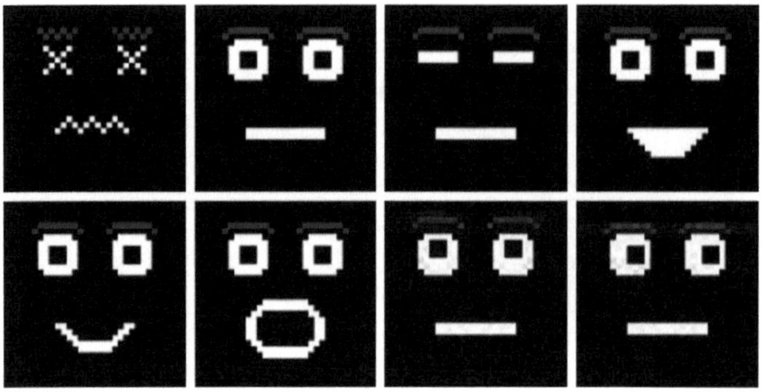

Fig. 3 Facial expression samples of the robot

- **Feedback**: There are two parts for the feedback; loud speaker and head. The loud speaker is used for speech synthesis, which is also installed behind the acoustic clothes. Head is composed by face, necklace and ears, which are used to indicate the robot is listening, thinking and talking and also show different facial expressions. The hardware of head consists of Arduino (Mega 2560) for control, one 32 × 32 RGB matrix LED panel for face, and three Pololu RGB LED strips for two ears and one necklace. The robot can display different facial expressions based on the results from the speech recognition and person identification modules, according to predefined scripts or potentially using data-driven approaches [9]. Figure 3 shows some facial expressions that can be displayed by this system. Furthermore, by switching between two facial expressions at a low frequency, we are able to make the robot appear to be talking, listening etc. This makes the interaction with people more natural and dynamic.
- **Power module**: The base and Kinect are powered by the battery inside the base. The head is powered by an extra battery while the speaker and laptop are powered by their built-in battery separately.

3 Software

The iSocioBot software is developed to run on The Robot Operating System (ROS),[5] which runs on top of the standard OS (e.g., Ubuntu). Having a robot which runs ROS offers many advantages such as:

- ROS offers a great framework for each module/function to communicate (e.g., fusing speaker recognition with face recognition) in real-time.

[5]http://www.ros.org/.

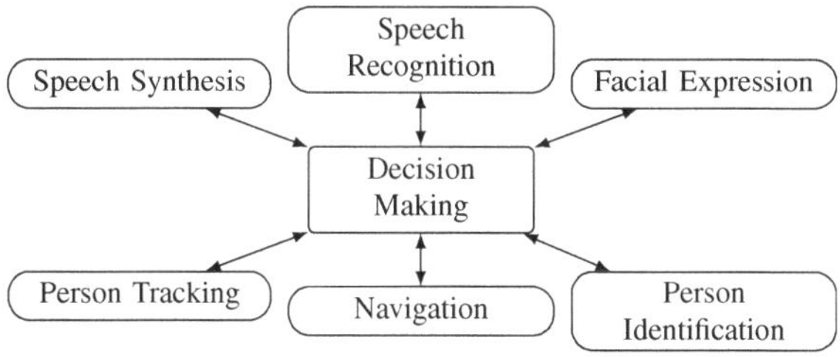

Fig. 4 Illustration of the iSocioBot software. *Arrows* indicate communication

- ROS is widely used thus much high-quality software is available. This makes it easier and faster to integrate various sensors (e.g., laser scanner, camera, arm etc.) without much effort.
- ROS is open-source which means that everyone can use it for free, thus making our research as reproducible as possible.
- ROS software can be written in Python or C++, thus allowing for fast development of code using Python and computationally efficient code using C++.

The high-level structure of the iSocioBot code framework is shown in Fig. 4. All decisions and actions are made in *Decision Making* based on the input from all other modules. All modules except for *Decision Making* are written in a generic way in terms of input-output, such that only *Decision Making* needs to be re-written if a new application is desired. This also has the benefit of allowing for easy substitution of a module without having to change in other modules, e.g., if one wanted to test different face detection algorithms. It is noted that the iSocioBot code partly relies on existing code and external libraries such as OpenCV, ALIZE, and partly on our own work [11, 12].

4 Evaluation

4.1 Public Demonstrations

iSocioBot has been an attraction at three different public events where it has demonstrated different applications. At the official opening of "Research Days of Denmark" in Aalborg and the opening of "Safe 7" in Nibe, the robot took part in on-stage scripted dialogues with the presenters and also performed a Q/A session, where it was supposed to answer a closed set of question and turning towards the speaker. iSocioBot also participated in "Culture Night" at the Danish Ministry of Higher Education and Science, where people were asked to take part in a dialogue with the robot and afterwards answer a questionnaire about the behaviour and appearance of iSocioBot. A total of 97 persons (mainly kids at the age of

6–12 years) participated and one conclusion was, that people liked the appearance of iSocioBot compared to robots with a more mechanical appearance.

4.2 Our Own Experience

Based on the participation at the above mentioned events and running tests, we have gained some valuable experience about what should be improved for the next version of iSocioBot.

- **Choice of base**: The base of the robot has turned out to be a weak link in several ways; it can only carry a payload of 5 kg which is too little when using an onboard laptop, and it is furthermore very sensitive to weight distribution resulting in unequal traction at the two active wheels which degrades the navigation and in practice makes it unable to drive around freely, thus limiting it to rotation in-place. The weight distribution on the current design combined with the current base also forces the robot to drive over bumps etc. at very low speeds in order not to tip over.
- **Power Consumption**: When all software modules are running the laptop is only capable of running for $\approx$1 h with a 9-cell 97 WH battery, which is a limiting factor for many applications. It should be noted that no effort has been made to optimizing the code or the choice of sensors with respect to power consumption.

4.3 Future Improvements

The issues regarding the base can be solved by either redistributing the weight or simply replacing it with a more powerful base such as Pioneer 3DX.[6] The first option can be achieved by making iSocioBot smaller, however this will degrade the appearance and thus the human-robot interaction, which is not desirable. The second option is more attractive from a mechanical point of view, however the price for good alternatives are considerably higher than for the current base.

One way of minimizing the power consumption is to shift all the heavy computational tasks from the on-board laptop to a server. This will also lower the weight of the robot, since only a less powerful computer is needed to handle communication with the server, the sensors and the actuators. This solution comes at the expense of being dependent on having a WiFi available. Another drawback of this solution is the introduction of an increase in the complexity of developing software for iSocioBot.

[6]http://www.mobilerobots.com/ResearchRobots/PioneerP3DX.aspx.

5 Conclusions

This work has presented the first iteration of designing and implementing the social robot called iSocioBot. Based on the implementation some shortcomings have been identified and possible solutions have been presented. One direction of future work is to make a second iteration to correct these shortcomings. Other directions of research will focus more on user modelling and the application of iSocioBot.

Acknowledgments The authors would like to thank Ben Krøyer and Peter Boie Jensen for constructing the iSocioBot and Trine Skjødt Axelgaard for making the mechanical design of the iSocioBot.

References

1. Baddoura, R., Venture, G.: Social vs. useful hri: experiencing the familiar, perceiving the robot as a sociable partner and responding to its actions. Int. J. Social Robot. **5**(4), 529–547 (2013)
2. Beňuš, Š.: Social aspects of entrainment in spoken interaction. Cogn. Comput. **6**(4), 802–813 (2014)
3. Fortenberry, B., Chenu, J., Movellan, J.: Rubi: A robotic platform for real-time social interaction. In: Proceedings of the International Conference on Development and Learning (ICDL04), The Salk Institute, San Diego (2004)
4. Kanda, T., Ishiguro, H., Imai, M., Ono, T.: Development and evaluation of interactive humanoid robots. Proc. IEEE **92**(11), 1839–1850 (2004)
5. Kober, J., Peters, J.: Reinforcement learning in robotics: a survey. In: Reinforcement Learning, pp. 579–610. Springer (2012)
6. Manohar, V., Crandall, J.: Programming robots to express emotions: interaction paradigms, communication modalities, and context. IEEE Trans. Hum. Mach. Syst. **44**(3), 362–373 (2014)
7. Metta, G., Sandini, G., Vernon, D., Natale, L., Nori, F.: The iCub humanoid robot: an open platform for research in embodied cognition. In: Proceedings of the 8th Workshop on Performance Metrics for Intelligent Systems, pp. 50−56. ACM, New York (2008)
8. Salichs, M.A., Barber, R., Khamis, A.M., Malfaz, M., Gorostiza, J.F., Pacheco, R., Rivas, R., Corrales, A., Delgado, E., Garca, D.: Maggie: a robotic platform for human-robot social interaction. In: IEEE Conference on Robotics, Automation and Mechatronics, pp. 1–7 (2006)
9. Shepstone, S.E., Tan, Z.H., Jensen, S.H.: Using audio-derived affective offset to enhance tv recommendation. IEEE Trans. Multimedia **16**(7), 1999–2010 (2014)
10. Springer: International Journal of Social Robotics. http://www.springer.com/about+springer/media/pressreleases?SGWID=0-11002-6-805078-0 . Accessed 19 Dec 2014
11. Tan, Z.H., Lindberg, B.: Low-complexity variable frame rate analysis for speech recognition and voice activity detection. IEEE J. Sel. Top. Sign. Process. **4**(5), 798–807 (2010)
12. Thomsen, N.B., Tan, Z.H., Lindberg, B., Jensen, S.H.: Improving robustness against environmental sounds for directing attention of social robots. In: Proceedings of the 2nd Workshop on Multimodal Analyses Enabling Artificial Agents in Human-Machine Interaction, Singapore (2014)

Portable Design and Range of Motion Control for an Ankle Rehabilitation Mechanism Capable of Adjusting to Changes in Joint Axis

A. Szigeti, Y. Takeda and D. Matsuura

Abstract This paper proposes improvements to a simple spatial rehabilitation mechanism for the human ankle, aiming to prevent exertion of forces on the joint, and control the range of flexion it experiences. Such functions would additionally provide information on recovery status in terms of degree of mobility regained, and the force and torque on the affected joint. This information would both provide support for physiotherapists in their work and help to establish self-rehabilitation on the part of the patients themselves. The mechanism presented is built upon previous research of an extended Oldham's coupling mechanism combined with a driving four-bar linkage. A complete rehabilitation device is designed and produced to be lightweight, affixed only to the patient's leg with no additional supporting structure, and able to drive motion along a variety of total flexion angles by changing link lengths using a removable 10 mm pin and pre-drilled holes.

Keywords Rehabilitation robotics · Mechanism design · Angle of flexion control · Self-rehabilitation · Passive adaptation to the spatial eccentricity of a human joint

1 Introduction

As Japan moves towards an aged society, the need for more efficient rehabilitation mechanisms becomes increasingly apparent. Limbs weaken with age, which increases susceptibility to leg injuries that can render a person bedridden. Ankle injuries in particular can take a long time to recover due to the complexity of the

A. Szigeti · Y. Takeda (✉) · D. Matsuura
Department Mechanical Sciences and Engineering, Tokyo Institute of Technology, Tokyo, Japan
e-mail: takeda@mech.titech.ac.jp

A. Szigeti
e-mail: szigalex@gmail.com

S. Bai and M. Ceccarelli (eds.), *Recent Advances in Mechanism Design for Robotics*, Mechanisms and Machine Science 33,
DOI 10.1007/978-3-319-18126-4_12

human ankle joint, which lacks a constant axis of rotation [1, 2], in actuality consisting of two separate joints acting as one complex [3]. This in turn results in a therapeutic process that is especially difficult to apply both promptly and effectively, being carried out personally by physiotherapists (PTs). At the present there is no simple and cost-effective mechanism developed that is capable of aiding swifter ankle rehabilitation, while also adjusting to a variety of users for self-rehabilitation. To solve this problem, concepts have been proposed ranging from a 6-DOF closed-loop mechanism [4], to combining grooved cams and non-circular gears [5]. Although effective, these mechanisms cannot dynamically adapt to a variety of users and must be custom-fit to match different body types. This issue can be solved with features such as a combination of prismatic joints [6], but they contain more DOF than the interacting joints, and rapidly become heavy, complicated, and expensive, while being unable to adapt to spatial motion.

In regards to this problem, the Tokyo Institute of Technology's Mechanical Systems Design Laboratory proposed an ankle rehabilitation mechanism based on a one-DOF Oldham's coupling mechanism that is capable of adjusting to changes in the joint axis [7]. This type of coupling is particularly known for being a compact mechanism designed to transmit rotational motion from one axis to another despite misalignment. By altering the design of such a coupling to expand upon these principles, any potentially dangerous transfer of forces to the user's ankle are avoided by restricting the rehabilitation motion to only the ranges of dorsiflexion (DF) and plantar flexion (PF), and compensating for all other motions using the mechanism's structure. However, a mechanism capable of either varying its motion range, or having a portable and adaptable design intended to be fit to a user's leg without an external support structure were not considered. Once these requirements are fulfilled, a practical, ready-to-use device can be built and tested. Proving the validity of such a low-cost and adaptable mechanism would make it an attractive proposition for production, while its ability to reduce the labor required of PTs without sacrificing quality of rehabilitation would lead to its ultimate adoption by hospitals and clinics. Thus the mechanism proposed in this paper seeks to improve upon the Oldham's coupling design by adding both range of motion control and portability, as well as serving as a testable prototype to aid in further refinement of the concept.

2 Mechanism Objectives, Features and Analysis

The mechanism concept is illustrated in Fig. 1, where θ_{in} and θ_{out} represent the input and output angles of the four-bar linkage OABC, and the XYZ and $x_a y_a z_a$ coordinate systems represent the locations of the driving motor, and human ankle joint at origins O and o_a, respectively. The mechanism joints vary between being cylindrical (C) or revolute (R) connections. In this paper, by combining the "driving

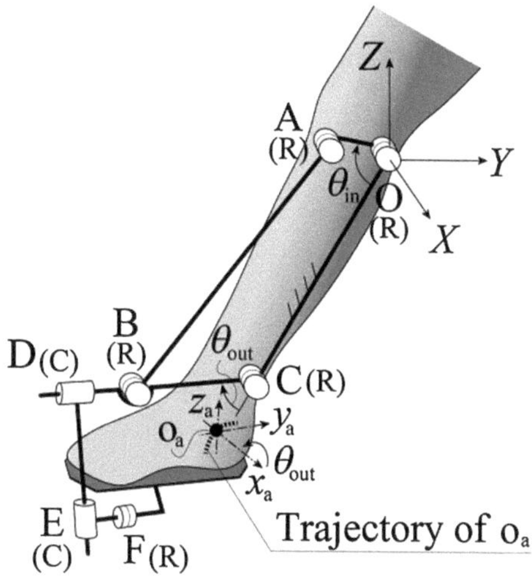

Fig. 1 Representation of proposed mechanism structure with ROM control capability

mechanism" consisting of the four-bar linkage OABC, with the original spatial, single DOF rehabilitation mechanism $CDEFo_a$ derived from Oldham's mechanism [7], we proposed a mechanism capable of providing Range of Motion (ROM) control over both DF and PF. Therefore by taking the problem background into consideration, the following necessary conditions were derived:

(1) Facilitate user self-rehabilitation by providing a system to easily adjust the length of key links, and consequently ROM, via a removable pin.
(2) Provide a lightweight and portable device usable regardless of body type.

For the purpose of meeting these requirements, a series of analyses have been carried out to improve understanding of the mechanism and provide a means of checking experimental results.

2.1 Geometric Analysis of the Driving Mechanism

As most mechanism joints passively adapt to the spatial motion of the human ankle [7], the four-bar linkage is separated for analysis in order to achieve ROM control. This will be accomplished through the Total Flexion Angle (TFA), or the combination of the complete PF and DF ranges of motion. In the mechanism, it consists of the angle centered on the output link, (length c) of the four-bar mechanism when it is perpendicular to the fixed link (length d) during a cycle of movement, or the envelope ϕ of the output link's motion. This concept is illustrated in Fig. 2.

The two "extreme positions" of this mechanism can be defined as the points where the output link is at either end of its envelope of motion, or the maximum and

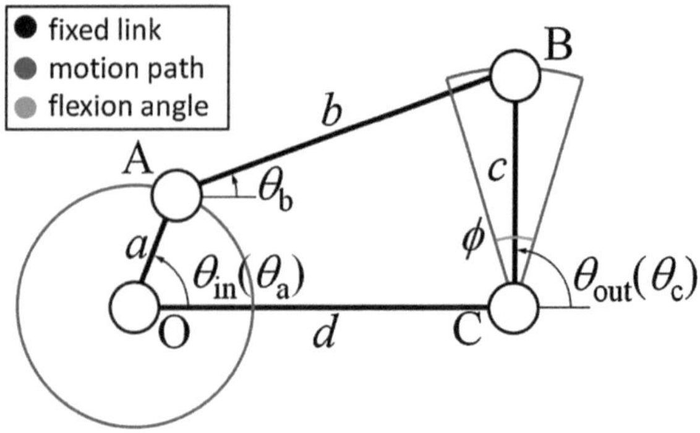

Fig. 2 Planar overview of driving/driven link ranges of motion and TFA of the mechanism

minimum values for θ_{out}. At each of these points the 4-bar mechanism forms one of two simple enclosed triangles, allowing the following relationships between links to be derived using trigonometric identities:

$$c = \sqrt{\frac{a^2 - d^2}{1 - \frac{d^2}{a^2}\sin^2\left(\frac{\phi}{2}\right)}},\ b = \frac{d}{a}\sin\left(\frac{\phi}{2}\right)c \tag{1}$$

where a, b, and c are the lengths of the moving links, d is the length of the fixed link, and ϕ is the TFA. For the purpose of creating a unified design, assume that a and d are constant values.

2.2 Kinematic Analysis of the Driving Mechanism

To better understand the driving mechanism and verify results, a kinematic analysis is performed on the four-bar linkage. By Freudenstein's Equation [8], the following relationship between the angular displacements θ_a and θ_c holds that

$$K_1 \cos\theta_c - K_2 \cos\theta_a + K_3 = \cos(\theta_c - \theta_a). \tag{2}$$

where,

$$K_1 = d/a,\ K_2 = d/c,\ K_3 = (d^2 + a^2 - b^2 + c^2)/(2ca). \tag{3}$$

Equation (2) can then be resolved to:

$$\theta_c = 2\arctan\left\{(-B \pm \sqrt{B^2 - 4AC})/(2A)\right\}. \tag{4}$$

where,

$$A = \cos\theta_a(1 - K_2) + K_3 - K_1, \tag{5}$$

$$B = -2\sin\theta_a, \tag{6}$$

$$C = -\cos\theta_a(1 + K_2) + K_1 + K_3. \tag{7}$$

By further independent analysis of the system using the relationships described in this section, it can also be found that

$$\theta_b = \arcsin\{(\sin\theta_c - a\sin\theta_a)/b\}. \tag{8}$$

The angular relationships determined in Eqs. (4) and (8) for the mechanism consequently allow for movement onto the design of a tangible prototype.

3 Mechanism Design and Testing

Once the design parameters have been determined, a prototype was constructed and evaluated based on a series of design considerations explained in the following sections.

3.1 Design Specifications

By using Eq. (1) and varying the TFA ϕ, a range of parameters is determined as displayed in Table 1. Here, Δb and Δc are the changes in link length caused by switching forward to the angle in the current row, away from the angle in the previous row, and the shaded values are results dismissed as unusable as explained below. The resulting lengths of b and c outside the range of 90–400 mm were dismissed for being exceedingly long, thin links that would hinder the ease of use necessary for self-rehabilitation. Similarly, values for Δb and Δc exceeding 200 mm in magnitude would prove impractical as this is the distance a pin would need to be switched to readjust the mechanism. The pin must also fit into a hole approximately 10 mm in diameter in order to provide a stable connection, so all values of Δb and Δc approximately at or below 10 mm in magnitude are also rendered unusable. Lastly, values of ϕ over 45 % are undesirable both for the risk of causing painful joint overextension.

As can be seen in Table 1, the two only completely usable rows that appear led to the selection of TFAs of 25° and 30° for their appealing qualities in regards to design. By skipping over 35° and 40°, the angle of 45° was also successfully added

Table 1 Table driven by a range of TFA values for selecting design specifications (link lengths *a* and *d* are fixed at 34.9 and 300.0 mm respectively)

ϕ (°)	*b* (mm)	*c* (mm)	Δb (mm)	Δc (mm)
15	657.1	585.6	X	X
20	401.3	268.9	−255.8	−316.7
25	353.3	189.9	−48.0	−79
30	333.6	149.9	−19.7	−40
35	323.1	125.0	−10.5	−24.9
40	316.9	107.8	−6.2	−17.2
45	312.8	95.1	−4.1	−12.7

Table 2 Final design parameters

X	Minimum range setting	Middle range setting	Maximum range setting
ϕ (°)	25	30	45
a (mm)	34.9	34.9	34.9
b (mm)	353.3	333.6	312.8
c (mm)	189.9	149.9	95.1
d (mm)	300.0	300.0	300.0
K_1	8.60	8.60	8.60
K_2	1.58	2.00	3.15
K_3	0.186	0.229	0.364

to the mechanism's range, fixing the issue of its acceptable *b* and *c* yet impractical Δb and Δc values. The final parameters resulting from this process and Eq. (1) are presented in Table 2.

3.2 Resulting Mechanism Design

For a labeled picture of the mechanism's finished structure, see Fig. 3.

As link 1 (disc) is required to interface with the DC motor by an especially rigid connection in order to drive the mechanism, aluminum was selected to provide this rigidity while still being fairly lightweight and low-cost. For the remainder of the links, ultra-high density polyethylene (UPE) was selected for its high strength, low weight qualities. Such a selection helps the mechanism to avoid any unexpected influences from gravity. To properly represent all of the degrees of freedom, two cylindrical joints were designed to allow for smooth translation and rotation.

Link 1 was specifically converted to a disc shape to capitalize on the 4-bar mechanism design. This structure benefits from following the same range of motion while allowing for the user to rest their leg on a flat surface like a chair without snagging as seen in Fig. 4, as the disc spins freely against most surfaces. Such a

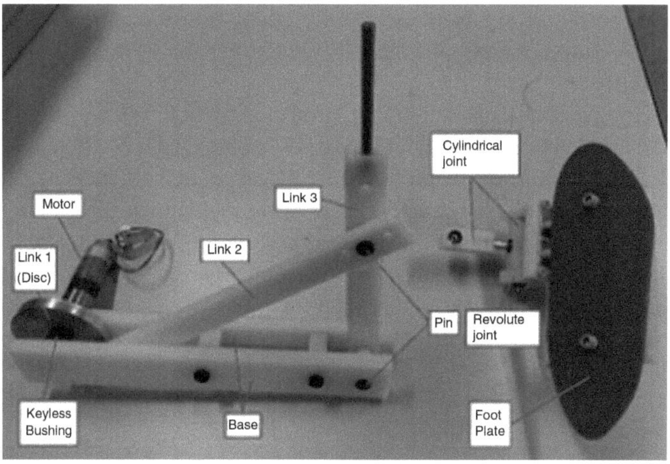

Fig. 3 Labeled image of basic, disassembled mechanism design

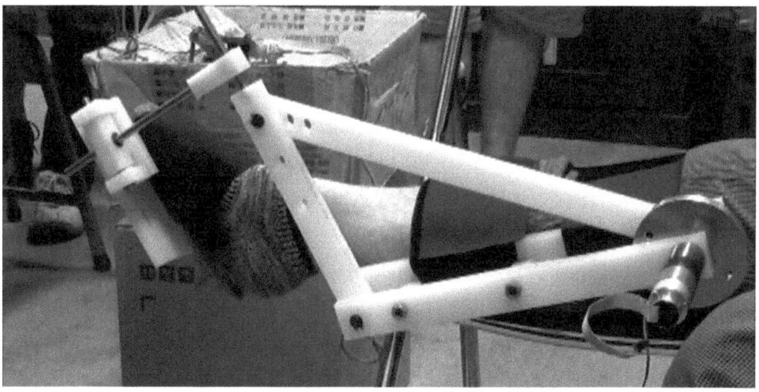

Fig. 4 Complete mechanism attached to user

feature renders the mechanism easy to use in many different locations. The foot plate is made of a sheet of laser-cut acrylic to achieve the efficient, custom shape visible in Fig. 3, and multiple identical sheets have been cut out such that increasing the strength of the plate can be achieved by inserting additional sheets with longer bolts. All bolts assembled onto the foot plate are made of resin to further reduce weight.

The links are connected to each other by pins, with link 1's pin being threaded directly into the aluminum disc. A series of three holes each along links 2 and 3 allows the TFA to be changed by effectively modifying the lengths of these two links while holding all other parameters constant, as seen in Eq. (1). The pins attached to link 3 and the base also serve to support the mechanism via a resin spacer within which it can spin freely, creating an effectively fixed column between links. Therefore the mechanism specifically adapts to different stages of rehabilitation by providing a means of gradually increasing the TFA as the user regains ankle functionality.

The moving parts of the assembly are protected on both sides by the base, which is connected along the middle by two square columns. One half of the base provides a stable mounting point for the motor while the opposite half attaches to the user's leg by a pair of wide Velcro straps. The smooth, rectangular plate that is the base also helps to prevent snagging on the user's clothing. As the assembly fixes the user's leg as the "ground" link, it also causes unexpected forces generated during use to be transmitted to the leg, rather than the more vulnerable ankle joint. Such a feature helps to further protect against unexpected misuse scenarios during self-rehabilitation.

3.3 Range of Motion and Testing

The completed mechanism was tested for range of motion as well as ability to adapt to changes in axis of rotation. It was then set to run independently through its ROM under each pin setting, and the ROM was confirmed as properly decreasing or increasing respective to each setting. However, more sophisticated testing is required to verify the exact angles produced by the practical mechanism. The mechanism was affixed to the leg of several volunteers ranging in height from 170 to 200 cm and set to run for 1 min in order to test adaptability. The process of attaching the mechanism was simple enough for the user to carry out in a comfortable and safe manner. Furthermore, the mechanism is constructed such that once it was affixed and the user's leg, set on a flat surface, and switched on, there was no special adjustment necessary for it to run automatically along its controlled motion. This supports the objective of furthering self-rehabilitation as a user could either rehabilitate without a PT's aid, or potentially even completely independently. All users reported the mechanism as comfortable to use, without any forces imparted to the foot save for the constrained DF and PF motion. However, it should be noted that these tests serve as a relatively simple proof-of-concept and cannot be empirically substantiated without more thorough testing such as inserting a force gauge into the foot plate to take data while in use.

3.4 Safety of Use in Rehabilitation

Although the mechanism was designed with safety and stability as primary concerns, it is important to address potential misuse scenarios during rehabilitation. As the device was designed with a feature that allows its constrained range of motion to be changed via pin placement, proper instruction must be given to users that the device can cause potentially harmful discomfort if used at ROM settings that exceed what their current degree of ankle functionality should tolerate. This result would be the same as if an inexperienced PT had over-flexed a patient's ankle, however it should be noted that the mechanism's ranges of motion are not large enough to harm an uninjured human ankle. Here, risk can be reduced by the addition of an "emergency" switch for the user that immediately returns the mechanism to its zero-flexion positon and shuts it off. As seen in Fig. 3 the pin connecting links 2 and 3 is connecting the middle holes in a series of three, but in Fig. 4 it connects the top holes. This system allows for switching of the range of motion between the three available settings, but is only intended to be used with the respective pairs of top, middle, or bottom holes in the series, and if mismatched will create unintended ROM that could cause the user physical discomfort. Therefore, in any final product care must be taken to label proper connection of the movable pin.

4 Conclusions

This paper has discussed the design and analysis of a simple rehabilitation mechanism capable of adapting to fluctuation in both rotation axis and load. Based on the constructed prototype mechanism, experiments will be carried out to further validate the functions of adaptation to joint axis and adjustment to loading force. The results obtained can be summarized as follows.

(1) Geometric analysis of the driving 4-bar linkage was carried out for the purpose of instituting control over the TFA without changing the orientation of the user's foot. Using a removable pin and matched to a series of holes, the stability of the orientation as perpendicular to the static link while varying TFA size was successfully achieved, and further experimentation will verify the exact accuracy of the TFAs.
(2) The device was designed with safety, adaptability, and portability in mind, in order to further self-rehabilitation by users. The device can be strapped onto one's leg without aid, and is easy to manipulate due to its compact, lightweight structure of primarily aluminum and UPE. Misuse scenarios can be avoided by clear explanation of use, and due consideration of the user's ankle functionality when using the mechanism.
(3) General validation of the desired flexion characteristics and usability of the proposed prototype form a basis from which we will further investigate its characteristics by empirical testing using a force gauge. This will lay the foundation for further additions for improving safety and functionality such as the introduction of shock-absorbing springs in the system.

References

1. Malosio, M., et al.: Analysis of elbow-joints misalignment in upper-limb exoskeleton. IEEE International Conference on Rehabilitation Robotics, (2011)
2. Gregorio, R., et al.: Mathematical models of passive motion at the human ankle joint by equivalent spatial parallel mechanisms. Med. Biol. Eng. Compu. **45**(3), 305–313 (2007)
3. Dai, J.S., et al.: Sprained ankle physiotherapy based mechanism synthesis and stiffness analysis of a robotic rehabilitation device. Auton. Robot. **16**(2), 207–218 (2004)
4. Jin, D., et al.: Kinematic and dynamic performance of prosthetic knee joint using six-bar mechanism. J. Rehabil. Res. Dev. **40**(1), 39–48 (2003)
5. Terada, H., et al.: Developments of a knee motion assist mechanism for wearable robot with a non-circular gear and grooved cams. Mech. Mach. Sci. **3**(2), 69–76 (2012)
6. Steinen, A.H.A., et al.: Self-aligning exoskeleton axes through decoupling of joint rotations and translations. IEEE Trans. Robot. **25**(3), 628–633 (2009)
7. Matsuura, D., et al.: Design of ankle rehabilitation mechanism using a quantitative measure of load reduction. In: Ceccarelli, M., Glazunov, V.A. (eds.) Advances on Theory and Practice of Robots and Manipulators, Springer, pp. 27–36 (2014)
8. Freudenstein, F.: An analytical approach to the design of four-link mechanisms. Trans. ASME **76**(3), 483–492 (1954)

Dynamic Modeling and Torque Distribution of a Climbing Hexapod Robot

Dong Liu, Weihai Chen, Zhongcai Pei, Jianhua Wang and Xingming Wu

Abstract This paper deals with the kinematics, dynamics and joint torque distribution of a novel hexapod robot. In order to climb over large obstacles or high steps, a neck joint has been installed between the front and central part of the body. The formulation of dynamics is performed by the Lagrange's equations, using the robot screw theory and product of exponential method. The torque distribution model is settled based on the inverse dynamics and force distribution of the tip point. The analysis has been verified by simulation and experiments to further improve the design and control of the hexapod robot.

Keywords Dynamics · Torque distribution · Screw theory · Hexapod robot

1 Introduction

It has long been recognized that legged robots can provide a superior mobility platform for complex environment to wheeled or tracked ones. This is a result of the abilities of legs to use discrete footholds to contact with the ground and adapt to various terrains. Legged robots include single-legged hoppers, two-legged humanoid robots, quadruped robots, hexapod robots, and octopod robots. In recent years, hexapods have gained the interest of international research community, as they possesses the preferred all-round performance in the field of robotics for their better static stability, more flexibility and terrain adaptability, compared to other types [1–3]. However, hexapods are still unable to traverse highly fractured and unstable terrain, e.g., large obstacles and high steps. In most cases, they detect the

D. Liu (✉) · W. Chen · Z. Pei · J. Wang · X. Wu
School of Automation Science and Electrical Engineering,
Beihang University (BUAA), Beijing 100191, China
e-mail: dong.liu@epfl.ch

S. Bai and M. Ceccarelli (eds.), *Recent Advances in Mechanism Design for Robotics*, Mechanisms and Machine Science 33,
DOI 10.1007/978-3-319-18126-4_13

surroundings with multiple sensors and execute obstacle avoidance behaviors [4]. This would largely restrict the applications of hexapod, especially climbing stairs, which is a fundamental and also challenging behavior for small robot [5].

Several attempts have been made by various research groups to solve this problem. RHex had been designed with small size, compliant legs and one actuator per leg in a reliable manner, which was the first to be able to climb full-scale stairs [6]. RiSE [7] was constructed later to be the first legged robot capable of walking on diverse terrain in both the ground and scansorial regimes, without the use of suction, magnets, or adhesives. A novel hexapod named 'ASTERISK' [8] was proposed for ladder climbing. By alternately changing its posture and distributing the weight on supporting legs based on its force margins, ASTERISK could achieve a stable climbing behavior.

In contrast to the aforementioned configurations, rectangular hexagonal ones are more widely used. Our previous work presented a multiple chaotic central pattern generators model for the hexapod locomotion, with learning mechanism to handle some malfunction compensation [9]. For this kind of hexapod, an additional body joint has been added for climbing over obstacles. Bartsch et al. [10] presented their Space Climber with the body joint to enable the robotic system to adapt to the surface structure. The design of the body joint was based on its leg actuators. AMOS-WD06 was proposed by Manoonpong et al. [11] with 19 active degrees of freedom (DoFs) to realize sensor-driven omni-directional walking. Neural control and learning mechanisms were performed in the six-legged prototype without modeling of this machine or dynamic analysis of its climbing behaviors. Besides, the most frequent sequence used by cockroaches when climbing high obstacles was to first elevate their front of the body upward, and then raise their center of mass (COM) to the level of the top of the obstacle. This had been verified by movement observation of cockroaches conducted by Watson et al. [12] and Harley et al. [13]. During climbing attempts, cockroaches bend their body to swing their front leg and redirect their middle legs to successfully reach the top of the block or misses.

Inspired by biology observations and previous works, here in this paper, an addition neck joint was integrated into the body between the front and rear compartments. To achieve the torque mathematic model, especially the moment curve of the neck joint, full kinematics and dynamics were analyzed in this paper. Conventional method was to use free body diagrams to analysis the torque distribution of six-legged robot, provided that the robot remains stationary or keeps moving at the same speed [14, 15]. Then Lagrange-Euler formulation, together with Denavit-Hartenberg (D-H) description of links, was used to calculate the dynamics of the robot. Least square method and minimization of norm joint torques method were proposed in [16, 17]. Compared with these works, a compact form of dynamics of hexapod was presented in this contribution, using the robot screw theory and exponential product equations. As a superior alternative to the use of D-H method [18], product of exponentials (POEs) representation, combined with the geometric significance of the twists, were used here to calculate the joint torques of our climbing robot. Torque distributions were achieved in a convenient, compact and systematic form by deriving the description of kinetic and potential energy.

The sum of the squares of each joint torque was computed as the index of energy consumption, which was straightforward and determined. Torque distribution and tip-point distribution were then compared to obtain a more energy efficiency for the hexapod performance.

The remaining part of the paper is organized as follows: Sect. 2 describes the walking platform which would execute the climbing tasks. Kinematics, dynamics model and torque distribution analysis are presented in Sect. 3. Simulation results and experimental performances are stated and discussed in Sect. 4. Finally, conclusions and future works are listed in Sect. 5.

2 Walking Platform

A six-legged walking machine was developed to realize field experiments. The body is made up of two parts, connected with an extra neck joint, which can be activated to rotate around the lateral axis (maximum 30°). The hexapod has six identical legs, each of which consists of three linkages. The six legs are defined as leg j (j = 1, 2, …0.6), as shown in Fig. 1a. Considering the natural counterpart, the three links are named coxa, femur and tibia, respectively. Correspondingly, the three rotation joints in each leg are denominated as TC (the thoraco-coxal joint enables forward and backward movements), CTr (the coxa-trochanteral joint, enables elevation and depression of the leg), and FTi (the femur-tibia joint enables extension and flexion of the tibia). A single leg serves as a three-revolute kinematical chain, with the tarsus segment ignored. Besides, a spring-sensor system was installed on each tibia to absorb the impact force when the tip point touches down the ground. With the force sensor, the contact force could be measured and serves as feedback to the control loop, which would be discussed in our future work but it is not the focus here. The central control board is based on an ARM (*Cortex—M*3, *LPC*1752) processor and all the signals from the sensors are fused in the central control module.

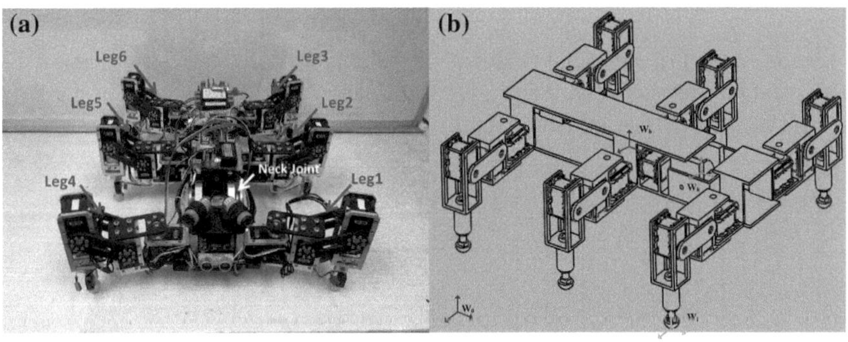

Fig. 1 **a** Prototype of the hexapod robot; **b** reference frames assigned at the robot

3 Kinematics, Dynamics and Torque Distribution

In Fig. 1b, the hexapod robot was at its initial position, with the body paralleling to the ground. To describe the position and orientation, several reference coordinate systems were set: the absolute reference frame (the inertial frame) W_0, body frame W_b (attached to the geometric center of the body), neck joint frame W_{k} (located in the cone vertex of the straight bevel gears), and single leg frame W_j, which was attached to the tip point of *j*th leg. Let point *p* be an arbitrary point on the robot, according to the theory of rigid body transformations,

$$p_{W_0} = p_{W_0 W_b} + R_{W_0 W_b} p_{W_b} \tag{1}$$

where p_{W_0}, p_{W_b} are the coordinates of an relative to frames W_0 and W_b, respectively, and $R_{W_0 W_b}$ is the rotation matrix. By (1), the problem is transformed from global frame to local coordinate system. As the relative position between W_b and W_k remains unchanged during the locomotion, W_k serves as the basic frame rather than W_b to simplify the description of the transmission system on the neck joint.

3.1 Kinematics and Dynamics

The twist parameters for the revolute joint axes are given by

$$\omega_1 = \begin{pmatrix} 0 \\ -1 \\ 0 \end{pmatrix} \quad \omega_2 = \begin{pmatrix} 0 \\ 0 \\ 1 \end{pmatrix} \quad \omega_3 = \begin{pmatrix} 1 \\ 0 \\ 0 \end{pmatrix} \quad \omega_4 = \begin{pmatrix} 1 \\ 0 \\ 0 \end{pmatrix} \tag{2}$$

and four axis points are chosen as

$$q_1 = \begin{pmatrix} 0 \\ 0 \\ 0 \end{pmatrix} \quad q_2 = \begin{pmatrix} x_0 \\ y_0 \\ 0 \end{pmatrix} \quad q_3 = \begin{pmatrix} 0 \\ y_0 + l_1 \\ z_0 \end{pmatrix} \quad q_4 = \begin{pmatrix} 0 \\ y_0 + l_1 + l_2 \\ z_0 \end{pmatrix} \tag{3}$$

where x_0, y_0 and z_0 are initial positions of the axis with the values (in mm) of 76, –76.5 and 6.8, respectively. And $l_1(41)$, $l_2(60)$ and $l_3(123)$ are lengths of the three links. The joint twists are then given by

$$\xi_1 = \begin{bmatrix} 0 \\ 0 \\ 0 \\ 0 \\ -1 \\ 0 \end{bmatrix} \quad \xi_2 = \begin{bmatrix} y_0 \\ -x_0 \\ 0 \\ 0 \\ 0 \\ 1 \end{bmatrix} \quad \xi_3 = \begin{bmatrix} 0 \\ z_0 \\ -l_1 - y_0 \\ 1 \\ 0 \\ 0 \end{bmatrix} \quad \xi_4 = \begin{bmatrix} 0 \\ z_0 \\ -l_1 - l_2 - y_0 \\ 1 \\ 0 \\ 0 \end{bmatrix} \tag{4}$$

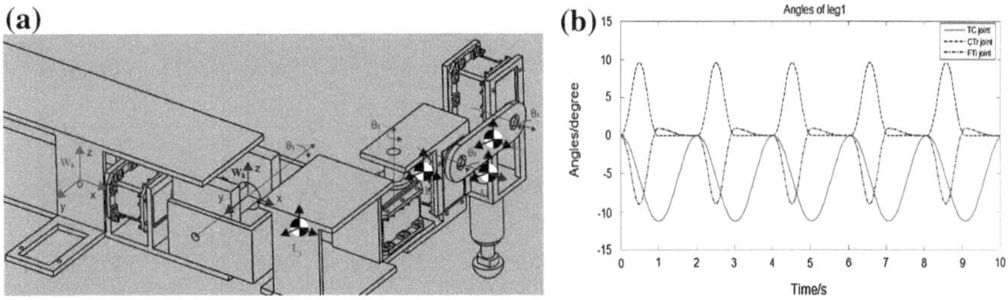

Fig. 2 **a** Kinematics and dynamics of leg 1; **b** joint angles of coxa, femur and tibia

The forward kinematics follows the POE formula, taking leg 1 for analysis as shown in Fig. 2a and

$$g_{W_kj}(\theta_j) = e^{\widehat{\xi_1}\theta_1} e^{\widehat{\xi_2}\theta_2} \ldots e^{\widehat{\xi_4}\theta_4} g_{W_kj(0)} \tag{5}$$

Equation (5) maps the variables from the joint space to the workspace. For the inverse kinematics of the fore legs, the Paden-Kahan sub-problems were introduced here to solve the joint track from the pre-defined tip point trajectories. Rotating about certain axis can be solved as,

$$\theta_i = \arctan 2\left(w_i^T\left(u_i' \times v_i'\right), u_i'^T v_i'\right) \tag{6}$$

where u_i' and v_i' are projected vectors from the rotating points about its axis. The inverse kinematics problem here corresponds to four times rotating a point about one axis each, which could be resolved by substituting these parameters into (6). And the results are described in Fig. 2b.

To each link, a coordinate frame L_i is attached to the COM of ith link as shown in Fig. 2a, with the same rotation matrix as the TC-joint frame.

$$g_{W_kL_i}(0) = \begin{bmatrix} \mathrm{I} & \begin{pmatrix} r_{i1} \\ r_{i2} \\ r_{i3} \end{pmatrix} \\ 0 & 1 \end{bmatrix} \tag{7}$$

where r_{i1}, r_{i2} and r_{i3} have been given in Table 1. The body velocity of the center of mass of the ith link is given by $V^b_{W_kL_i} = J^b_{W_kL_i}(\theta_i)\dot{\theta}_i$, where $J^b_{W_kL_i}$ is the body Jacobian corresponding to $g_{W_kL_i}$, with the form of

$$J_i = J^b_{W_kL_i}(\theta_i) = \left[\xi_1^+ \ldots \xi_i^+ \; 0 \ldots 0\right] \tag{8}$$

Table 1 Physical parameters of the hexapod

Parameters		Body1 (i = 1)	Coxa (i = 2)	Femur (i = 3)	Tibia (i = 4)
Weight (kg)	m_i	0.474	0.193	0.014	0.189
Position of the COM (mm)	r_{i1}	65.487	76.002	98.5	76.002
	r_{i2}	0.770	−112.126	−147.5	−177.500
	r_{i3}	−6.101	−1.342	6.8	-8.403
Moment of inertia (kg mm^2)	$^iI_{xx}$	1484.195	127.769	7.910	335.213
	$^iI_{yy}$	468.958	105.442	0.797	344.161
	$^iI_{zz}$	1586.043	71.314	7.150	38.567
Product of inertia (kg mm^2)	$^iI_{xy}$	−8.391	−0.002	0	0
	$^iI_{xz}$	−31.668	0.024	0	−0.017
	$^iI_{yz}$	0.782	−2.282	0	0

And ξ_i^+ is the instantaneous joint twist relative to the ith link frame. Each body Jacobian is calculated separately as follows,

$$\xi_j^+ = Ad^{-1}_{e^{\hat{\xi}_j\theta_j}\cdots e^{\hat{\xi}_i\theta_i}g_{W_kL_i(0)}} \xi_j \; (j \le i) \tag{9}$$

where Ad_g^{-1} is inverse adjoint transformation associated with g, which maps one coordinate system into another. Ad_g^{-1} is defined as, $Ad_g^{-1} = \begin{bmatrix} R^T & -R^T\hat{p} \\ 0 & R^T \end{bmatrix}$, where $\hat{p}$ is the cross product operation as $\hat{p} = \begin{bmatrix} 0 & -p_3 & p_2 \\ p_3 & 0 & -p_1 \\ -p_2 & p_1 & 0 \end{bmatrix}$.

The inertia tensor $\Im$ is expressed in the body frame, and it is a symmetric matrix ($\Im \in \Re^{3\times3}$), as defined by $\Im_i = \begin{bmatrix} ^iI_{xx} & ^iI_{xy} & ^iI_{xz} \\ ^iI_{yx} & ^iI_{yy} & ^iI_{yz} \\ ^iI_{zx} & ^iI_{zy} & ^iI_{zz} \end{bmatrix}$. These parameters are also concluded in Table 1. The generalized inertial matrix M is given by $\mathrm{M}_i = \begin{bmatrix} m_i\mathrm{I} & 0 \\ 0 & \Im_i \end{bmatrix}$.

Next, the inertial matrix for the robot leg is calculated by

$$M(\theta) = \begin{bmatrix} M_{11} & M_{12} & M_{13} & M_{14} \\ M_{21} & M_{22} & M_{23} & M_{24} \\ M_{31} & M_{32} & M_{33} & M_{34} \\ M_{41} & M_{42} & M_{43} & M_{44} \end{bmatrix} = J_1^T\mathrm{M}_1J_1 + J_2^T\mathrm{M}_2J_2 + J_3^T\mathrm{M}_3J_3 + J_4^T\mathrm{M}_4J_4 \tag{10}$$

which results in 16 components of $M(\theta)$. The coriolis and centrifugal forces are computed from the 16 elements in inertial matrix via the formula $C_{ij}(\theta, \dot{\theta}) = \sum_{k=1}^{n} \Gamma_{ijk}\dot{\theta}_k$ where n is the degree of freedom ($n = 4$), and $\Gamma_{ijk} = \frac{1}{2}(\frac{\partial \mathrm{M}_{ij}}{\partial \theta_k} + \frac{\partial \mathrm{M}_{ik}}{\partial \theta_j} - \frac{\partial \mathrm{M}_{kj}}{\partial \theta_i})$.

The effects of gravitational forces on each leg are computed as

$$V(\theta) = m_1 g h_1(\theta_1) + m_2 g h_2(\theta_2) + m_3 g h_3(\theta_3) + m_4 g h_4(\theta_4) \tag{11}$$

where V is the potential energy of the leg, and h_i is the height of the COM for ith link, which can be calculated by kinematics above mentioned. By deriving $M(\theta)$, $C(\theta, \dot{\theta})$ and $N(\theta, \dot{\theta})$, the inverse dynamics model of our hexapod can be achieved by substituting all the parameters into the following equation,

$$\tau_j = \left[M(\theta)\ddot{\theta} + C(\theta, \dot{\theta})\dot{\theta} + V(\theta)\right]_j \tag{12}$$

3.2 Torque Distribution

Torque distribution for climbing can be achieved by inverse dynamics and trajectory planning. The results will be shown in next section. For flat terrain walking, the hexapod can be treated as quasi-static rigid system. It is assumed here leg $j, \ldots r$ are in swing phase and the rest in stance phase are expressed as leg s. The general matrix form of force/moment equation can be written as,

$$A_s F_s + B_c W = 0 \tag{13}$$

where A_s is a coefficient matrix representing the position of supporting legs and F_s is the force vector of supporting legs. A_s and F_s meet the conditions of matrix multiplication. A_s is determined by the duty factor(gait) of hexapod with the form of $A_s = \begin{bmatrix} I_3 \ldots I_3 \\ \hat{t}_j \ldots \hat{t}_s \end{bmatrix}$, where $\hat{t}_j$ is the cross production form of tip point of jth leg and I_3 is a 3×3 unit matrix. B_c is a coefficient matrix representing the position of COM with the form of $B_c = \begin{bmatrix} I_3 & O_3 \\ \hat{t}_{COM} & I_3 \end{bmatrix}$ and W is the global body force/moment vector as $W = \begin{bmatrix} F_x & F_y & F_z & M_x & M_y & M_z \end{bmatrix}^T$. By solving Eq. (13), force distribution of tip point can be achieved when the hexapod walking on flat ground. By substituting $F_s = J_s \tau_s$ into this formula, torque distribution of the hexapod during walking on even terrain can be obtained and used for further analysis, where J_s is a diagonal matrix with the inverse of body Jacobian J_j^T as the elements of the principal diagonal.

4 Results and Discussion

The results of torque calculation and torque distribution analysis were presented in this section. The torque of the neck joint was described in Fig. 3, with the output of its actuator (Dynamixel RX-28). In our first version of the design, all joints were driven by this digital actuator with a maximum torque of 28.3 kgf cm as shown by the red lines in Fig. 3, which could hardly raise the fore part of the robot. Based on the analysis of dynamic modeling, RX-64 is then used for neck joint with a torque of 64 kgf cm and the other joints remain the previous actuators. Using model-based method here, our hexapod can climb a high step up to 75 % of its tibia length to realize climbing behavior for outdoor application of hexapod. The driving torques for each rotational joint were estimated in Fig. 4. TC joint provides relatively smaller output torque during climbing, because it only needs to keep rotating along vertical axis while CTr joint and FTi joint provide higher output joint with almost the same values with opposite rotations. Simulations in Webots [19] (Fig. 5) and experimental tests (Fig. 6) were conducted to verify our design of the climbing hexapod.

In this paper, to compute the actuator torques, the equations of motion were derived first. The kinematic variables of the hexapod were assumed to be given. The dynamics of the hexapod was described based on the robot screw theory, which made explicit use of POE equations and joint twist for representing a Lagrange derivation. It only needed to calculate the kinetic and potential energies rather than all the forces acting on the robot to reduce modeling errors. Body velocity of the COM of each link (Fig. 7d) was computed for the torque distribution analysis

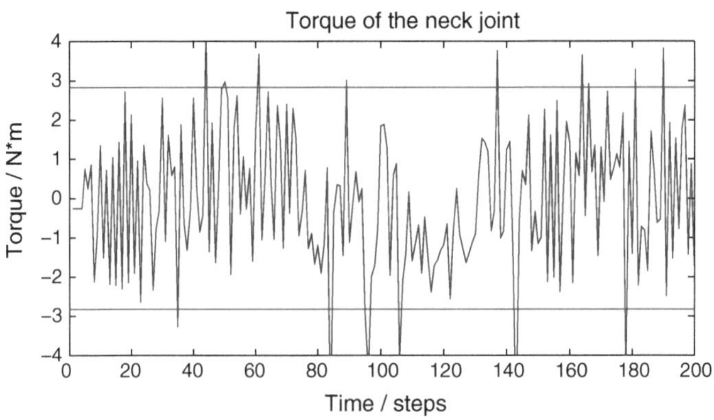

Fig. 3 Torque curve of the neck joint during climbing a high steps about 75 % of its leg length

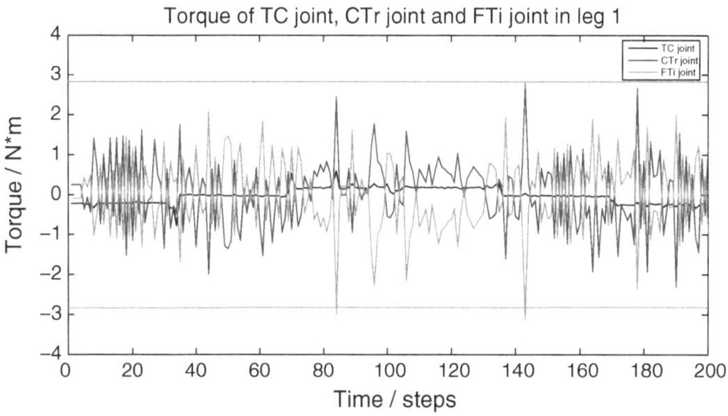

Fig. 4 Comparison of torque values of the TC, CTr and FTi joints of the hexapod while the neck joint rotate for 30° to reach the *top* of a stair

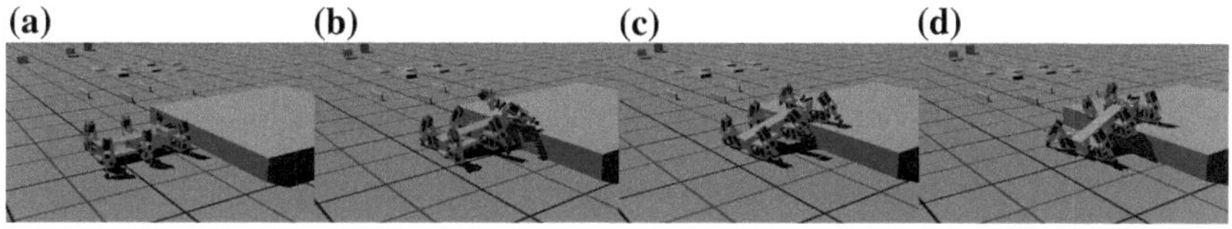

Fig. 5 Snapshots of the climbing behavior for the hexapod in simulation environment

Fig. 6 Snapshots of the hexapod prototype with climbing behavior: **a** Start climbing. **b** Front legs on the stair. **c** Climb entirely onto the stair

(Fig. 7a–c). Force and torque of the tip point were calculated by the force/moment equation. Compared with tip point forces, torque distribution was a more energy-efficient approach for hexapod control, as it could make full use of frictions and inter-actions between each joint.

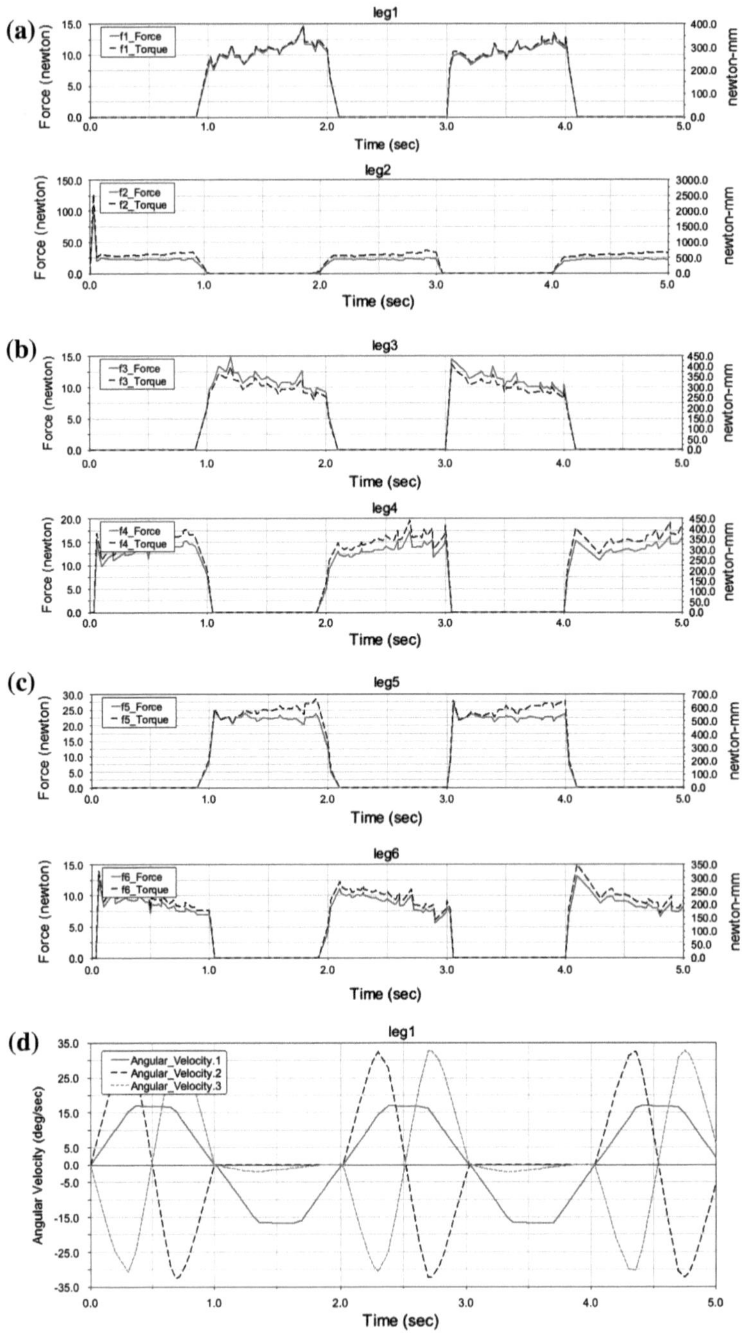

Fig. 7 Torque distribution and force analysis in tip points for each leg: **a** Leg 1 and leg 2. **b** Leg 3 and leg 4. **c** Leg 5 and leg 6. **d** Body velocity of each joint during walking

5 Conclusions

In this paper, a hexapod with a neck joint for climbing over high obstacles were presented. Screw theory and Lagrange method were used to calculate the kinematics, dynamics and torque distribution. Improvement was made based on the analysis and estimation of joint torques. Force and torque distribution of tip point were calculated for a more energy-efficient way. Future works include optimization of the trajectory and torque distribution for a more flexible climbing behavior.

Acknowledgments This work was supported in part of the National Natural Science Foundation of China under Grant No. 61175108, and in part of the Beijing Natural Science Foundation under Grant No. 4142033.

References

1. Quinn, R.D., Nelson, G.M., Bachmann, R.J., Kingsley, D.A., Offi, J.T., Allen, T.J., Ritzmann, R.E.: Parallel complementary strategies for implementing biological principles into mobile robots. Int. J. Robot. Res. **22**(3–4), 169–186 (2003)
2. Steingrube, S., Timme, M., Wörgötter, F., Manoonpong, P.: Self-organized adaptation of a simple neural circuit enables complex robot behaviour. Nat. Phys. **6**(3), 224–230 (2010)
3. Roy, S.S., Pratihar, D.K.: Dynamic modeling, stability and energy consumption analysis of a realistic six-legged walking robot. Robot. Comput. Integr. Manuf. **29**(2), 400–416 (2013)
4. Siciliano, B., Khatib, O.: Springer Handbook of Robotics. Springer Science and Business Media, New York (2008)
5. Campbell, D., Buehler, M.: Stair descent in the simple hexapod RHex. In: Proceedings of the 2003 IEEE International Conference on Robotics and Automation, ICRA 2003, vol. 1, pp. 1380–1385. Taipei, Taiwan, 2003
6. Moore, E.Z., Buehler, M.: Stable stair climbing in a simple hexapod robot. Technical report, DTIC Document (2001)
7. Spenko, M.J., Haynes, G.C., Saunders, J.A., Cutkosky, M.R., Rizzi, A.A., Full, R.J., Koditschek, D.E.: Biologically inspired climbing with a hexapedal robot. J. Field Robot. **25**(4–5), 223–242 (2008)
8. Fujii, S., Inoue, K., Takubo, T., Mae, Y., Arai, T.: Ladder climbing control for limb mechanism robot ASTERISK. In: Proceedings of the 2008 IEEE International Conference on Robotics and Automation, pp. 3052–3057. 2008
9. Ren, G., Chen, W., Dasgupta, S., Kolodziejski, C., Wörgötter, F., Manoonpong, P.: Multiple chaotic central pattern generators with learning for legged locomotion and malfunction compensation. Inf. Sci. **294**, 666–682 (2015)
10. Bartsch, S., Birnschein, T., Römmermann, M., Hilljegerdes, J., Kühn, D., Kirchner, F.: Development of the six-legged walking and climbing robot space climber. J. Field Robot. **29**(3), 506–532 (2012)
11. Manoonpong, P., Parlitz, U., Wörgötter, F.: Neural control and adaptive neural forward models for insect-like, energy-efficient, and adaptable locomotion of walking machines. Front. Neural Circuits **7**, 1–28 (2013)
12. Watson, J.T., Ritzmann, R.E., Zill, S.N., Pollack, A.J.: Control of obstacle climbing in the cockroach, blaberus discoidalis. i. kinematics. J. Comp. Physiol. A. **188**(1), 39–53 (2002)
13. Harley, C.M., English, B.A., Ritzmann, R.E.: Characterization of obstacle negotiation behaviors in the cockroach, blaberus discoidalis. J. Exp. Biol. **212**(10), 1463–1476 (2009)

14. Marhefka, D.W., Orin, D.E.: Gait planning for energy efficiency in walking machines. In: Proceedings of the 1997 IEEE International Conference on Robotics and Automation, vol. 1, pp. 474–480. IEEE 1997
15. Erden, M.S., Leblebicioglu, K.: Torque distribution in a six-legged robot. Robot. IEEE Trans. **23**(1), 179–186 (2007)
16. Roy, S.S., Pratihar, D.K.: Kinematics, dynamics and power consumption analyses for turning motion of a six-legged robot. J. Intell. Robot. Syst. **74**(3–4), 663–688 (2014)
17. Kar, D.C., Issac, K.K., Jayarajan, K.: Minimum energy force distribution for a walking robot. J. Robot. Syst. **18**(2), 47–54 (2001)
18. Murray, R.M., Li, Z., Sastry, S.S., Sastry. S.S.: A Mathematical Introduction to Robotic Manipulation. CRC Press, Baton Rouge (1994)
19. Kramer, J., Scheutz, M.: Development environments for autonomous mobile robots: a survey. Auton. Robot. **22**(2), 101–132 (2007)

Development and Simulation of an Automated Twistlock Handling Robot System

C. Liang, Y. Zou, I.-M. Chen and M. Ceccarelli

Abstract In this paper, an automated twistlock handling robot system is proposed with the aim to perform tedious and danger works with safe efficient operations. The proposed robot system is composed of an unmanned ground vehicle, a human friendly lightweight robot manipulator, a robot gripper with three fingers, and a 3D Kinect sensor for object recognition. A segmentation algorithm has been developed and 3D models have been elaborated for different twistlocks. Then, a library has been built for automatic object recognition and registration purposes. An architecture design of the software module is introduced and a full cycle of twistlock handling operation is illustrated. Simulations are reported in order to evaluate feasibility and operation performances of the system.

Keywords Machine design · Container twistlock handling · 3D object scanning · Mobile manipulation

1 Introduction

In modern container terminals, most of container logistic tasks are automated by using Automated Ground Vehicles (AGV), container handling cranes and other heavy machines [1]. However, in the process of container loading and discharging from ocean vessel, the handling of twistlock (coning and deconing) are still performed as manual tasks that are performed over a billion times in container terminals worldwide per year. This is a heavy burden for container terminal operation

C. Liang (✉) · Y. Zou · I.-M. Chen
School of Mechanical and Aerospace Engineering, Nanyang Technological University, Singapore, Singapore
e-mail: chliang@ntu.edu.sg

M. Ceccarelli
University of Cassino and South Latium, Cassino (Fr), Italy
e-mail: ceccarelli@unicas.it

S. Bai and M. Ceccarelli (eds.), *Recent Advances in Mechanism Design for Robotics*, Mechanisms and Machine Science 33,
DOI 10.1007/978-3-319-18126-4_14

companies. The burden consists of operational costs to hire a large number of dockworkers working 24 h/7 days to place and remove these heavy twistlocks. Additionally, the twistlock handling typical takes place in a container terminal where dense traffic takes place and dockworkers work closing to moving containers. It is therefore often a source of injuries or casualties [2]. Hence, shortage of labor, increasing costs, and workplace safety are the problems faced by terminal operation companies. Moreover, productivity is also a big concern for a busy container terminal like Singapore PSA transshipment port where 33.55 million TEUs of containers are handled in 2014 [3]. Therefore, productivity can also be increased when the twistlock handling tasks are automated.

There are attempts to develop automated twistlock robot handling systems. In Refs. [4, 5], a mobile robot manipulator has been proposed to handle twistlocks. Twistlock pose estimation and grasping strategies are developed as based on CAD model of twistlock. The main issue in this work is focusing on the 3D irregular object recognition method in an unconstructed environment. A method of 3D object recognition using Kernel PCA with depth information for twistlock grasping is presented in Ref. [6]. In Ref. [7], an electro-hydraulic manipulator has been designed for loading and unloading container twistlock and design issues of the control system for the manipulator are also addressed with solutions.

In industry frames, there have been also many efforts in developing automated twistlock handling stations. Several commercial products are available like the ROMMA ALP/ALS automated twistlock handling machine system and RAM spreader 4000 series PinSmart machine [1, 8]. These systems are electro-hydraulic operated heavy machines which are capable of fully automate the coning and deconing of twistlocks. Although several tests have been carried out, it has not led to a large-scale application due to the high cost and its limited capabilities. The speed of the automated procedure and extra step of putting the container into and getting out of it makes the system not suitable for the transshipment port, which requires high throughput rate. Additionally, these machines are not capable of handling all kinds of twistlocks.

In this paper, an automated twistlock handling robot system is proposed with the aim for performing tedious manual twistlock handling task in a transshipment port. This paper is organized as following: In Sect. 2, a brief description of twistlock handling process is described to analyze the operation characteristics and a design solution of robot system is proposed. In Sect. 3, a system software is described for module design and a fully cycle of twistlock handling flow is illustrated. In Sect. 4, 3D models of different solutions for twistlocks are discussed and a library is presented for recognition purposes. Finally in Sect. 5, simulation results are reported for evaluation purposes in feasibility and operation performances.

2 Operation Characteristics for Robot Systems

Figure 1 shows dockworkers manually handling twistlocks in the corner casting after a container has been discharged from a cargo vessel. In coning and deconing operations, twistlocks are mounted to and dismounted from the four corners casting of the container manually by two experienced dockworkers. The mounting and dismounting actions are mechanical actions and take about a couple of seconds for one single twistlock. A bin nearby the container loading and unloading area will store the unused twistlocks. Dockworkers need to handle the twistlocks in and out of the bin. Thus, the overall time in handling the coning or deconing operation for one container will be about 1 or 2 min. Since the twistlock is made of cast iron weighing from 5 to 10 kg each, the overall manual operation is very tiresome and dockworkers need to work in shift at peak hours of the container loading and unload operation.

It is observed that there are two fundamental challenges in automating the coning and deconing operations: the wear-and-tear of twist locks due to heavy loading and impact, and the variety of twist locks. The wear-and-tear of the twist locks makes the operation not 100 % accomplishable by human operators. Automated machines may not be able to resolve such mechanical jamming. Secondly, the twist locks can be provided with different shapes and sizes. To handle all of them automatically requires a universal robot gripper and a library of 3D models of twistlocks for automated object recognition.

Figure 2 shows a robot system that is proposed for automated robot twistlock handling. The robot system a mobile manipulator Grizzly from Clearpath Robotics®, which is composed of a Unmanned Ground Vehicle, a human friendly lightweight robot manipulator UR10, a gripper with three fingers from Robotiq® and a low-cost kinect sensor for Microsoft® [9, 10]. The robot manipulator has a maximum payload capability of 10 kg while mobile platform has a speed of 5 m/s.

Fig. 1 Examples of dockworkers manual deconing twistlocks on container corner castings

Fig. 2 A mobile manipulator system Grizzly® for automated twistlock handling [9]

3 An Architecture of Software Module

A control software of the twistlock handling robot system is developed on ROS (Robot Operating System) in Ubuntu Linux environment. Separated function modules can be linked together through messages subscribing and publishing mechanisms. This feature makes it is very flexible when the module is integrated suitable in a whole complex robot system with a vision guided mobile manipulator as reported in this paper. Figure 3 shows an architecture scheme of the developed software modules, which consists of seven modules as described in the following:

(1) The system initialization and calibration module is responsible for calibrating position of the robot and stereo camera system in order to ensure a precisely positioning and robot grasping. (2) Task planner is a module for planning twistlock handling sequences and motion trajectories of the UGV by receiving execution results from other modules. (3) Perception module is composed of three

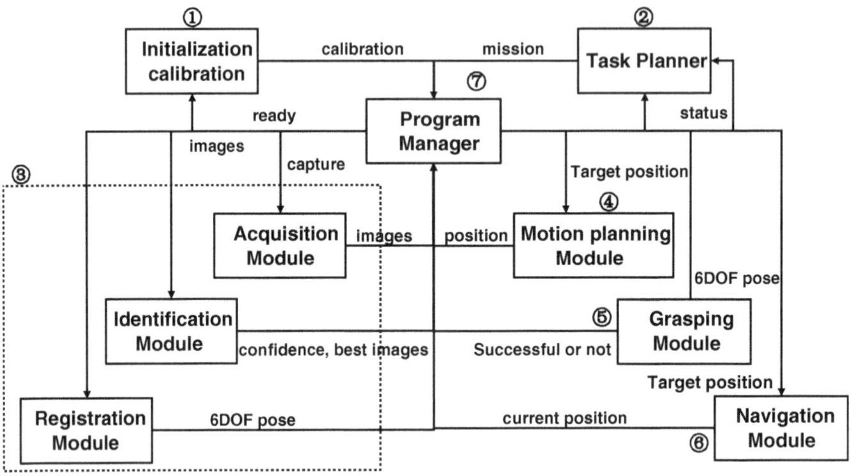

Fig. 3 A scheme illustrates design architecture of the system software module

sub-modules, namely acquisition module, identification module and registration module on the left side in Fig. 3. Acquisition module captures RGB-D images and point clouds data and it generates mask images. Identification module is used to distinguish a twistlock by indexing and comparing with the registered object RGB images in a library. Registration module is to compute position and orientation information of the target twistlock by analyzes point clouds data and pre-registered 3d twistlock model in a library. (4) Motion planning module is responsible of generating real-time obstacle avoidance trajectories for the lightweight robotic arm. (5) Grasping module includes an offline library of dynamic grasping strategies for coning and deconing movements. (6) The function of navigation module is to guide the UGV to target position while avoiding obstacles. (7) Program manager module monitors the whole execution sequences and decides which task the robot is going to perform by communicating with other modules.

Figure 4 shows a flowchart of a full cycle of automated twistlock deconing procedure, which can be described in following steps:

Step 1: Once a container is discharged from an ocean vessel, the robot moves to a predefined position that is nearby the discharged container. Onboard laser scanner system detects the surrounding environment and helps the robot staying in a safe region that will not be hitted by the moving container.

Step 2: RGB images of twistlock and corner casting are obtained from stereo camera, the perception module detects the target twistlock and its position with respect to the base frame of robot manipulator is computed.

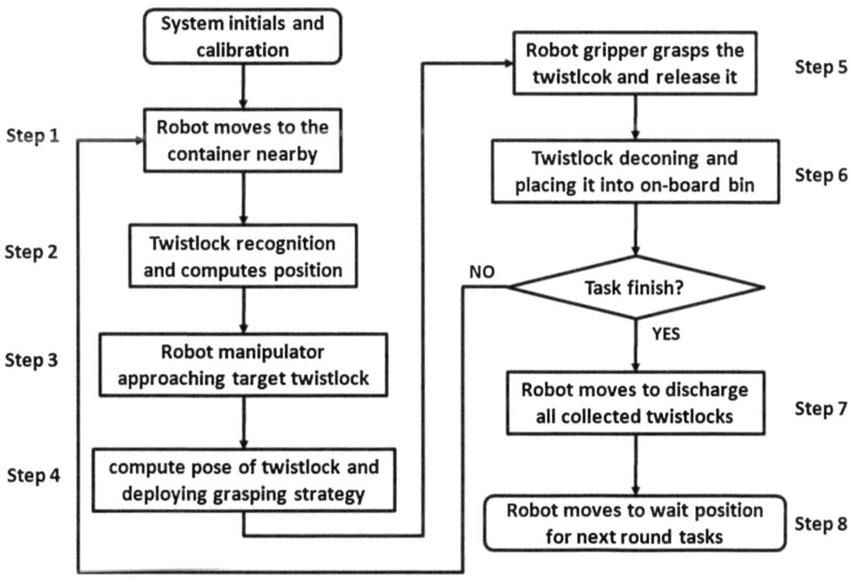

Fig. 4 A flowchart for a full cycle of automated robot twistlock deconing procedure

Step 3: Collision avoidance motion trajectories are generated. Lightweight robot manipulator moves towards the target twistlock for grasping.

Step 4: A short ranger stereo camera (creative Senz3D) mounted on the robot captures several images in different orientation angles. Registration module computes 6DOF pose of the target twistlock, which helps for a successful grasping.

Step 5: Robot gripper with tactile sensors starts the twistlock grasping. An offline grasping strategy is used which guarantees a quick and successful grasping.

Step 6: Robot gripper rotates grasped twisttlock for a certain degrees so that it can be released from the container corner casting. Then, robot manipulator moves downward until the grasped twistlock is taken out from the corner casting.

Step 7: Robot manipulator placing the grasped twistlock into on-board twistlock storage bin. Task planner module decides if the robot moves to another twistlock.

Step 8: Robot moves to the twistlock storage bin and discharges all collected twistlocks. If the task is finished, the robot moves to the wait position and waits for another discharged container from ocean vessel.

4 Twistlock 3D Model Library

It is estimated that there are more than 20 types of twistlocks that are currently used in the container terminal port around the world [1, 2]. Since the twistlock is an irregular shaped object, a library that includes all 3D twistlock models has been built in this work for recognition and registration purposes. As shown in Fig. 5, a 3D irregular object scanning platform has been built. The scanning system is composed of a rotation platform actuated by a step motor, a kinect sensor mounted on a frame. Moreover, Microsoft Kinect fusion® is used to get a complete 3D model of the scanned twistlock that is placed on the rotation platform. Additionally, RGB-D images from different viewpoints are collected. These two data will help in accurately estimating the pose of the scanned twistlock. The second figure in Fig. 5 shows a raw 3D model of one scanned twistlock. An algorithm has been elaborated for object segmentation between the twistlock and rotation platform.

Fig. 5 An experiment setup for 3D model scanning of the twistlock and a raw 3D model

Fig. 6 3D models of different twistlocks as part of a library for automated object perception

The obtained 3D model of twistlocks can be used for simulation, as well as for robot recognition and grasping strategy development purposes. For the recognition and registration, appropriate feature extraction and matching algorithms are elaborated to detect and extract the twistlock using the object library. 2D-3D registration is implemented to align the sensing data to its corresponding reference model in the library. Figure 6 show several examples of the built library with 3D models twistlocks. At the moment, the built library includes 6 types of twistlocks.

5 Simulation Results

Simulations are carried out in order to evaluate feasibility and operation performance of the robot system. The robot system is simulated in Gazebo, which is a dynamic simulation environment running in Linux OS. Control algorithms have been elaborated specifically and overall operation performance of the software modules can be easily evaluated before deploying a real robot platform. In the simulation model, robot structures are imported by using urdf file (unified robot description format) with the same geometry sizes of real robots. Twistlock models are obtained from a 3D object scanning system as presented in Sect. 4. An example is reported in Fig. 7, the left figure shows the simulation in Rviz.

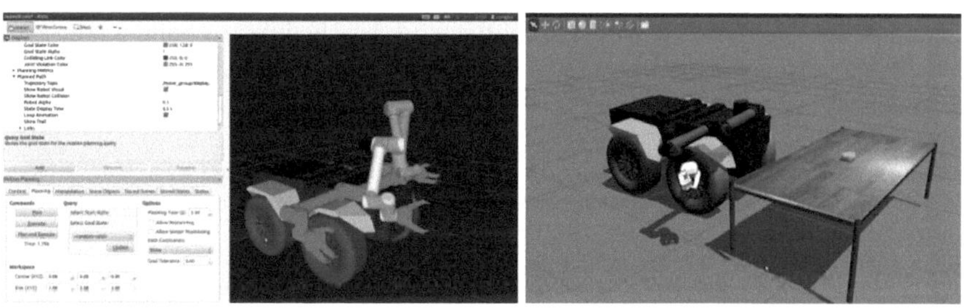

Fig. 7 Simulated automated twistlock handling robot system in Rviz and Gazebo environment

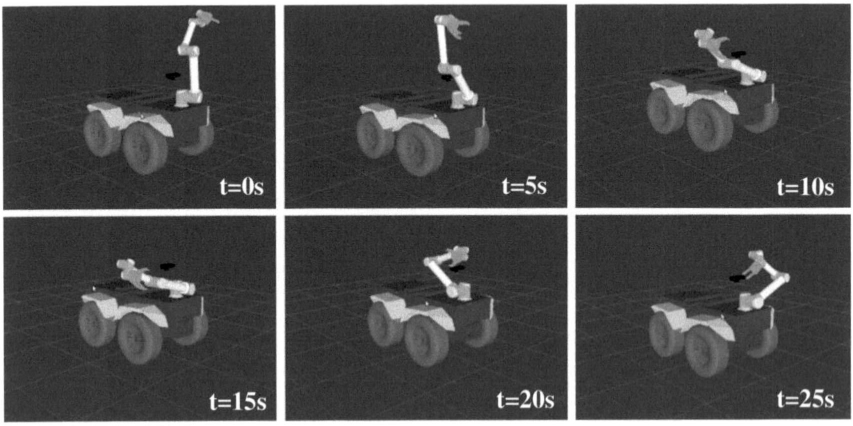

Fig. 8 Snapshots of simulation for a full cycle of twistlock deconing task

Figure 8 shows snapshots of simulation results for a motion planning of the robot manipulator. It can be noted the robot first moves to the wait position. The robot is capable of reaching the target position with a proper orientation of the gripper. Then, robot gripper grasps the twistlock and a rotation movement to deconing the twistlock. This simulation uses OMPL (Open Motion Planning Library) in Moveit! and it takes about 25 s to finish a full cycle of simulation.

6 Conclusions

This paper presents a design for an automated twistlock handling robot system, which is aimed to perform tedious manually handling task in a very busy container transshipment port. The proposed solution is based on a mobile manipulator that is combined with automated object recognition system and grasping capability end-effectors. A library of 3D models of different twsitlocks has been built for automated perception and registration. In order to evaluate feasibility and verify the developed control software, simulations are carried out in Gazebo environment. Simulation results show that the robot system is capable of performing the twistlock deconing task successfully. Next step of the work is scheduled to build a prototype in the laboratory and experiment tests will be carried out on site in order to evaluate the practical operation performance.

Acknowledgments This work has been partially supported by the Singapore Agency for Science, Technology and Research (A*star) under SERC industrial robotics research program Project No. 1225100008. The authors would like to thank the technical supports from Ms. Liu lili, Mr. Teguh Santoso Lembono and Dr. Qilong Yuan.

References

1. Stahlbock, R., et al.: Operations research at container terminals: a literature update. OR Spectrum **30**(1), 1–52 (2008)
2. Saanen, Y.: An operations perspective on new twistlock handling in terminals, port technology international. PT53–21_11 (2012)
3. Maritime and Port Authority of Singapore (MPA): The world busiest seaport, Available via MPA official website. http://www.mpa.gov.sg/sites/pdf/infokit2.pdf
4. Zhang, L., et al.: Twist-lock pose estimation and grasping based on CAD model. In: Proceedings of the 3rd IFTOMM International Symposium on Robotics and Mechatronics, ISRM2013, Singapore, pp. 739–747 (2013)
5. Shuang, M. et al.: 3D irregular object recognition for twist-lock handling system. In: IEEE—2014 26th Chinese Control and Decision Conference, CCDC2014, Harbin, pp. 2729–2734 (2014)
6. Shuang, M. et al.: 3D object recognition using kernel PCA based on depth information for twist-lock grasping. In: IEEE International Conference on Robotics and Biomimetics, ROBIO2013, Shenzhen, pp. 2667–2672 (2013)
7. Ding, J., et al.: Design of manipulator controlling system for loading and unloading container twist lock. J. Wuhan Univ. Technol. **28**(1), 33–35 (2004)
8. Bromma: Automatic twistlock handling solutions. Available via Bromma official website. http://www.bromma.com/source.php/1531385/BROMMA_ALP_low.pdf
9. Clearpath Robotics Inc: Datasheet and technical specifications of Grizzly mobile manipulator, Available via Clearpath robotics Inc. website. http://clearpath.wpengine.netdna-cdn.com/wp-content/uploads/2014/11/Grizzly-Mobile-ManipulatorV1.pdf
10. Ostergaard, E.H.: Light weight robot for everybody. IEEE Robot. Autom. Mag. **19**(4), 17–18 (2012)

Grinding Methodology Research for Helical Drill Based on the Biglide Parallel Grinder

P. Zou, R. Lv, X.J. Gao and F. Li

Abstract A new grinding method for the helical drill point based on a biglide parallel grinder was introduced in this paper. According to the mechanism description and kinematics analysis about the biglide parallel grinder, the mathematical models for the flank surface near the chisel edge was established in Cartesian coordinate frame and parametric form respectively. Moreover, the simulation of the flank surface was conducted. It was also discussed the effects of the velocity of the two glider and the distance between two gliders on determining the flank surface distribution near the chisel edge. Experiments have been implemented on the biglide parallel grinder and the helical drill point can be realized based on optimizing grinding parameters of the biglide parallel grinder. Therefore, the grinding helical drill can be obtained with the optimized parameters.

Keywords Biglide parallel grinder · Helical drill point · Flank surface · Parallel machine tool

1 Introduction

In recent years there has been significant development in the application of parallel kinematic mechanisms to machine tools. As parallel mechanisms and parallel kinematic machines (PKM) have advantages of high stiffness, more

P. Zou (✉) · R. Lv · X.J. Gao · F. Li
School of Mechanical Engineering & Automation, Northeastern University, Shenyang, China
e-mail: pzou@mail.neu.edu.cn

R. Lv
e-mail: lvruiahut0421@163.com

X.J. Gao
e-mail: gxj7976@126.com

F. Li
e-mail: ysulf@163.comu

S. Bai and M. Ceccarelli (eds.), *Recent Advances in Mechanism Design for Robotics*, Mechanisms and Machine Science 33,
DOI 10.1007/978-3-319-18126-4_15

agile, better accuracy and simple mechanical structure, etc. over serial mechanisms, optimal architecture design of parallel mechanisms, therefore, has been paid attention in recent years [1, 2]. The motions of a parallel machine tool are provided by the legs which bear only axial load. Due to its parallel structure, the deflection of the load is thus reduced significantly and legs with small cross section can be used. Moreover, the parallel structure does not accumulate the moving errors of its legs. As a result, the parallel machine tools are more accurate and can move much faster than that of the conventional machine tools. However, some disadvantages of the parallel structure such as small workspace and complex forward kinematics also limit their applications. But due to that tool grinding process does not need a big workspace and grinding force is relatively small. It is very helpful to using parallel mechanism to develop new type grinder to realize curing tools grinding automatically and numerically [3, 4].

Helical drill points require a special drill grinder which can produce three axis simultaneous motions when helical drill points are realized. Therefore, a new parallel machine tool grinder was developed and the helical drill is realized by the parallel mechanism motion. In this paper, a new type of biglide parallel grinder (as shown in Fig. 1) is developed in Northeastern University [5]. During helical drill point grinding process with this biglide parallel grinder, only two axes simultaneous motions are required instead of three axis simultaneous motions of conventional tool grinders. Combining the homogenous coordinate transformation and conjugate surface theory, the mathematical model in different formats is developed to facilitate the design of the helical drill point. Then, the flank surfaces of helical drill point for different grinding velocities are described. As a result, the helical drill points are realized based on optimization of grinding parameters by the helical drill point grinder.

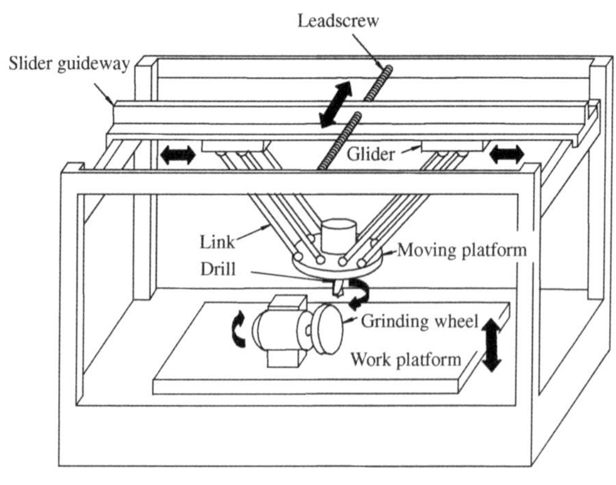

Fig. 1 Schematic drawing of the biglide parallel grinder

2 Kinematics of the Biglide Parallel Grinder

As shown in Fig. 1, the biglide parallel grinder combines the rigidity of a conventional machine tool and flexible movement of a robot. It has two platforms: One work platform on which a grinding wheel is mounted can move along vertical directions, the other mobile platform on which a tool holder is attached can move in workspace. The mobile platform is connected to a fixed-base through two parallel subchains. Each subchain consists of four parallel constant length links that are connected to revolute joints at both top end and bottom end. These rotary joints at the top end are then attached to a slide guideway through two gliders. The slide guideway that carries the two gliders is driven by a rotary servomotor and lead-screw, Which transforms rotations into translations.

As shown in Fig. 2, the global coordinate system (X, Y, Z) is built with the origin O located on the fixed-base frame, where the X-axis direction parallel to the slide guideway and the Z-axis direction perpendicular to the platforms. Thus, the Y-axis direction is defined by the right-hand-rule from Z-axis to X-axis. The kinematics problem of the biglide parallel grinder can be stated as follows: Give the positions of (x_1, y_1, z_1) of glider 1 and (x_2, y_2, z_2) of glider 2 respectively, and then the coordinate of drill tip P (x_p, y_p, z_p) can be given.

$$\begin{cases} x_p = \dfrac{x_1 + x_2}{2} \\ y_p = y_1 = y_2 \\ z_p = \dfrac{\sqrt{4l^2 - (x_2 - x_1 - 2r)}}{2} \end{cases} \tag{1}$$

where l is the length of the constant length links, and r is the radius of moving platform.

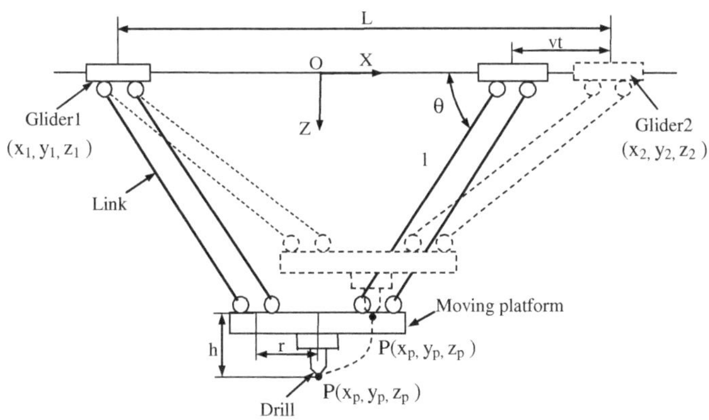

Fig. 2 Kinematic diagram of the biglide parallel grinder

Differentiating Eq. (1) with respect to time and then the velocity equation of point P can be obtained as follows:

$$k\begin{bmatrix} v_x \\ v_y \\ v_z \end{bmatrix} = \begin{bmatrix} \frac{1}{2} & \frac{1}{2} & 0 \\ 0 & 0 & 0 \\ \frac{x_2 - x_1 - 2r}{2\sqrt{4l^2 - (x_2 - x_1 - 2r)^2}} & \frac{-(x_2 - x_1 - 2r)}{2\sqrt{4l^2 - (x_2 - x_1 - 2r)^2}} & 0 \end{bmatrix} \begin{bmatrix} v_1 \\ v_2 \\ v_{z_1}(v_{z_2}) \end{bmatrix} \tag{2}$$

In the real grinding practice, it is required that only glider 2 moves while glider 1 stays stationary. Thus, Eq. (2) is rewritten as:

$$\begin{cases} v_x = -\dfrac{1}{2}v \\ v_z = \dfrac{v(L - vt - 2r)}{2\sqrt{4l^2 - (L - vt - 2r)}} \end{cases} \tag{3}$$

where, L is the distance from glider 1 to glider 2.

3 Flank Surface Modeling Based on the Biglide Parallel Grinder

In the drill sharpening operation, the generatrix of the major flank face is the common tangent of the flank and the cylindrical surface of the grinding wheel. When the generatrix moves helically around the guiding cylinder, the flank surface is generated [6, 7]. However, if the helical surface is formed only using a single line, no surface can be generated within the guiding cylinder. In the real grinding process, in fact, a fragment of the wheel peripheral, rather than only a single line, takes part in the formation of the drill point surface and thus generates the surface within the guiding cylinder. Namely, a part of a circle on the grinding wheel's end-face is the generatrix of the drill point near the chisel edge and the surface near the chisel edge is a circular helical surface.

As shown in Fig. 3, the global coordinate system {O-XYZ} is fixed on the drill blank with the origin O located on the centre of drill tip and the Z-axis coinciding with drill axis. The moving coordinate system {O_1-$X_1Y_1Z_1$} is established on the drill blank. Initially, it coincides with the global system and then will move helically along with the drill blank, namely, rotating around Z-axis with the angular velocity w, translating along Z-axis with velocity v_x and traversing along X-axis with velocity v_y. Another coordinate system {O_2-$X_2Y_2Z_2$} is attached to the grinding wheel with the origin O_2 located on the center of the wheel's end-face, X_2-axis parallel to the wheel's geometric axis and inclined angle θ between the geometric axis and drill axis [8, 9].

In the grinding process, the grinding wheel keeps stationary except for high-speed rotation. The blank moves helically along the positive Z and traverses along the

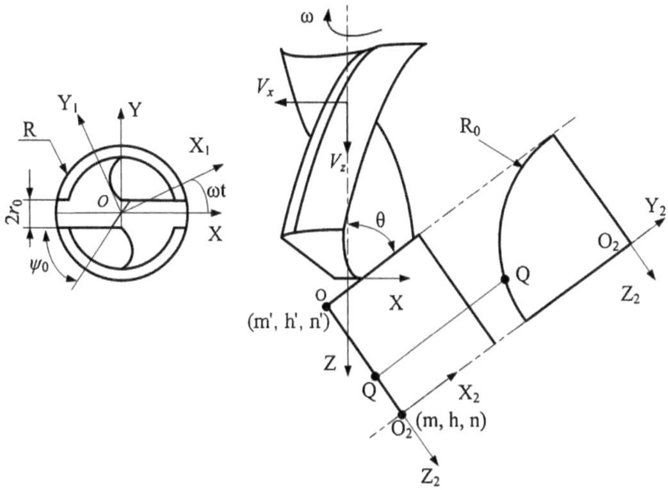

Fig. 3 Schematic diagram of the helical drill point grinding

negative X direction. Conversely, in terms of relative motion, considering the drill blank as stationary while the wheel traverses along the positive X-axis and moves helically around the negative Z-axis. The circular center of the wheel's end-face with homogenous coordinate $[x_r\ y_r\ z_r\ 1]^T$ can be given by the homogeneous coordinate transformation.

It (m, h, n) is the initial position of the circular center of the wheel's end-face and h is the half of the drill web thickness, namely R_0, shown in Fig. 3.

$$\begin{bmatrix} x_r \\ y_r \\ z_r \\ 1 \end{bmatrix} = \begin{bmatrix} \cos\omega t & \sin\omega t & 0 & 0 \\ -\sin\omega t & \cos\omega t & 0 & 0 \\ 0 & 0 & 1 & -v_z t \\ 0 & 0 & 0 & 1 \end{bmatrix} \begin{bmatrix} 1 & 0 & 0 & v_x t \\ 0 & 1 & 0 & 0 \\ 0 & 0 & 1 & 0 \\ 0 & 0 & 0 & 1 \end{bmatrix} \begin{bmatrix} m \\ h \\ n \\ 1 \end{bmatrix} \tag{4}$$

Equation (4) is rewritten as,

$$\begin{cases} x_r = m\cos\omega t + h\sin\omega t + v_x t\cos\omega t \\ y_r = -m\sin\omega t + h\cos\omega t - v_x t\sin\omega t \\ z_r = n - v_z t \end{cases} \tag{5}$$

The tangent vector of the axis of the grinding wheel can be obtained as Eq. (6).

$$\begin{cases} \tau_x = \sin\theta\cos\omega t \\ \tau_y = -\sin\theta\sin\omega t \\ \tau_z = -\cos\theta \end{cases} \tag{6}$$

An arbitrary point Q on the circle of the wheel's end-face is denoted as (x, y, z). Obviously, the coordinate (x, y, z) is subject to the following constraints:

$$(x - x_r)^2 + (y - y_r)^2 + (z - z_r)^2 = R_0^2 \tag{7a}$$

$$\tau_x(x - x_r) + \tau_y(y - y_r) + \tau_z(z - z_r) = 0 \tag{7b}$$

Equation (7a) indicates that the wheel's radius R_0 is equivalent to the distance from an arbitrary point on the circle to the circular center of the wheel's end-face. Equation (7b) indicates the circle is located in the normal plane which is perpendicular to the axis of the grinding wheel and passes through the circular center.

Substituting Eqs. (5) and (6) into Eqs. (7a) and (7b), the mathematical model of flank surface near the chisel edge in the Cartesian coordinate frame can be obtained expressed by Eq. (8).

$$\begin{cases} [x - (m\cos\omega t + h\sin\omega t + v_x t\cos\omega t)]^2 \\ + [y - (-m\sin\omega t + h\cos\omega t - v_x t\sin\omega t)]^2 + [z - (n - v_z t)]^2 = R_0^2 \\ \sin\theta\cos\omega t[x - (m\cos\omega t + h\sin\omega t + v_x t\cos\omega t)] \\ - \sin\theta\sin\omega t[y - (-m\sin\omega t + h\cos\omega t - v_x t\sin\omega t)] - \cos\theta[z - (n - v_z t)] = 0 \end{cases} \tag{8}$$

4 Simulation for the Flank Surface and Experiments

Based on the analysis above and according to different desired drill geometry, the relative parameters of grinding helical drill point for different moving velocity under the condition of constant angular velocity (w = 2 rad/s, R_0 = 62.5 mm, R = 6 mm, r = 0.925 mm) has been obtained, as shown in Table 1. According to Eq. (3), the parameters (v_1, v_2, L) of the biglide parallel grinder were solved. The simulation results demonstrate that it is negative correlation between the chisel rake angle and the traversing speed v_x, namely, the chisel rake decreasing with the increasing of the speed v_x. In addition, the speed v_x has more influence on the flank surface shape than speed v_z, thus, the speed v_z is supposed to be controlled more precisely. Then, the flank surface helical drill point simulation top view of grinding helical drill point was described as shown in Fig. 4.

Table 1 Parameters of grinding helical drill point

Number	v_x (mm/s)	v_z (mm/s)	v_1 (mm/s)	v_2 (mm/s)	L (mm)	m' (mm)	w (rad/s)	θ (rad)
1	1.0604	1.8271	0	2.1208	901.3	−1.1	2	1.0297
2	1.5635	2.2519	0	3.1272	924.7	−1.6	2	1.0297
3	2.0129	3.4844	0	4.0158	978.0	−2.16	2	1.0297

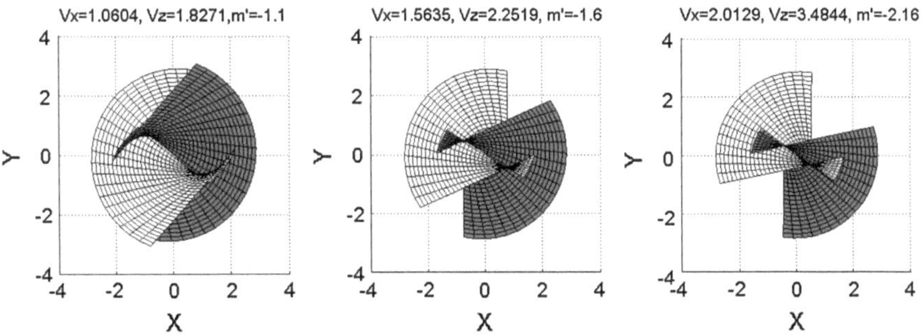

Fig. 4 The *top view* of the flank surface simulation diagram

From Fig. 4 it may be seen that two small curved surfaces form a chisel edge and the chisel edge has presented "S" shape. Several surfaces with grid in Fig. 4 are formed from the wheel highest generatrix and the end face of the grinding wheel. According to the data of the three groups in Table 1, v_x and v_z are changing, so the motion parameters of the machine have an important influence on the flank surface. It can be seen from Fig. 4, the lateral edges of the "S" shaped degree becomes larger gradually enhanced with the increase of v_x and v_z. So the slide speed value of the biglide parallel grinder can be magnified in order to get a clear "S" type of chisel edge.

Experiments have been implemented on the biglide parallel grinder. Grinding is an experiment based on the established interface of parameters solution. Finally, helical drill point can be realized based on the above table grinding parameters of the biglide parallel grinder in Fig. 5. Experimental results about the flank surface helical drill point show good correlation with the simulated results. By comparing

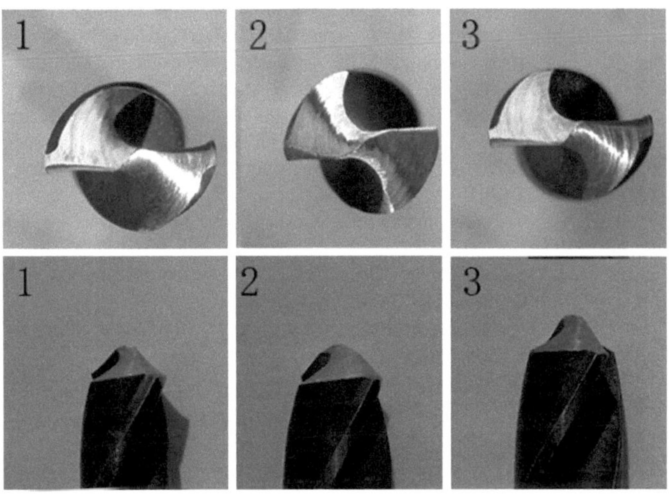

Fig. 5 Drill points in the parameters of grinding helical drill point

the drill geometry with the simulated draw, the clearance angle presents the gradual increment trend, which is in conformity with the given design parameter. Comparing grinding drill with theoretical drill in geometric parameters, the parameters solution and mathematical model are validated.

5 Conclusions

This paper presents a new method for the helical drill grinding using the biglide parallel grinder which combines the rigidity of a conventional machine tool and flexible movement of a robot. A simpler grinding control is required for this new grinder under the condition of only two axes simultaneous motions required instead of three axis simultaneous motions. The kinematic equations of the biglide parallel grinder are derived. According to the kinematic equations, the paper presents the mathematical model of the flank surface near the chisel edge in two different forms, namely, rectangular coordinate form and parametric form respectively. In addition, the simulation of the flank surface is conducted. It is also discussed the effects of the velocity of glider and the distance between two gliders on determining the flank surface distribution near the chisel edge. Therefore, grinding parameters of helical drill point can be optimized.

References

1. Dasgupta, B., et al.: The Stewart platform manipulator: a review. Mech. Mach. Theory **35**, 15–40 (2000)
2. Bing, L. et al.: Analysis and simulation for a parallel drill point grinder. Part I: kinematics, workspace and singularity analysis. Int. J. Manuf. Technol. **31**, 915–925 (2007)
3. Bing, L. et al.: Analysis and simulation for a parallel drill point grinder. Part 2: grinding kinematic modeling and simulation. Int. J. Manuf. Technol. **30**, 221–226 (2006)
4. Fei, T. et al.: Practical and reliable carbide drill grinding methods based on a five-axis CNC grinder. Int. J. Manuf. Technol. **73**, 659–667 (2014)
5. Zou, P. et al.: Study on helical drill point grinding with a biglide parallel grinder. Adv. Mater. Res. **97–101**, 2119–2122 (2010)
6. Tsai, W.D., et al.: A mathematical model for drill point design and grinding. Int. J. Mach. Tools Manuf. **19**, 95–108 (1979)
7. Lan, Y., et al.: A practical optimization design of helical geometry drill point and its grinding process. Int. J. Manuf. Technol. **64**, 1387–1394 (2013)
8. Wei, Z., et al.: Parameterized geometric design for complex helical drill point. J. Manuf. Sci. Eng. **127**, 319–327 (2005)
9. Jung, F.H.: Mathematical model for helical drill point. Int. J. Mach. Tools Manuf. **45**(7), 967–977 (2005)

Part III
Actuators and Control

Towards Safe Robotic Surgical Systems

C. Sloth and R. Wisniewski

Abstract A proof of safety is paramount for an autonomous robotic surgical system to ensure that it does not cause trauma to patients. However, a proof of safety is rarely constructed, as surgical systems are too complex to be dealt with by most formal verification methods. In this paper, we design a controller for motion compensation in beating-heart surgery, and prove that it is safe, i.e., the surgical tool is kept within an allowable distance and orientation of the heart. We solve the problem by simultaneously finding a control law and a barrier function. The motion compensation system is simulated from several initial conditions to demonstrate that the designed control system is safe for every admissible initial condition.

Keywords Safety · Barrier certificates · Surgical robotics · Control

1 Introduction

In the past decades, there has been an immense development of robotic surgical systems. This has led to regular use of robotic assistance in surgery of human patients.

In cardiac bypass surgery several advantages exist if the procedure can be carried out without the use of extracorporeal circulation. However, this means that the delicate coronary artery anastomosis has to be performed on a beating heart. This is a very technical demanding task due to the pulsating movement of the heart.

C. Sloth (✉) · R. Wisniewski
Automation and Control, Department of Electronic Systems, Aalborg University, Aalborg, Denmark
e-mail: ces@es.aau.dk

R. Wisniewski
e-mail: raf@es.aau.dk

S. Bai and M. Ceccarelli (eds.), *Recent Advances in Mechanism Design for Robotics*, Mechanisms and Machine Science 33,
DOI 10.1007/978-3-319-18126-4_16

It was proposed in [9] to provide a video image in the robotic console to the surgeon that looks as if the heart is static (although it still is beating), and move the surgical instruments in a similar pulsating motion as the heart is beating. Then the surgeon can perform the procedure using a real time image where the heart as well as the surgical instruments look static. This makes subsequent coronary artery anastomosis much easier for the surgeon to perform.

In this paper, we focus on the design of the required control system for moving the surgical instruments. The addition of automatic control introduces a risk for the surgical robot to cause trauma to patients, without human intervention [7], as experienced with the Therac-25 [5]. To alleviate this potential issue, the safety (correct behavior) of control systems must be formally proved before they are employed in surgical systems.

In the past two decades, lots of methods for designing safe dynamical systems have been developed [3]. However, these methods have not been frequently applied to robotic surgical systems. An exception is [8] that presents a verification of high level plans for composing procedures in robotic surgery in relation to puncturing. Formal verification has not yet been used much in robotic surgery, since methods based on reachability can only handle systems with a few continuous states [3], and other more scalable methods do not apply to the nonlinear dynamics of the robotic system. Recently, attention has been directed towards the development of methods for improving the scalability of verification methods by using assume-guarantee reasoning [4] and compositional computational techniques [13].

To design a safe motion compensation control system for beating-heart surgery, we use the barrier certificate method, which was developed for safety verification [10], and extended to the design of safe controllers [17]. We use an experimentally generated model from [12] to describe the movement of the heart as a combination of respiratory motion and heart beating. The safe control system is designed for a 6° of freedom robot based on the end-effector in [16]. To make the controller design method applicable to high dimensional systems, we exploit ideas from the computational techniques in [4, 13] to allow a sequential calculation of controllers for each joint of the robot. This is possible since the safety requirement is decomposed into separate requirements for the tracking of each joint angle. To find the control laws, we derive a method for realizing the control law similar to [15, 17]. To delimit the content of this paper, we abstract the important image processing away and assume that the position and orientation of the heart are measured [11]. To make the study realistic, we introduce uncertainties in the model. A detailed description of the utilized models is provided in [14].

This paper is organized as follows. Section 2 presents a model of the considered system and states the addressed problem, and Sect. 3 presents the designed control system, where the safety of the system is guaranteed by a barrier certificate. Section 4 shows simulated trajectories satisfying the specification, and Sect. 5 comprises conclusions.

2 Problem Formulation

The purpose of this section is to present a mathematical model of the interrelation between the beating heart and the robotic arm that compensates for the movement of the heart. Additionally, we define safety and finalize by stating the objective of the paper in Problem 1.

2.1 Model of Beating Heart

It is chosen to base this work on an existing model of a beating heart presented in [12] that describes the movement of the coronary arteries. This model has been experimentally generated. The principle of the model is to have two independent motions: respiratory motion and beating of the heart. Both motions are combinations of rotation and translation, and as the diaphragm moves the heart, the resulting movement of a point on the heart surface is a combination of both movements. Figure 1 illustrates the arrangement of reference frames used in the modeling.

The objective of this paper is to design a controller that can control a surgical tool to a point of interest on the heart surface at a given orientation. To accomplish this, we model the physiological motion using time-varying coordinate transformations [2].

For a point $p \in \mathbb{R}^3$ and a coordinate frame Ψ_A, let ${}^{A}p$ denote the coordinates of p in the coordinate frame Ψ_A. For coordinate frame Ψ_A and Ψ_B, the coordinates ${}^{A}p$ and ${}^{B}p$ of the point p are related by

$$\begin{bmatrix} {}^{A}p \\ 1 \end{bmatrix} = \underbrace{\begin{bmatrix} {}^{A}_{B}R & {}^{A}_{B}p \\ 0 & 1 \end{bmatrix}}_{{}^{A}_{B}T} \begin{bmatrix} {}^{B}p \\ 1 \end{bmatrix}, \tag{1}$$

where ${}^{A}_{B}R$ is a 3×3 rotation matrix and ${}^{A}_{B}p$ is a 3×1 vector translating the origin of coordinate frame Ψ_A to the origin of coordinate frame Ψ_B.

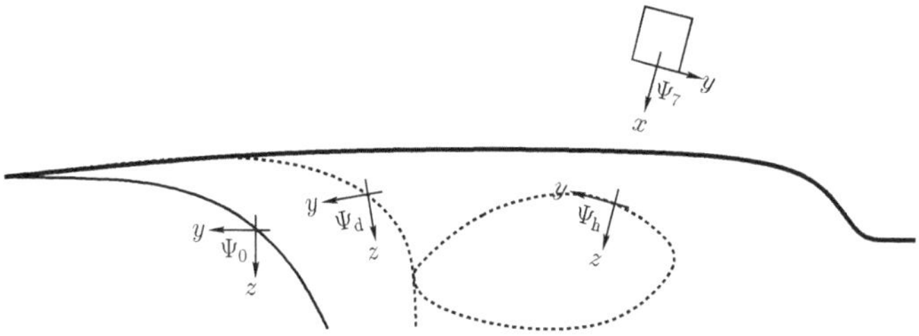

Fig. 1 Out of scale illustration of coordinate frame in the y, z-plane. The reference frame is denoted by Ψ_0, the frame on the diaphragm is denoted by Ψ_d, the frame on the heart is denoted by Ψ_h, and the frame on the surgical tool is denoted by Ψ_7

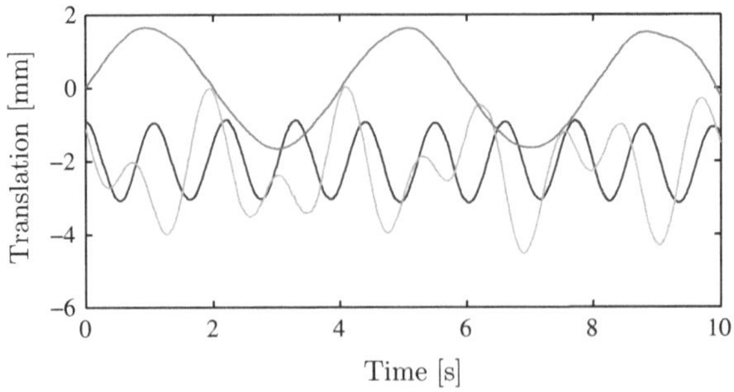

Fig. 2 Translation of the point of interest on the heart surface along the *x*-axis (*blue*), *y*-axis (*green*), and *z*-axis (*red*) of the reference frame Ψ_0

We use this formalism to describe the behavior of the diaphragm and the heart. It is chosen to let Ψ_0 and Ψ_d coincide at time 0, i.e., ${}^0_\mathrm{d}R(0) = I$ and ${}^0_\mathrm{d}p = 0$. The diaphragm (Ψ_d) rotates approximately 2° around the *x*- and *z*-axis of Ψ_0 in Fig. 1 and is translated along the *y*- and *z*-axis of Ψ_0. The behavior is periodic with a time period of 4 s. The point of interest on the heart surface (Ψ_h) rotates around the *x*- and *y*-axis of Ψ_d and is translated along the *x*- and *y*-axis of Ψ_d. This behavior has a period of 1.1 s. The resulting movement of the point of interest on the heart surface is described in the reference frame Ψ_0 using the following time-dependent transformation

$$ {}^0_\mathrm{h}T(t) = {}^0_\mathrm{d}T(t)\, {}^\mathrm{d}_\mathrm{h}T(t). \tag{2} $$

A patient's heart movement is nondeterministic; hence, we add a disturbance to the system that represents a variation in the time period of the heart and diaphragm.

Figure 2 shows the translation of the heart with respect to the reference frame Ψ_0. The elements of the transformation matrices are given by the solution of the system of polynomial differential equations describing the heart model [14].

2.2 Model of Robotic Arm

We consider a 6° of freedom robotic arm composed of a da Vinci Surgical end-effector [16], and a 3° of freedom prismatic base robot. The robotic arm is described using the same formalism as the heart, but as the length of the links of the robot are fixed, only the rotations can vary. A kinematic model of the robotic arm is detailed in [14]. The angles of the 6 joints must be controlled to steer the tool frame to the point of interest on the heart surface. Therefore, we need an inverse kinematic description of the robot that, for a desired position and orientation of the tool frame, provides the six angles that realize the desired pose. The inverse kinematic description is provided in [14].

The model of the robot can be augmented with the heart model to form a system of ordinary differential equations (affine in control and disturbance)

$$\dot{x} = f(x) + g(x)u + h(x)d, \tag{3}$$

where $x \in \mathbb{R}^n$ is the state, $u \in \mathbb{R}^m$ is the control input, and $d \in D \subseteq \mathbb{R}^p$ is the disturbance input. In this work, we control the angular velocities of the robotic arm; hence, $\dot{\theta}_i = u_i$, where u_i is the control input.

2.3 Safety of the System

We consider a control system given by $\Gamma = (f, g, h, X, X_0, X_\mathrm{u}, D)$, where $f : \mathbb{R}^n \to \mathbb{R}^n$, $g : \mathbb{R}^n \to \mathbb{R}^{n \times m}$, and $h : \mathbb{R}^n \to \mathbb{R}^{n \times p}$ are continuous, $X \subseteq \mathbb{R}^n$, $X_0 \subseteq X$, $X_\mathrm{u} \subseteq X$, and $D \subseteq \mathbb{R}^p$ is convex. The system is controlled via the continuous map $k : \mathbb{R}^n \to \mathbb{R}^m$ defining the closed-loop behavior

$$f_\mathrm{cl} : x \mapsto f(x) + g(x)k(x). \tag{4}$$

The closed-loop system is denoted by $\Gamma_\mathrm{cl} = (f_\mathrm{cl}, h, X, X_0, X_\mathrm{u}, D)$. For a measurable and essentially bounded disturbance function $\bar{d} : \mathbb{R}_{\geq 0} \to D$, we denote the solution of the Cauchy problem for the closed-loop system with $x(0) = x_0$ on an interval $[0, T]$ by $\phi_{x_0}^{\bar{d}}$, i.e.,

$$\frac{d\phi_{x_0}^{\bar{d}}(t)}{dt} = f_\mathrm{cl}\left(\phi_{x_0}^{\bar{d}}(t)\right) + h\left(\phi_{x_0}^{\bar{d}}(t)\right)\bar{d}(t) \tag{5}$$

for almost all $t \in [0, T]$. We denote the set of solutions from all initial conditions x_0 in X_0 by $\phi_{X_0}^{\bar{d}}$.

In the safety verification, we only consider trajectories initialized in X_0 that are contained in the set X. We verify if there exists a trajectory that can reach an unsafe set X_u. The safety of a system Γ_cl is defined as.

Definition 1 (*Safety*) Let $\Gamma_\mathrm{cl} = (f_\mathrm{cl}, h, X, X_0, X_\mathrm{u}, D)$ be a control system. A trajectory $\phi_{X_0}^{\bar{d}} : [0, T] \to \mathbb{R}^n$ with disturbance $\bar{d}$ is unsafe if there exists a time $t \in [0, T]$, such that $\phi_{X_0}^{\bar{d}}([0, t]) \cap X_\mathrm{u} \neq \emptyset$ and $\phi_{X_0}^{\bar{d}}([0, t]) \subseteq X$.

We say that the system Γ_cl is safe if there are no unsafe trajectories.

The surgical robotic system is safe if the relative rotation and position between the tool frame and the heart frame is kept within certain bounds.

The objective of this work is to solve the following problem.

Problem 1 Given a system $\Gamma = (f, g, h, X, X_0, X_u, D)$, design a control law $k: \mathbb{R}^n \to \mathbb{R}^m$ such that the system $\Gamma_{cl} = (f_{cl}, h, X, X_0, X_u, D)$, where f_{cl} is given by (4), is safe.

3 Control Algorithm

The purpose of this section is to present the method used for designing the safe control system. To ease the controller design, we initially decompose the specification into sub specifications.

To allow the design of a safe control system for a complex system, it is appropriate to decompose the specification into a specification for each subsystem. This enables the design of separate controllers for each subsystem, based on the assumption that each of the controllers comply with their partial specifications [4]. We use this principle in the design of the controller for the surgical robot. The requirement to the tracking of the position and orientation of the surgical tool is decomposed into requirements on the tracking of joint angles of the robot. The decomposed requirements are $|\theta_i - \theta_{i,\mathrm{ref}}| \leq \theta_{i,\mathrm{e}}$, where $\theta_{i,\mathrm{e}}$ is the maximum allowed tracking error of the ith joint angle.

To satisfy the specification, we design a control system consisting of six controllers, one per joint of the robot. The barrier certificate method used for designing the controller is similar to [17]; however, we define a control barrier function in a less restrictive manner. Furthermore, we allow unknown but bounded disturbance inputs. The addition of disturbances is crucial in medical applications, where the system (the patient) is described by an uncertain model.

In the definition of a control barrier function, we denote the complement of X by X^c, and use the notion of contingent cone [1]. Let K be a nonempty subset of a space X and let x belong to K. The contingent cone to K at x is the set

$$T_K(x) = \left\{ v \in X | \liminf_{h \to 0+} \frac{d_K(x + hv)}{h} = 0 \right\}, \tag{6}$$

where $d_K(y)$ denotes the distance of y to K, defined by

$$d_K = \inf_{z \in K} ||y - z||. \tag{7}$$

A control barrier function is defined as

Definition 2 Given a control system Γ. A continuously differentiable function $B: X \to \mathbb{R}$ satisfying

$$X_0 \subset B^{-1}((-\infty, 0]) \subset X_u^c, \text{ and} \tag{8a}$$

there exists $u \in \mathbb{R}^m$ such that for any $d \in D$ and $x \in B^{-1}(0)$

$$f(x) + g(x)u + h(x)d \in T_{B^{-1}((-\infty,0])}(x) \tag{8b}$$

is called a control barrier function.

The intuition of a control barrier function B is to separate the initial set X_0 and the unsafe set X_u by the zero level set of B. This is ensured by (8a). On the zero level set, it must be possible to control the vector field to point into the zero sublevel set; hence, all solutions avoid the unsafe set X_u that is in the complement of the zero sublevel set. This is ensured by (8b).

Given a control barrier function, a control law must be found that ensures the safety of the system. A selection of such a control law is provided in the following, inspired by [15]. Let g_i, h_i denote the ith column of g respectively h, and denote the Lie derivative of B along f by

$$L_f B = \sum_{i=1}^{n} \frac{\partial B}{\partial x_i} f_i.$$

Proposition 1 *Let Γ be a control system, let B be an associated proper control barrier function, and let $L_g B(x) \neq 0$ for $x \in B^{-1}(0)$. There exists a pair of real numbers (γ_1, γ_2) with $0 < \gamma_1 < \gamma_2$ such that the control*

$$k = -\xi(||b||) \frac{a + \alpha + \sqrt{(a+\alpha)^2 + \kappa^2 b^T b}}{b^T b} b, \tag{9}$$

where $\kappa > 0$, $a \equiv L_f B$,

$$b^{\mathrm{T}} \equiv L_g B = \left[L_{g_1} B \ldots L_{g_m} B\right],$$

$$c^{\mathrm{T}} \equiv L_h B = \left[L_{h_1} B \ldots L_{h_p} B\right],$$

$$\alpha(x) \equiv \sup_{d \in D} c^{\mathrm{T}}(x) d, \tag{10}$$

and $\xi : \mathbb{R} \to [0, 1]$ defined by

$$\omega(z) \equiv \begin{cases} 0 & \text{if } z \leq 0 \\ \exp(-1/z) & \text{if } z > 0 \end{cases} \tag{11}$$

$$\xi(z) \equiv \xi_{(\gamma_1, \gamma_2)}(z) \equiv \frac{\omega(z - \gamma_1)}{\omega(z - \gamma_1) + \omega(\gamma_2 - z)}, \tag{12}$$

is continuous and ensures safety for the closed-loop system Γ_{cl}.

Proof The safety of Γ_{cl} is proved by showing that the Lie derivative of the control barrier function is negative for all $x \in B^{-1}(0)$; hence, by Nagumo's Theorem $B^{-1}((-\infty, 0])$ is positively invariant [1]. The Lie derivative of the control barrier function for the closed-loop system is

$$L_{f_{\mathrm{cl}}}B = L_f B + L_g Bk + L_h Bd \tag{13a}$$

$$L_{f_{\mathrm{cl}}}B = -\xi(||b||)(\alpha + \sqrt{(a+\alpha)^2 + \kappa^2 b^{\mathrm{T}} b}) + c^{\mathrm{T}} d. \tag{13b}$$

The function B is proper, thus $B^{-1}(0)$ is compact. By compactness of $B^{-1}(0)$ and since $L_g B(x) \neq 0$ for $x \in B^{-1}(0)$, there exists (γ_1, γ_2) such that $\xi(||b(x)||) = 1$ for all $x \in B^{-1}(0)$ (ξ is a bump function [6]). It is seen that $-\sqrt{(a+\alpha)^2 + \kappa^2 b^T b} < 0$, as $\kappa^2 b^{\mathrm{T}} b > 0$. From (10), it is seen that $c^{\mathrm{T}} d - \alpha \leq 0$ for all $x \in B^{-1}(0)$. This implies that $L_{f_{\mathrm{cl}}}B < 0$ for all $x \in B^{-1}(0)$.

By Sect. 6.5 in [1], α is continuous since D is convex, c is continuous, and α is linear in d for each x. Thus the controller k is continuous, as it is the product of the continuous function ξ and a second function that is continuous at any point except at $b = 0$. Since $\xi(||b||) = 0$ in a neighborhood of $b = 0$, the product is continuous for all x.

Corollary 1 *The properness of* B *imposed in* Proposition 1 *can be replaced with compactness of* $\{x \in X | b(x) = 0\}$.

To design the control system for the surgical robot, we find one control barrier function per joint of the robot, and use it in the design of a controller. It is possible to completely separate the design of the six controllers, as the reference signal $\theta_{i,\mathrm{ref}}$ for the ith joint angle is generated independent of the other joint angles. Therefore, all control barrier functions are of the following form

$$B_i(x) = (\theta_i - \theta_{i,\mathrm{ref}}(x))^2 - \theta_{i,\mathrm{e}}^2. \tag{14}$$

It is seen that the value of B_i is nonpositive when the tracking error is within the specified bound. Since the dynamics of the robot is given as integrators $\dot{\theta}_i = u_i$ for $i = 1, \ldots, 6$, B_i is a control barrier function when $\theta_{i,\mathrm{e}} > 0$. Therefore, we can directly calculate the control laws using Proposition 1 and the designed controllers are safe by design. This finalizes the controller design.

4 Simulation Results

The purpose of this section is to demonstrate how well the designed controllers track the movement of the heart, and keep the tracking error within the specified bounds.

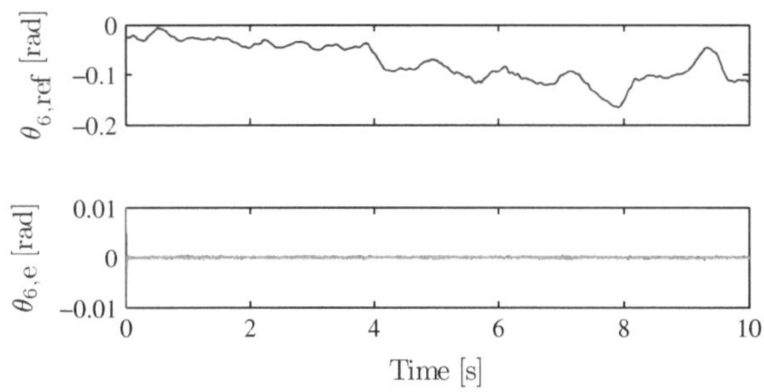

Fig. 3 The *upper* subplot shows the reference signal $\theta_{6,\mathrm{ref}}$, and the *lower* subplot shows the tracking error of θ_6 from different initial conditions

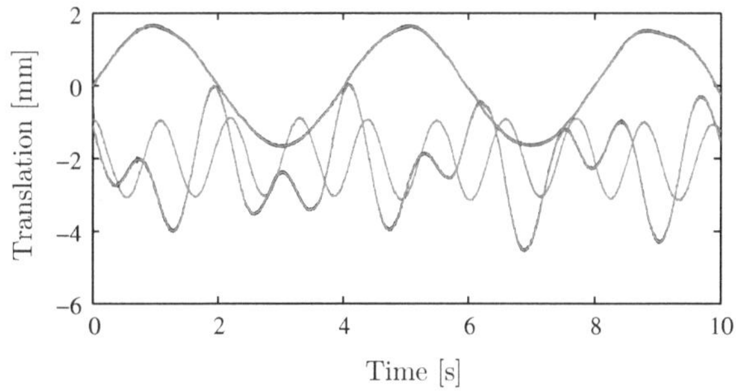

Fig. 4 Position of the tool (origin of Ψ_7, *blue lines*) and position of the point of interest on the heart surface (origin of Ψ_{h}, *green lines*) in the reference frame Ψ_0

The control barrier functions are chosen to obtain a tracking error less than 0.02 rad of any joint angle. Figure 3 shows 5 simulations of the motion compensation with different initial conditions and disturbances to the model. Note that we initialize the system with a nonzero tracking error.

All trajectories are within the safety bound of 0.02 rad.

The tracking of the position of the tool is shown in Fig. 4. It demonstrates that the controller puts the surgical tool at a constant relative position with respect to the heart.

The tracking error of the position is also within the required bounds. We have omitted a visualization of the tracking of the orientation to save space.

5 Conclusion

In this paper, it was demonstrated that it is possible to derive a safety proof for a robotic surgical system, despite its high complexity. The key to constructing the proof of safety is to decompose the safety problem into small subproblems that can be solved independently. This allows the generation of multiple control barrier functions, allowing the design of individual controllers for each subsystem.

We have designed and simulated a safe controller for a 6° of freedom robot that is capable of compensating for the physiological movements experienced in beating-heart surgery.

The use of safety verification may lead the prevail of autonomous control in robotic surgery, as guaranteed safety will increase the acceptance of sophisticated computer aids in robotic surgery.

References

1. Aubin, J.P.: Viability Theory. Birkhäuser, Boston (1991)
2. Craig, J.J.: Introduction to Robotics: Mechanics and Control, 3rd edn. Pearson Prentice Hall, New Jersey (2005)
3. Ding, J., Gillula, J.H., Huang, H., Vitus, M.P., Zhang, W., Tomlin, C.J.: Hybrid systems in robotics. IEEE Robot. Autom. Mag. **18**(3), 33–43 (2011). doi:10.1109/MRA.2011.942113
4. Kerber, F., van der Schaft, A.: Compositional analysis for linear control systems. In: Proceedings of the 13th ACM International Conference on Hybrid Systems: Computation and Control, pp. 21–30. ACM, New York (2010). doi:10.1145/1755952.1755957
5. Leveson, N., Turner, C.: An investigation of the Therac-25 accidents. Computer **26**(7), 18–41 (1993). doi:10.1109/MC.1993.274940
6. Madsen, I.H., Tornehave, J.: From Calculus to Cohomology: De Rham Cohomology and Characteristic Classes. Cambridge University Press, Cambridge (1997)
7. Moustris, G.P., Hiridis, S.C., Deliparaschos, K.M., Konstantinidis, K.M.: Evolution of autonomous and semi-autonomous robotic surgical systems: a review of the literature. Int. J. Med. Robot. Comput. Assist. Surg. **7**(4), 375–392 (2011). doi:10.1002/rcs.408
8. Muradore, R., Bresolin, D., Geretti, L., Fiorini, P., Villa, T.: Robotic surgery: formal verification of plans. IEEE Robot. Autom. Mag. **18**(3), 24–32 (2011). doi:10.1109/MRA.2011.942112
9. Nakamura, Y., Kishi, K., Kawakami, H.: Heartbeat synchronization for robotic cardiac surgery. IEEE Int. Conf. Robot. Autom. **2**, 2014–2019 (2001). doi:10.1109/ROBOT.2001.932903
10. Prajna, S., Jadbabaie, A., Pappas, G.J.: A framework for worst-case and stochastic safety verification using barrier certificates. IEEE Trans. Autom. Control **52**(8), 1415–1428 (2007). doi:10.1109/TAC.2007.902736
11. Richa, R., Bo, A.P.L., Poignet, P.: Towards robust 3d visual tracking for motion compensation in beating heart surgery. Med. Image Anal. **15**(3), 302–315 (2011). doi:10.1016/j.media.2010.12.002
12. Shechter, G., Ozturk, C., Resar, J., McVeigh, E.: Respiratory motion of the heart from free breathing coronary angiograms. IEEE Trans. Med. Imaging **23**(8), 1046–1056 (2004). doi:10.1109/TMI.2004.828676
13. Sloth, C., Pappas, G.J., Wisniewski, R.: Compositional safety analysis using barrier certificates. In: Proceedings of Hybrid Systems: Computation and Control, pp. 15–23 (2012). doi:10.1145/2185632.2185639
14. Sloth, C., Wisniewski, R., Larsen, J., Leth, J., Poulsen, J.: Model of Beating-Heart and Surgical Robot. Technical report, Aalborg University, Aalborg (2012)

15. Sontag, E.D.: A 'universal' construction of Artstein's theorem on nonlinear stabilization. Syst. Control Lett. **13**(2), 117–123 (1989). doi:10.1016/0167-6911(89)90028-5
16. Sun, L.W., Van Meer, F., Schmid, J., Bailly, Y., Thakre, A.A., Yeung, C.K.: Advanced da Vinci surgical system simulator for surgeon training and operation planning. Int. J. Med. Robot. Comput. Assist. Surg. **3**(3), 245–251 (2007). doi:10.1002/rcs.139
17. Wieland, P., Allgöwer, F.: Constructive safety using control barrier functions. In: Proceedings of the 7th IFAC Symposium on Nonlinear Control Systems, pp. 462–467 (2007). doi:10.3182/20070822-3-ZA-2920.00076

A Comprehensive Analytical Model and Experimental Validation of Z-shaped Electrothermal Microactuators

Zhuo Zhang, Weize Zhang, Qiyang Wu, Yueqing Yu, Xinyu Liu and Xuping Zhang

Abstract This paper presents a comprehensive analytical model for a Z-shaped electrothermal microactuator operating in air condition. The model provides the estimate of the tip displacement of the microactuator directly based on a voltage difference applied on the both anchors of the actuator. In an attempt to improve the accuracy of modeling, the impact of the shuttle and multiple pairs of beams are included in the model. The numerical simulations with a Finite Element (FE) model are conducted using the commercialized FE software ANSYS to verify the analytical model. The experimental testing is performed as well to validate the analytical model. The analytical results based on the proposed model agree well with both the FE modeling and experimental results.

Keywords Z-shaped electrothermal microactuator · Analytical model · Thermal-mechanical · MEMS

Z. Zhang · Y. Yu
School of Mechanical and Electrical Engineering,
Beijing University of Technology, Beijing, China
e-mail: zhangzhuozz123@126.com

Y. Yu
e-mail: yqyu@bjut.edu.cn

W. Zhang · Q. Wu · X. Liu
Department of Mechanical Engineering, McGill University, Montreal, Canada
e-mail: weize.zhang@mail.mcgill.ca

Q. Wu
e-mail: qiyang.wu@mail.mcgill.ca

X. Liu
e-mail: xinyu.liu@mcgill.ca

X. Zhang (✉)
Department of Engineering and Aarhus School of Engineering,
Aarhus University, Aarhus, Denmark
e-mail: xuzh@ase.au.dk

S. Bai and M. Ceccarelli (eds.), *Recent Advances in Mechanism Design for Robotics*, Mechanisms and Machine Science 33,
DOI 10.1007/978-3-319-18126-4_17

1 Introduction

Electrothermal actuation in micro-scale devices has attracted deep attention in recent years, as it has been demonstrated to be a compact, stable and high-force actuation technique [8]. Various configurations of electrothermal microactuators have been developed for applications of in-plane motion. Among them, the V-shaped and U-shaped microactuators [2, 4, 6, 19] have been received extensive studies involving modeling [3, 5, 10, 13, 14], design [15, 16], material properties [8] etc. The Z-shaped microactuator developed in [8, 17] provides a complementary alternative to the U-shaped and V-shaped actuators as it offers the wide range of stiffness and output force. Investigation on the analytical models for predicting the tip deflection of the Z-shaped actuators has been studied in [8]. However, the current models only apply to the actuators operating in vacuum conditions or the SOI (slilicon-on-insulator) devices in which the silicon substrates are totally etched and the heat conduction through the air layer beneath the actuator can be ignored. For the actuators fabricated using the PolyMUMPs process, the heat conduction through the air layer between the structure and the substrate cannot be ignored as the size of the air gap between the polysilicon and the nitride layer is very thin (typically 2 μm) [9]. Moreover, in the current analytical models of the Z-shaped microactuators [8], the impact of the shuttle is not included. Omitting the impact of shuttle is acceptable for the 'vacuum' condition, but it may leads to poor estimates for the 'in air' condition. The reasons is that the shuttle area is large, which results in that its heat conduction to the substrate through the air gap cannot be ignored. The other necessary improvements in the current models include: (1) Current analytical models only applies to a single pair of beams. However in practical applications, multiple beams are used in the microactuators as multiple pairs of beams increase the overall current flowing through the structure, and the stiffness in reaction to the external force. (2) Some assumptions are always made to simplify the derivations. For example, the length of the two long beams was assumed to be the same and the thermal expansion of the central beam was small enough to be omitted. Such assumptions narrow the space to optimize the device performance.

To achieve analytical modeling with better accuracy, this paper develops a comprehensive model for the Z-shaped actuator operating in air conditions with including the impact of the shuttle and multiple pairs of beams. First, the electro-thermal model is built to derive the average temperature increase along the beams. Second, the thermal-mechanical model is established to compute the tip displacement with the known average temperature increase of the beam given by the electro-thermal model. Finally, the electro-thermal and thermal-mechanical models are combined to formulate the system model. The established model can be used to directly estimate the tip displacement based on the voltage difference applied on the anchors. Finite-element simulations and experiments are performed to verify the developed analytical model with comparisons of the tip displacements as a function of voltages.

2 Analytical Modeling

In this work, modeling the electrothermal microactuators is first divided two sub-modeling: the electro-thermal modeling on the temperature change caused by the voltage applied to the anchors, and the thermal-mechanical modeling on the tip displacement driven by the temperature change. The system model is then established based on the combination of the two sub-models.

The schematic diagram of a Z-shaped electrothermal microactuator is shown in Fig. 1. b_0, h_0 and L_b are the thickness, width and length of the beam respectively; b_1, h_1 and L_s denote the thickness, width and length of the shuttle respectively. The half span of the beam, i.e., distance between the shuttle and the anchor, is defined as L_0, and total length of the beam is expressed as L_b (i.e., $L_b = L_0 + L_2 = L_1 + L_2 + L_3$).

2.1 Electro-Thermal Modeling

The electro-thermal analysis is conducted to derive the formulation predicting the steady-state temperature distribution and the average temperature increase along the beams with the applied voltage on the microactuator. The heat is transferred through conduction, convection, and radiation. Convection where heat is transferred via a bulk, microscopic motion of a fluid (liquid or gas) is small and negligible [7]. The heat losses by radiation can also be ruled out based on the fact that radiation becomes significant only at high temperatures (>1000 °C) [18]. The electro-thermal model therefore only includes the heat conduction to the anchors across the beams and the substrate via the air layer between the structure and the substrate. With the assumption of a slender beam, the heat conduction analysis can be treated as a 1-D problem. The 1-D form of Fourier's Law of heat conduction [7] is expressed as

$$q_x = -kA\frac{dT}{dx} \tag{1}$$

where q_x is the rate at which heat is transferred from the hot side to the cold side of one differential element of length dx, k the thermal conductivity, A the cross-sectional area

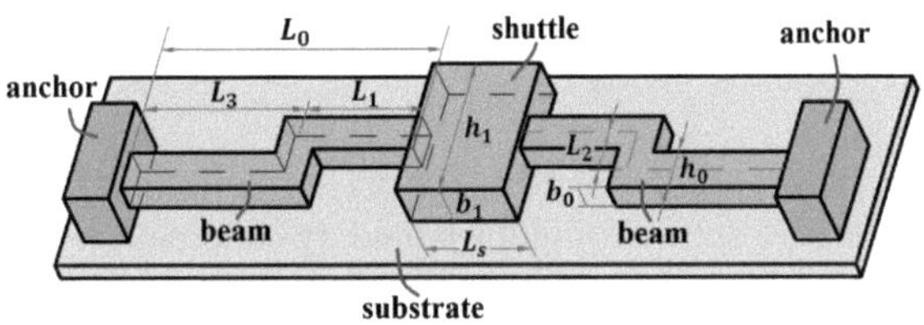

Fig. 1 Schematic diagram of a Z-shaped microactuator

of the element, and T the temperature. The heat balance in the differential element of length dx can be described as [7]

$$k\frac{d^2T}{dx^2}-\frac{k_a(T-T_0)}{bh_a}+\dot{q}=0 \tag{2}$$

where k_a is the thermal conductivity of the air, T_0 the reference temperature, $\dot{q}$ the heat (thermal energy) per unit volume per unit time generated by electric current, we call it the electrical power density, b and h_a the thickness of the structure (i.e., the beams and the shuttle) and the air layer, respectively. Note that the thickness of the beams b_0 and the shuttle b_1 are usually the same due to fabrication limits, i.e. $b=b_0=b_1$. The first and the second term on the left side of Eq. (2) describe the heat conduction in the structure (i.e., beams and shuttle) and the air gap respectively. In practice, the shuttle is usually supported by multiple pairs of beams to increase the stiffness and output force of the actuator. In this case, we first calculate the current passing through the beams I_b and the shuttle I_s, respectively, and then the electrical power density of the beams $\dot{q}_b$ and the shuttle $\dot{q}_s$ are derived as

$$\dot{q}_b=\frac{I_b^2R_b}{V_b}=\frac{A_s^2}{\hat{\rho}(2L_bA_s+nL_sA_b)^2}U^2 \tag{3}$$

$$\dot{q}_s=\frac{I_s^2R_s}{V_s}=\frac{n^2A_b^2}{\hat{\rho}(2L_bA_s+nL_sA_b)^2}U^2 \tag{4}$$

where R_b and R_s are the electric resistance of the beam and the shuttle respectively ($R_b=\hat{\rho}\frac{L_b}{A_b}$ and $R_s=\hat{\rho}\frac{L_s}{A_s}$); V_b and V_s the volume of the beam and the shuttle respectively, $\hat{\rho}$ the electric resistivity of the material, A_b and A_s the cross-sectional area of the beam and the shuttle respectively, n the number of pairs of beams, and U the voltage difference applied on the anchors. The total resistance of the actuator is expressed as $R_{total}=\frac{2R_b}{n}+R_s=\hat{\rho}\left(\frac{2L_b}{nA_b}+\frac{L_s}{A_s}\right)=\hat{\rho}\frac{2L_bA_s+nL_sA_b}{nA_bA_s}$. $I_s=\frac{U}{R_{total}}$, and $I_b=\frac{I_s}{n}$. Equation (2) is rewritten for the beams and the shuttle as

$$\frac{d^2T_b}{dx^2}+A_1T_b=-B_1 \tag{5}$$

$$\frac{d^2T_s}{dx^2}+A_2T_s=-B_2 \tag{6}$$

where $A_1=-\frac{k_a}{kb_0h_a}, B_1=\frac{k_aT_0}{kb_0h_a}+\frac{\dot{q}_b}{k}, A_2=-\frac{k_a}{kb_1h_a}, \textit{and } B_2=\frac{k_aT_0}{kb_1h_a}+\frac{\dot{q}_s}{k}$. The linear differential equations are solved using the boundary conditions: (1) the anchors and substrate are considered to remain at room temperature T_0 throughout actuation; (2) the temperature of the beams and the shuttle at $x=L_b$ and $x=L_b+L_s$ are the same. For the thermal analysis, the direction of x axis is along the direction of current passing through the actuator and the origin is located on the left end of the structure; (3) in addition, the rate of heat conduction at $x = L_b$ and $x = L_b + L_s$ should be the same, which yields $nq_b = q_s$. Solving the equations, the temperature distributions along the left beam T_{b1}, the shuttle T_s, and the right beam T_{b2} can be obtained, and the average temperature increase along the beam ΔT_{b_avg} is expressed as

$$\Delta T_{b\_avg} = \frac{1}{L_b}\int_0^{L_b}(T_{b1} - T_0)dx = Q_a \cdot T_{a1} \cdot U^2 \tag{7}$$

where $Q_a = \frac{1}{k_a \hat{\rho}}$, $T_{a1} = \frac{(1-s_1)^2}{L_b r_1 s_1} O_1 + \left(\frac{1-s_1}{L_b r_1 s_1} + 1\right) J_1$,
$O_1 = \frac{nA_b s_1 r_1 (s_4+1)\left[2s_1 J_1 + (s_1^2+1)(J_1 - J_2)\right]}{\left(s_1^2-1\right)\left[A_s R_1 (s_4-1)\left(s_1^2-1\right) + nA_b r_1 (s_4+1)\left(s_1^2+1\right)\right]} - \frac{s_1(J_1-J_2)+J_1}{s_1^2-1}$, $J_1 = \frac{A_s^2 b_0 h_a}{(2L_b A_s + nL_s A_b)^2}$,
$J_2 = \frac{n^2 A_b^2 b_1 h_a}{(2L_b A_s + nL_s A_b)^2}$, $r_1 = \sqrt{-A_1}$, $R_1 = \sqrt{-A_2}$, $s_1 = e^{r_1 L_b}$, $s_2 = e^{R_1 L_b}$, and $s_4 = e^{R_1 L_s}$.

In order to derive the thermal elongation of the central beam L_2 and the long beam L_0 $(= L_1 + L_3)$, the average temperature change along L_2 and L_0 are calculated as

$$\Delta T_2 = \frac{1}{L_2}\int_{L_3}^{L_2+L_3}(T_{b1} - T_0)dx = Q_a \cdot S_{a2} \cdot U^2 \tag{8}$$

$$\Delta T_{13} = \frac{\Delta T_{b\_avg} \cdot L_b - L_2 \cdot \Delta T_2}{L_0} = Q_a \cdot S_{a13} \cdot U^2 \tag{9}$$

where $S_{a2} = \frac{e^{r_1(L_2+L_3)} - e^{r_1 L_3} + e^{-r_1(L_2+L_3)} - e^{-r_1 L_3}}{r_1 L_2} O_1 + \left[\frac{e^{-r_1(L_2+L_3)} - e^{-r_1 L_3}}{r_1 L_2} + 1\right] J_1$ and $S_{a13} = \frac{L_b}{L_0} T_{a1} - \frac{L_2}{L_0} S_{a2}$. For the simplicity of expression, the following notations are made: $\beta_{a2} = \frac{\Delta T_2}{\Delta T'_{b\_avg}} = \frac{S_{a2}}{T_{a1}}$; $\beta_{a13} = \frac{\Delta T_{13}}{\Delta T'_{b\_avg}} = \frac{L_b}{L_0} - \frac{L_2}{L_0}\beta_{a2}$

2.2 Thermal-Mechanical Modeling

In this section, the formulation is derived to predict the tip displacement with the given average temperature increase of the beam ΔT_{b_avg} obtained in Sect. 2.1. With the consideration of the symmetry, the development of the analytical model is simplified by simply considering half of the structure without the shuttle and replacing the other half with the reaction forces and moments as shown in Fig. 2. The Castigliano's theorem is applied to establish this modeling based on the steady-state assumption.

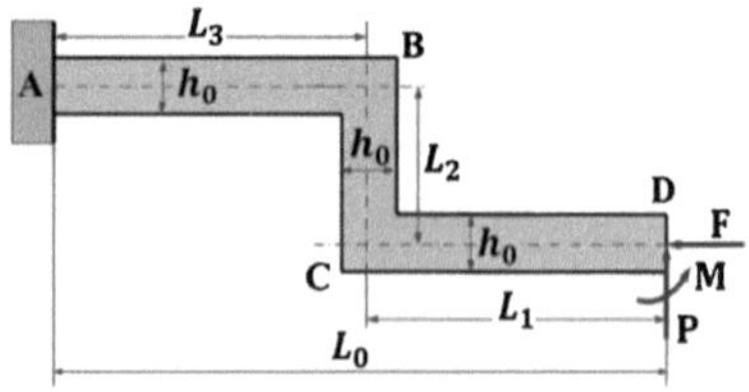

Fig. 2 One half of the Z-shaped beam

The bending moments applied on each segment of the Z-shaped beam are expressed as

$$M_1 = M + Px \qquad 0 < x < L_1 \tag{10}$$

$$M_2 = M + PL_1 - Fx \qquad 0 < x < L_2 \tag{11}$$

$$M_3 = M + P(L_1 + x) - FL_2 \qquad 0 < x < L_3 \tag{12}$$

Here, the axial forces on the beam are $N_1 = F$ and $N_2 = P$. According to the Castigliano's theorem, the tip displacement of the beam is written as

$$\begin{aligned} y = \Delta L_{T_2} - \int_0^{L_1} \frac{M_1}{EI}\frac{\partial M_1}{\partial P}dx - \int_0^{L_2} \frac{M_2}{EI}\frac{\partial M_2}{\partial P}dx - \int_0^{L_3} \frac{M_3}{EI}\frac{\partial M_3}{\partial P}dx \\ - \int_0^{L_2} \frac{N_2}{EA}\frac{\partial N_2}{\partial P}dx \end{aligned} \tag{13}$$

where ΔL_{T_2} is the elongation of L_2 due to thermal expansion. Note that $\Delta L_{T_2} = \alpha(\Delta T_2)L_2$, where α is called the coefficient of thermal expansion (CTE). It is clear that the beams are fixed to the shuttle and the beam cannot produce deflection in the direction along L_0 due to symmetry. Hence, the boundary conditions are given as

$$\delta_{\theta_D} = \int_0^{L_1} \frac{M_1}{EI}\frac{\partial M_1}{\partial M}dx + \int_0^{L_2} \frac{M_2}{EI}\frac{\partial M_2}{\partial M}dx + \int_0^{L_3} \frac{M_3}{EI}\frac{\partial M_3}{\partial M}dx = 0 \tag{14}$$

$$\begin{aligned} \Delta L_{T_{13}} - \int_0^{L_1+L_3} \frac{N_1}{EA}\frac{\partial N_1}{\partial F}dx - \int_0^{L_1} \frac{M_1}{EI}\frac{\partial M_1}{\partial F}dx - \int_0^{L_2} \frac{M_2}{EI}\frac{\partial M_2}{\partial F}dx \\ - \int_0^{L_3} \frac{M_3}{EI}\frac{\partial M_3}{\partial F}dx = 0 \end{aligned} \tag{15}$$

where $\Delta L_{T_{13}}$ is the total elongation of L_1 and L_3 due to thermal expansion, and $\Delta L_{T_{13}} = \alpha(\Delta T_{13})L_0$. Combining with the above boundary conditions, the tip displacement of the actuator can be calculated as

$$y = \begin{pmatrix} L_b B_1 & -\frac{L_2}{A_b} B_2 \end{pmatrix} \begin{pmatrix} \alpha & \frac{1}{E} \end{pmatrix} \begin{pmatrix} \Delta T_{b\_avg} & P \end{pmatrix} \tag{16}$$

where $B_1 = \beta_{a2}\omega_2 + \beta_{a13}\omega_{13}G_1$, $B_2 = 1 + G_2$, $\omega_2 = \frac{L_2}{L_b}$, $\omega_{13} = \frac{L_0}{L_b}$, $G_1 = \frac{3L_2\left[2L_1^2L_3+L_1^2L_2+2L_1L_3^2+L_2L_3^2\right]}{h_0^2L_0L_b+L_2^2\left[L_2^2+12L_1L_3+4L_1L_2+4L_2L_3\right]}$, and $G_2 = \frac{36L_2\left[(2L_3+L_2)L_1^2+(2L_1+L_2)L_3^2\right]^2}{\left(-4h_0^2\right)L_b\left\{h_0^2L_0L_b+L_2^2\left[L_2^2+12L_1L_3+4L_1L_2+4L_2L_3\right]\right\}}$ $+\left(\frac{4}{h_0^2L_2}\right)\left[L_1^3 + 3(L_2 + L_3)L_1^2 + 3L_1L_3^2 + L_3^3\right] - \frac{3\left[L_1^2+2L_1L_2+2L_1L_3+L_3^2\right]^2}{h_0^2L_2L_b}$.

2.3 System Modeling

In Sect. 2.2, P is the external force acting at one half of the beam while the external force acting at the whole structure should be $2P$. Henceforth, we denote P as the external force applied on the whole structure. Combining the electro-thermal model with thermal-mechanical model, the tip displacement for the Z-shaped actuator is derived as

$$y = \begin{pmatrix} F_1 & -F_2 \end{pmatrix} \begin{pmatrix} U^2 & P \end{pmatrix} = \begin{pmatrix} W_1 & -W_2 \end{pmatrix} \begin{pmatrix} \alpha Q_a & \frac{1}{E} \end{pmatrix} \begin{pmatrix} U^2 & P \end{pmatrix} \tag{17}$$

where $F_1 = \alpha Q_a W_1$, $F_2 = \frac{W_2}{E}$, $W_1 = L_b B_1 T_{a1}$, and $W_2 = \frac{L_2 B_2}{2nA_b}$. Apparently, F_1 and F_2 can be seen as the flexibility of the beams in response to the voltage U and external force P, respectively.

3 Simulations and Experimental Testing

The simulations have been conducted using ANSYS Multiphysics (version 14.0) to verify the derived analytical model in Sect. 2. In the simulations of ANSYS, A 3-D element type (SOLID227) is used, and a coupled-field analysis is conducted including electric, thermal, and mechanical fields.

To further validate the analytical model, we fabricated Z-shaped electrothermal actuators with different in-plane dimensions using the PolyMUMPs process, and compared the experimental calibration results with the theoretical and FE simulation results. Figure 3 shows the optical microscopy photograph of the Z-shaped actuators. We calibrated a Z-shaped actuator with dimensional parameters, listed in Table 1, on a probe station (MicroZoom, Signatone). A DC power supply is used to apply voltages of 0–9.5 V (0.5 V increments) to the Z-shaped actuator, and the resultant output displacements of the actuator are measured through an optical microscope (50 × objective, NA = 0.6) with a CMOS camera (A602f, Basler; image size: 640 × 480 pixels). The pixel size of the imaging setup is calibrated to be

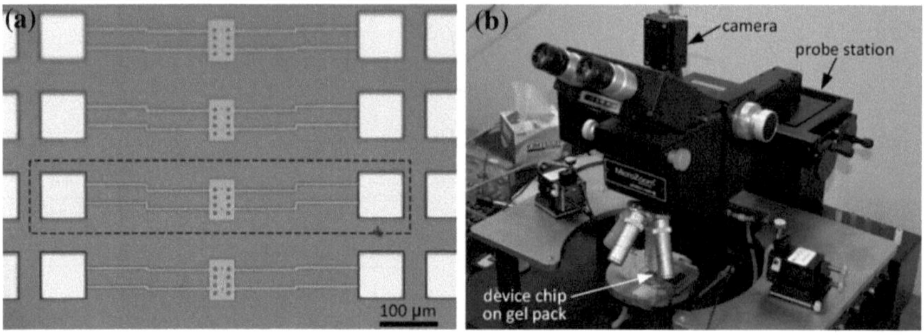

Fig. 3 **a** Z-shaped actuators with different dimensions. The *dashed block* indicates the device we tested for comparison with the theoretical/simulation results. **b** Experimental setup for testing the Z-shaped actuator

173.7 nm. A sub-pixel computer-vision algorithm is used to track the movable central shaft of the actuator at a tracking resolution of 0.08 pixels [11]. Therefore, the measurement resolution of the output displacements of the Z-beam actuator is 13.9 nm.

In this research, the temperature-dependent properties i.e., the coefficient of thermal expansion (CTE) α, the thermal conductivity of polysilicon k, the electric resistivity $\hat{\rho}$, and thermal conductivity of air k_a, are treated as constant for simplicity. In addition, the thermal expansion of the shuttle is neglected. The dimensions and material properties are listed in Tables 1, 2, respectively. The room temperature T_0 is 21 °C.

By virtue of the coupled-field simulation method in ANSYS, the temperature distributions along the structure are calculated and shown in Fig. 4a. The rectangular blue background area is the air gap. The curve in Fig. 4a represents the temperature distribution via analytical calculation. It was reported in [8] that the hottest part of the structure is the shuttle when the actuator operating in vaccum

Table 1 Structural dimensions of Z-shaped actuator (Unit: μm)

b_0	h_0	L_0	L_3	L_2	b_1	h_1	L_s	h_a
3.5	4	240	120	12	3.5	78	50	2

Table 2 Material properties

Material properties	Units	Value
Coefficient of thermal expansion (CTE): α	K^{-1}	2.5×10^{-6}[1]
Electric resistivity: $\hat{\rho}$	Ω μm	34 [16]
Young's modulus: E	μN μm^2	170×10^3 [19]
Thermal conductivity of polysilicon: k	W μm^{-1} K^{-1}	148×10^{-6} [9]
Thermal conductivity of the air: k_a	W μm^{-1} K^{-1}	0.026×10^{-6} [12]

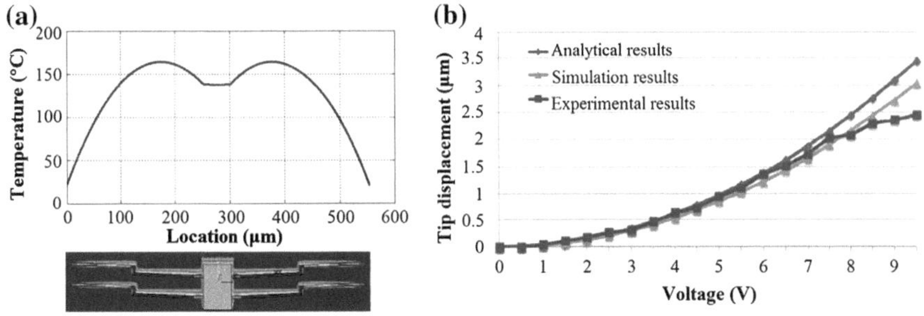

Fig. 4 **a** Temperature distributions. **b** Tip displacement of the structure

conditions. However, the hottest part is located on the beams operating in air conditions in this work. The reason is that heat is conducted more efficiently through the air gap from the shuttle to the substrate than the beams to the substrate.

The comparison of the tip displacement, as shown in Fig. 4b and Table 3, was made based the results from the analytical prediction, FE multiphysics simulation, and experiment testing. The Δ_1 and Δ_2 listed in most right two columns represent

Table 3 Tip displacement results

Voltage (V)	Analytical (μm)	Simulation (μm)	Experiment (μm)	Δ_1 (%)	Δ_2 (%)
0	0	0	0	0	0
0.5	0.009544705	0.0084203	1.13747E-07	13.3535	83911608
1	0.038178819	0.033681	0.04138427	13.3542	−7.7456
1.5	0.085902342	0.075782	0.112994108	13.3545	−23.9763
2	0.152715275	0.13472	0.176698023	13.3575	−13.5727
2.5	0.238617617	0.21051	0.270344408	13.3522	−11.7357
3	0.343609369	0.30313	0.315104822	13.3538	9.0461
3.5	0.46769053	0.41259	0.465857554	13.3548	0.3935
4	0.610861101	0.5389	0.623721203	13.3533	−2.0618
4.5	0.77312108	0.68204	0.717260755	13.3542	7.788
5	0.95447047	0.84203	0.930160656	13.3535	2.6135
5.5	1.154909268	1.0189	1.097340338	13.3486	5.2462
6	1.374437476	1.2125	1.352085148	13.3557	1.6532
6.5	1.613055094	1.423	1.513333198	13.3559	6.5896
7	1.870762121	1.6504	1.724355583	13.3520	8.4905
7.5	2.147558557	1.8946	2.024390457	13.3516	6.0842
8	2.443444402	2.1556	2.07698914	13.3533	17.6436
8.5	2.758419657	2.4335	2.28800269	13.3519	20.5602
9	3.092484322	2.7282	2.359863835	13.3526	31.0450
9.5	3.445638396	3.0397	2.452880409	13.3546	40.4732

the relative errors compared with the FE simulations and the experiments, respectively. The results show that the analytical predictions agree well with the simulation and experimental results particularly when the voltage is lower than 7.5 V. Relative larger errors occurs when voltage increases (resulting in temperature rises). The possible reason is that $\hat{\rho}$ and k_a increase with increased temperature resulted in weaker heat generation and greater heat dissipation to the substrate, despite the fact that k decreases and α increases with the increased temperature which resulted in more efficient elongation and less heat dissipation to the anchors. The comparison also shows that there exists the small deviation (around 13.35 %) between analytical and simulation results. It is mainly because the analytical model ignored the extra heat dissipation to the ambient air around the conducting path. This analytical model developed in this paper is targeted at the microactuators operating in air condition. However, the model can be used for microactuators operating in vacuum condition by simply setting k_a to be closed to zero or letting the air gap be larger.

4 Conclusions

A comprehensive analytical model for Z-shaped electrothermal microactuator operating in the air condition has been developed in this paper. It can be used to directly predict the tip displacement of the actuator for a given voltage. This model has been incorporated with the heat conduction through the air gap between the structure and the substrate, the impact of the shuttle, and multiple pairs of beams. FE simulations and experimental testing have been conducted to verify the analytical model. The analytical results agree well with the FE multiphysics simulations and experiments testing results, especially at voltages lower than 7.5 V where the assumption of the constant material properties is reasonable. The derived analytical model provides the insights and guidance to the design and control of the Z-shaped electrothermal microactuators.

Acknowledgments The work was partially supported by National Science Foundation of China (No. 51175006).

References

1. Aaron, A.G.: Electrothermal properties and modeling of polysilicon microthermal actuators. J. Microelectromech. Syst. **12**(4), 513–523 (2003)
2. Baker, M.S.: Final Report: Compliant Thermomechanical MEMS Actuators LDRD #52553, Sandia Report SAND20046635, Sandia National Laboratories, Albuquerque (2004)
3. Chiao, M.: Self-buckling of micromachined beams under resistive heating. J. Microelectromech. Syst. **9**(1), 146–151 (2000)

4. Eniko, E.T.: Analytical and experimental analysis of folded beam and V-shaped thermal microactuators. In: SEM 10th International Congress and Exposition, Costa Mesa, California (2004)
5. Eniko, E.T.: Analytical model for analysis and design of V-shaped thermal microactuators. J. Microelectromech. Syst. **14**(4), 788–798 (2005)
6. Girija, M.N.: Modelling and simulation of thermal actuator using polysilicon material. Int. J. Nanosci. Nanotechnol. **4**(2), 175–180 (2013)
7. Gregory, N.: Heat Transfer. Cambridge University Press, New York (2009)
8. Guan, C.: An electrothermal microactuator with Z-shaped beams. J. Micromech. Microeng. **20** (8), 85014–85022 (2010)
9. Jorge, V.: Polysilicon thermal microactuators for heat scavenging and power conversion. J. Micro-Nanolithog. MEMS and MOEMS, **8**(2), 023020 1–8 (2009)
10. Kushkiev, I.: Modeling the thermo-mechanical behavior of a "V"-shaped composite buckle-beam thermal actuator. In: Proceedings of the COMSOL Multiphysics User's Conference, Bostion (2005)
11. Liu, X.: A MEMS stage for 3-axis nanopositioning. J. Micromech. Microeng. **17**, 1796–1802 (2007)
12. Long, Q.: Bent beam electrothermal actuators-part 1: single beam and cascaded devices. J. Microelectromech. Syst. **10**(2), 247–254 (2001)
13. Lott, C.D.: Thermal modeling of a surface-micromachined linear thermomechanical actuator. In: Modeling and Simulation of Microsystems, pp. 370–373 (2001)
14. Lott, C.D.: Modeling the thermal behavior of a surface-micromachined linear-displacement thermomechanical microactuator. Sens. Actuators **101**, 239–250 (2002)
15. Mallick, D.: Design and simulation of MEMS based thermally actuated positioning systems, 2012. In: 5th International Conference on Computers and Devices for Communication (CODEC), Kolkata, pp. 1–4 (2012)
16. Suen, M.S.: Optimal design of the electrothermal V-beam microactuator based on GA for stress concentration analysis. Lect. Notes Eng. Comput. Sci. **2189**, 1264–1268 (2011)
17. William, N.: Effect of specimen size on young's modulus and fracture strength of polysilicon. J. Microelectromech. Syst. **10**(3), 317–326 (2001)
18. Zhu, Y.: A microelectromechanical load sensor for *insitu* electron and x-ray microscopy tensile testing of nanostructures. Appl. Phys. Lett. **86**, 013506 (2005)
19. Zhu, Y.: A thermal actuator for nanoscale *insitu* microscopy testing: design and characterization. J. Micromech. Microeng. **16**(2), 242–253 (2006)

Optimising Configurations of KUKA LWR4+ Manipulator for Calibration with Optical CMM

Sergey A. Kolyubin, Leonid Paramonov and Anton S. Shiriaev

Abstract This work is aimed at a comprehensive discussion of experiments and numerical procedures for the open-loop geometric calibration of the KUKA LWR4+ redundant robotic arm, when a full 6D end-effector's pose is measured using Nikon K610 optical coordinate measuring machine (CMM). The later includes a comparative analysis of three different conjugate-type and meta-heuristic iterative algorithms used for numerical optimization of two observability indexes associated with Jacobian properties of the manipulator kinematics. While the former is based an original LEDs fixture design, which geometry is important for organization of the experiment. To the best of our knowledge, such integrated efforts are new for the KUKA robot widely used in robotics research community.

Keywords Calibration · Kinematics · Optimization · Optical coordinate measuring machine (CMM)

1 Introduction

Absolute accuracy of a robot's tool positioning is one of the key robot characteristics. It matters for almost all applications of industrial manipulators, but precise positioning is even more important for such tasks as laser welding, riveting, or automated fibre placement in composite manufacturing as well as for contact operations like grinding or drilling.

S.A. Kolyubin (✉) · L. Paramonov · A.S. Shiriaev
Department of Engineering Cybernetics, Norwegian University of Science and Technology, NO-7491, Trondheim, Norway
e-mail: sergey.kolyubin@ntnu.no

L. Paramonov
e-mail: leonid.paramonov@ntnu.no

A.S. Shiriaev
e-mail: anton.shiriaev@ntnu.no

S. Bai and M. Ceccarelli (eds.), *Recent Advances in Mechanism Design for Robotics*, Mechanisms and Machine Science 33,
DOI 10.1007/978-3-319-18126-4_18

While the industrial manipulators mechanics and manufacturing techniques are constantly improving, there is still a crucial trade-off between price and quality, therefore manufacturing and assembly defects still exist and cause significant deviation of the real robot parameters from its nominal values specified on the design phase. Such offsets in parameters used by a motion planner or a motion controller can significantly degrade the overall system performance and even cause the equipment damage. A way to handle the problem and improve the robot accuracy just on a software level without changing the mechanical structure of the robot is the proper calibration.

The calibration problem was extensively developed in fundamental works of Hollerbach, Khalil, Born and Menq, Daney, and others [1, 8, 11], and still remains an active research area [9, 10]. A good review of the main calibration aspects and state-of-art approaches is given in [5, 7].

There are sources of robot positioning errors of different nature such as joint's and link's flexibility or gear backlash and friction, but according to [11], poor estimation of basic parameters defining robot geometry could be responsible for up to 90 % of these errors.

This work is aimed at a comprehensive discussion of experiments and numerical procedures for the open-loop geometric calibration of the KUKA LWR4+ redundant robotic arm, when a full 6D end-effector's pose is measured using Nikon K610 optical CMM. While there are a number of studies reporting successful results on its dynamic identification, we have not been able to identify any work focused on calibration of the robot geometrical parameters. Without doubts, this work is critical for performing various dedicated experiments with the robot. However, it is important to emphasize the fact that any calibration of the robot geometry cannot be organized without external measuring system, and the lack of such reports in open literature is partly explained the cost of necessary metrological equipment and software.

Even though an external positioning system is available, the calibration procedure is a non-trivial assignment and should be properly organized. Indeed, from the very beginning an engineer is required to choose a number of robot's configurations (poses) that will be most informative and insensitive to noise in further processing of measurement records. Such poses are typically found numerically via optimizing one or several aggregated criteria describing sensitivity of parameters used in solution of a forwards kinematics for a robot. The described settings is referred to as *open-loop calibration* meaning that the robot end-effector is not physically fixed in space, but its position is defined by an external measurement device.

As the main contributions of this note, we have provided the comparative analysis of three different numerical optimization algorithms and two observability indexes. The description of the design of additional tools necessary for experiments with optical CMM can be also valuable.

The paper is organized as follows. At first, we will describe the experimental setup, including LEDs fixture original design. Section 3 includes the formulation of the problem of selecting optimal calibration configurations. Utilized optimization algorithms and observability indexes will be introduced in Sect. 4. Optimization results will be presented in Sect. 4 as well. Finally, we will conclude the paper with comparative analysis and performance boundaries of implemented numerical procedures for the particular task of KUKA LWR4+ geometrical calibration.

2 Equipment

KUKA LWR4+ is a redundant robotic arm with 7 rotational joints. It offers unique features via its low-latency Fast Research Interface (FRI) developed with focus on research purposes [6]. At the same time, to the best of our knowledge, the only source of geometrical parameters' data for this robot is a CAD model, which provides quite rough estimations.

There are different kinds of metrological equipment suitable for calibration experiments, including laser trackers [9] and photogrammetric systems [10], which are considerably less expensive. We use the latter type Nikon K610 system. This optical CMM allows us localizing end-effector in 3D space with sampling rate up to 1 kHz and volumetric accuracy of 60 μm for measurement volume of 17 m^3. This way, having complete 6D pose measurements in calibration experiments we are able to identify the maximum number of robot's geometric parameters [5, 7]. Tracking movable objects is enabled through dynamic frames concept [2]. The system defines positions of infrared LEDs by means of three linear CCD cameras scanning in different planes. Diodes on/off phases are synchronized via strobers such that the camera 'distinguish' between different LEDs.

One more element, which is needed for calibration, is a fixture with LEDs that should be attached to a robot tool plate. Based on Nikon specifications, an LED is guaranteed to be visible only if an angle between a normal to its surface and a line of sight with the camera's central point does not exceed 30°. It gives us 'a visibility cone' of each LED. Thereby, a pyramid geometry was optimized based on criteria:

- all three 'visibility cones' have at least one common point on a distance of 1.5 m[1] from the pyramid top, i.e. there is no 'blind spot' in between
- pyramids' sides allows to inscribe a circle about a size of a LED

Geometry behind the calculation is illustrated on Fig. 1b, c. We have chosen a design with nine LEDs arranged in triples on sides of three similar pyramids, which are distributed on the plate surface (see Fig. 1a).

[1]The distance of 1.5 m was selected based on the relative location of the camera and the robot and parameters of the camera's field of view.

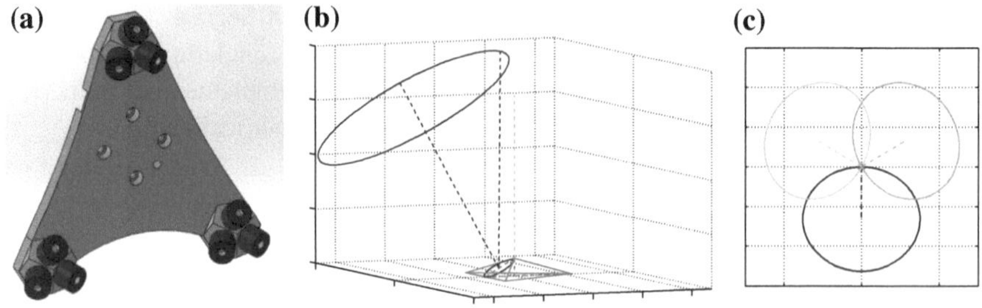

Fig. 1 LEDs fixture geometry. **a** Fixture CAD model. **b** Pyramid geometry. **c** LEDs' visibility cones *top view*

3 Problem Statement

An outcome of the geometric calibration is estimations of parameters describing relative location of neighbouring coordinate systems associated with robot links. Basically, it means identifying parameters of the forward kinematics equation

$$Y^l = F(\Phi, q)|_{q=q^l}, \tag{1}$$

where $Y^l = [x^l y^l z^l \phi^l \theta^l \psi^l]^T$ and $q^l = [q_1^l q_2^l \ldots q_n^l]^T$ are the end-effector's Cartesian pose and relative to the measurement device and the vector of joint angles for robot's configuration l respectively, n is the number of robot joints, and Φ is the vector of unknown geometric parameters.

It is clear that the calibration is feasible only if a total number of constraint Eq. (1) from all measurements $l = 1, 2 \ldots M$ is bigger than a number of estimated parameters. However, a good selection of the configurations is more important than its number to obtain good identification results.

As a quantitative measure of the configurations optimality, several so-called observability indices $O(Q)$, $Q = \{q^1, q^2 \ldots q^M\}$, were introduced (refer to [3, 5] for more details). However, all of these indexes rely on properties of the Jacobian evaluated in $l = 1, 2 \ldots M$ configurations

$$J(Q) = \left[J^1; J^2; \ldots; J^M\right], \tag{2}$$

where J^l in the Jacobian defined in the configuration q^l

$$J^l = \left.\frac{\partial F(\Phi, q)}{\partial \Phi}\right|_{q=q^l}.$$

Thus, the configurations optimization task can be formulated as a nonlinear constraint optimization problem

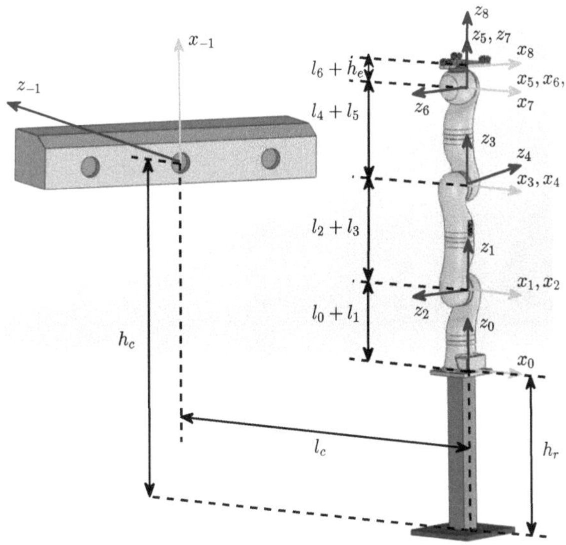

Fig. 2 Coordinate frames for camera, robot, and LEDs fixture

$$\max_{Q} O(Q), \quad \text{subject to} \tag{3}$$

$$\begin{cases} q_i^l \in (q_{i,\min}; q_{i,\max}), \\ Y^l \in (Y_{\min}; Y_{\max}), \end{cases} \quad \forall i = 1 \ldots n, l = 1 \ldots M \tag{4}$$

Figure 2 illustrates relations between different coordinate frames for the calibration setup. The most common way of composing the forward kinematics Eq. (1) is to use a set of modified Denavit-Hartenberg (mDH) parameters

$$\Phi = [\alpha_0, \ldots \alpha_{n+1}, d_0, \ldots d_{n+1}, \theta_0, \ldots \theta_{n+1}, r_0, \ldots r_{n+1}]$$

and a homogeneous transformation

$$^{j-1}T_j = Rot_{x,\alpha_j} Trans_{x,d_j} Rot_{z,\theta_j} Trans_{z,r_j}.$$

For more detailed explanation, please, refer to [8].

Nominal mDH parameters values for the considered system are in the Table 1.

In order to recursively calculate Jacobian (2) and simultaneously eliminate non-identifiable parameters, corresponding to linearly dependent Jacobian columns, we used equations introduced in [8]. Due to limited space, we skip these relations here. As a result, we determined that 4 parameters θ_0, r_0, θ_8, and r_8 are non-identifiable, which remains the set $|\bar{\Phi}| = 4(n+2) - 4 = 32$ of parameters to be calibrated and the modified Jacobian $\bar{J}_{[6M \times 32]}$.

Except for physical limitations on joint angles, we should impose additional constraints on the LEDs fixture Cartesian coordinates. Despite the fact that we can

Table 1 Nominal mDH parameters

Frame index	α_i (rad)	d_i (m)	θ_i (rad)	r_i (m)
0	0	0	$\gamma_z = \frac{\pi}{2}$	$b_z = -l_c = -4.39$
1	$\alpha_z = 0.48\pi$	$d_z = 0.07$	$q_1 + \theta_z = q_1 - \frac{\pi}{4}$	$l_0 + l_1 - (h_c - h_r) = -0.2$
2	$\frac{\pi}{2}$	0	q_2	0
3	$-\frac{\pi}{2}$	0	q_3	$l_2 + l_3 = 0.4$
4	$-\frac{\pi}{2}$	0	q_4	0
5	$\frac{\pi}{2}$	0	q_5	$l_4 + l_5 = 0.39$
6	$\frac{\pi}{2}$	0	q_6	0
7	$-\frac{\pi}{2}$	0	$q_7 + \gamma_e = q_7$	$b_e = 0$
8	0	0	$\frac{\pi}{4}$	$l_6 + h_e = 0.093$

While the robot parameters (i = 1…7) were taken from the CAD model, parameters of the transformation between the camera and the robot base frames (indexed with z) were calculated backwards from camera-based measurements, and parameters for the transformation between the robot last joint and LEDs fixture frames (indexed with e) were directly measured. In general, the Nikon camera allows direct measurements of the end-effector's pose relative to the robot base, therefore transformation between the camera and the robot base coordinate frames can be excluded from consideration

Table 2 Optimization constraints

Parameter	Constraint
Joint angles (rad)	$q_{1,3,5,7} \in [-\frac{17\pi}{18}, \frac{17\pi}{18}]$, $q_{2,4} \in [-\frac{2\pi}{3}, \frac{2\pi}{3}]$, $q_6 \in [-\frac{\pi}{2}, \frac{\pi}{2}]$
Fixture roll (rad)	$\phi_{x,LED} \in [-\frac{4\pi}{9}, \frac{7\pi}{18}]$
Fixture pitch (rad)	$\theta_{y,LED} \in [-\frac{\pi}{4}, \frac{\pi}{4}]$
Distance between LED TCP and robot base (m)	$\sqrt{{}^0x_8^2 + {}^0y_8^2} \geq 0.2$
z-coordinate of LED TCP (m)	${}^0z_8 \geq -0.1$

place the camera to cover the entire robot workspace, the end-effector pose can be recovered only if at least 3 LEDs remain visible. It gives us bounds on roll, pitch, and yaw angles of the LEDs fixture, expressed in the camera coordinate frame. Two more constraints were imposed to guarantee that the LEDs fixture will not collide with the robot base. All these constraints are summarized in Table 2.

4 Optimization

At the first step we randomly selected a number of robot configurations, satisfying (4), which were used as initial conditions for the optimization procedure. We distributed it uniformly across the robot's workspace, providing distinguishable variation in joint angles as well.

From the available variety of observation indices [5] we employed two. The first on is the inverse condition number of the Jacobian (2) $O_1 = \frac{\sigma_{\min}}{\sigma_{\max}}$, where $\sigma_{\min}$ and $\sigma_{\max}$ are Jacobian minimal and maximum singular values respectively (see [4]). The second index was introduced in [1] $O_2 = \frac{\sqrt[R]{\sigma_R \ldots \sigma_i \ldots \sigma_1}}{\sqrt{M}}$, where R is the Jacobian rank (in our case $R = 32$), M is the number of configurations, and σ_i, $i = 1 \ldots R$ are Jacobian singular values.

We tried three different algorithms to actually solve the optimization task.

The first one is conjugate-type deterministic algorithm similar to one described in [8]. It tries to optimize the entire set of configuration on every iteration. However, algorithms of this type have significant limitations: strong dependence on initial conditions and sticking in a local minimum [3].

To avoid these issues we implemented a modification of the previous algorithm utilizing multistart optimization, i.e. when the same algorithm starts N times (we selected $N = 50$) consequently from randomly uniformly distributed initial conditions.

As an alternative to conjugate-type algorithms, we tested an iterative meta-heuristic algorithm with the Tabu rule [3]. Instead of optimizing the entire configurations set, it implements a 2-stage procedure (see Fig. 3). We selected $L = 10$

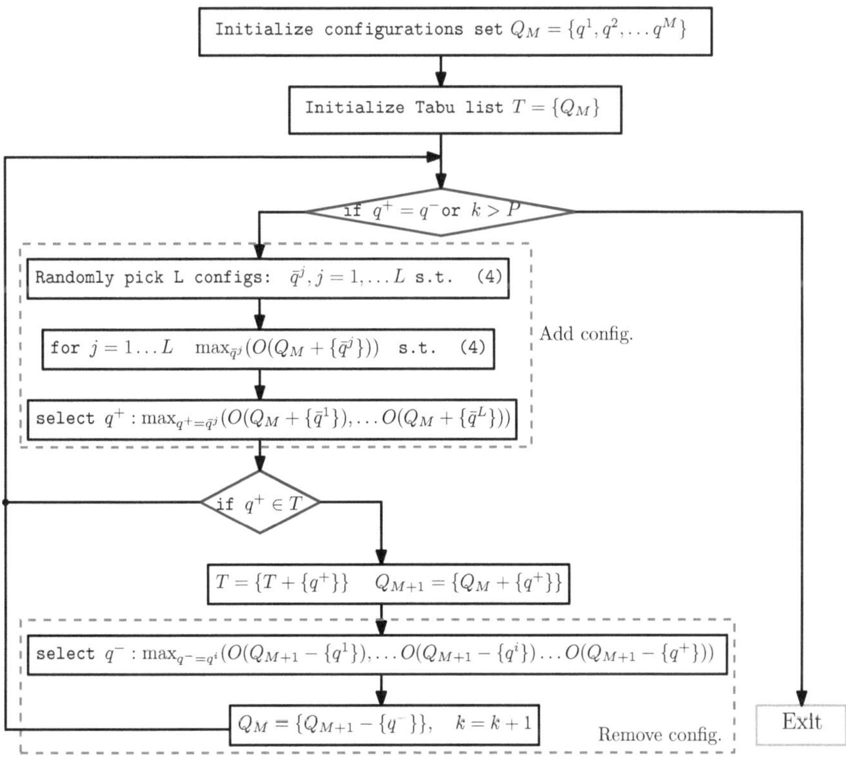

Fig. 3 Diagram for the iterative meta-heuristic algorithm algorithm with the Tabu rule

Table 3 Optimization results

Number of configurations	Algorithm	Observability index	
		O_1	O_2
$M = 10$	Non-optimized	0.000716	0.5685
	Conjugate-type deterministic	0.001535	0.8065
	Conjugate-type with multistart	0.001613	0.7955
	Randomized iterative with Tabu rule	0.001497	0.7904
$M = 20$	Non-optimized	0.000811	0.6435
	Conjugate-type deterministic	0.001618	0.8142
	Conjugate-type with multistart	0.001631	0.7973
	Randomized iterative with Tabu rule	0.001571	0.8091

and $P = 2M$. Compared to the conjugate-type algorithm, the optimization task takes only n variables instead of $n \times M$, but should be solved $L \times P$ times instead. In general, this approach is quite similar to genetic algorithms, when a stochastic behaviour is introduced during candidate-configuration selection.

We used a SQP (sequential quadratic programming) solver for all aforementioned algorithms.

Comparative results of the optimization for sets of 10 and 20 configurations for all three algorithms and two observability indices are given in Table 3. Figure 4 illustrates how joint angles change between initial and optimized configurations. Figure 5 visualizes positions of the robot end-effector before and after optimization in Cartesian space relative to the camera coordinate frame. And finally, Fig. 6 shows if the constraints imposed on the LEDs fixture position and orientation were matched.

5 Discussion

We tried 3 different algorithms and 2 observability indices solving the configurations optimization task. Comparative analysis of the obtained results leads to the following conclusions:

- Optimization algorithms gave tangible difference in terms of joint configurations and end-effector's poses allocation, but provided quite similar resulting observability indices. It reveals features of the cost function, which is highly nonlinear and has many local minimums.
- The observability index O_1 associated with the Jacobian condition number shows higher relative improvement between initially selected and optimized configurations compared to the Born-Menq index O_2.
- In practice, the iterative meta-heuristic algorithm with Tabu rule didn't ensure better results, while required longer computation time.
- The conjugate-type algorithm with multistart can be considered as the most effective one, since it provided the best observability index O_1 and had less tendency to configurations' clustering.

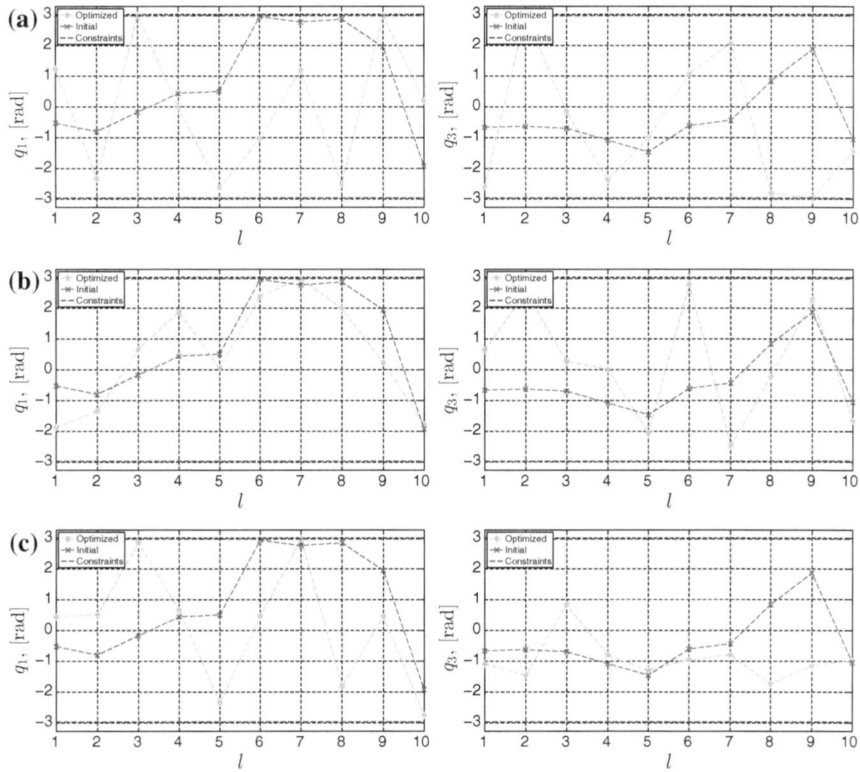

Fig. 4 Joint coordinates for $M = 10$ (only joints 1 and 3 are presented for illustration purposes). **a** Conjugate-type algorithm, O_1. **b** Conjugate-type algorithm, O_2. **c** Iterative algorithm, O_1

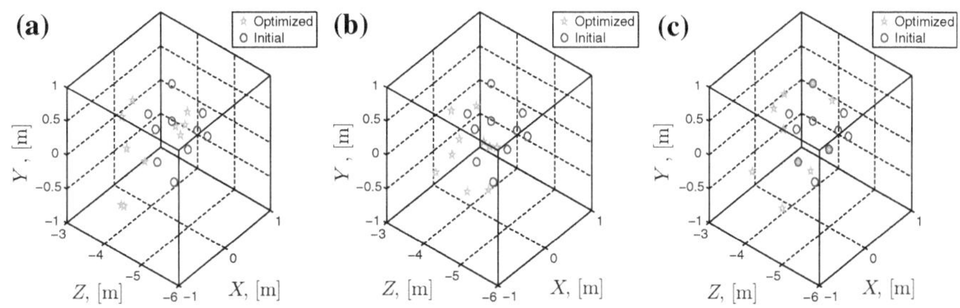

Fig. 5 Cartesian coordinates of the LED frame for $M = 10$. **a** Conjugate-type algorithm, O_1. **b** Conjugate-type algorithm, O_2. **c** Iterative algorithm, O_1

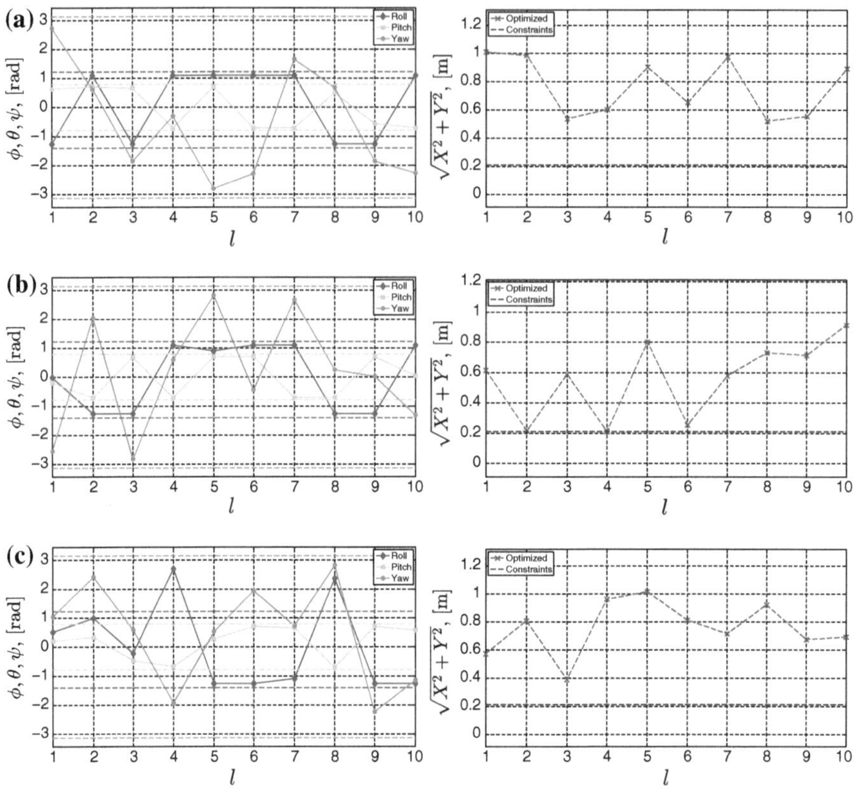

Fig. 6 Constraints for the LED frame orientation and its distance from the robot base for $M = 10$. **a** Conjugate-type algorithm, O_1. **b** Conjugate-type algorithm, O_2. **c** Iterative algorithm, O_1

On the other hand, we did not observe an order of magnitude increase in indices values before and after optimization, as it was reported, e.g. in [3, 8], that also can be an issue of kinematics features of the light-weight arm. Furthermore, such an aspect as an ease of calibration should be taken into account on the manipulator's design phase.

References

1. Borm, J.H., Menq, C.H.: Determination of optimal measurement configurations for robot calibration based on observability measure. J. Robot. Syst. **10**, 51–63 (1991)
2. Daney, D., Papegay, Y., Madeline, B.: Choosing measurement poses for robot calibration with the local convergence method and Tabu search. Int. J. Robot. Res. **24**, 501–518 (2005)
3. Driels, M.R., Pathre, U.S.: Significance of observation strategy on the design of robot calibration experiments. J. Robot. Syst. **7**, 197–223 (1990)

4. Hollerbach, J.M., Wampler, C.W.: The calibration index and taxonomy for robot kinematic calibration methods. Int. J. Robot. Res. **15**(6), 573–591 (1996)
5. Khalil, W., Besnard, S., Lemoine, P.: Comparison study of the geometric parameters calibration methods. Int. J. Robot. Autom. **15**(2), 56–67 (2000)
6. Khalil, W., Gautier, M., Enguehard, C.: Identifiable parameters and optimum configurations for robots calibration. Robotica **9**, 63–70 (1991)
7. Klimchik, A., Pashkevich, A., Wu, Y., Caro, S., Furet, B.: Design of calibration experiments for identification of manipulator elastostatic parameters. J. Mech. Eng. Autom. **2**, 531–542 (2012)
8. Marie, S., Courteille, E., Maurine, P.: Elasto-geometrical modeling and calibration of robot manipulators: application to machining and forming applications. Mech. Mach. Theory Elsevier **69**, 13–43 (2013)
9. Renders, J.-M., Rossignol, E., Becquet, M., Hanus, R.: Kinematic calibration and geometrical parameter identification for robots. IEEE Trans. Robot. Autom. **7**(6), 721–732 (1991)
10. Schreiber, G., Stemmer, A., Bischoff, R.: The fast research interface for the KUKA lightweight robot. IEEE ICRA 2010 Workshop on Innovative Robot Control Architectures (2010)
11. van den Bossche, A.: Procedure for determining the dynamic behavior of a vehicle on a test bench, US Patent 6,748,796 (2004)

Key Parameters Optimization of a Novel Tubular Double Excitation Windings Linear Switched Reluctance Motor

Liang Yan, Wei Li, Zongxia Jiao, Chin-Yi Chen and I-Ming Chen

Abstract This paper discussed the characteristic analysis of the novel tubular double excitation windings linear switched reluctance motor (LSRM), and the influence of this motor performance by different key parameters. Much of the work is accomplished by using finite element analysis method. The main contents include comparison of motor output force under the different condition of parameters, the flux linkage distribution and self induction electromotive force (EMF) when the motor running. In addition, the EMF can be used to apply sensorless technology into this machine.

Keywords Switched reluctance · Linear machine · Electromotive force · Electromagnetic force

1 Introduction

The linear machine is a device that converts electrical energy to linear mechanical energy. It has great value and applications, for example machining tools, precision drive system, transportation and so on. As a kind of linear motor, the linear

L. Yan (✉) · W. Li · Z. Jiao
The School of Automation Science and Electrical Engineering, Beihang University, Xueyuan Road, Beijing 100191, China
e-mail: Lyan1991@gmail.com

L. Yan
Research Institute of Beihang, University in Shenzhen, Shenzhen 518000, China

C.-Y. Chen
The Institute of Advanced Manufacturing Technology, Ningbo Institute of Material Technology and Engineering, Ningbo 315201, China

I-MingChen
The School of Mechanical and Aerospace Engineering, Nanyang Technological University, 50 Nanyang Avenue, Singapore 639798, Singapore

S. Bai and M. Ceccarelli (eds.), *Recent Advances in Mechanism Design for Robotics*, Mechanisms and Machine Science 33,
DOI 10.1007/978-3-319-18126-4_19

switched reluctance motor (LSRM) has achieved great development [1–4], LSRM has the advantages of simple structure, low starting current and high efficiency [5]. But it has a fatal drawback which is can't provide a stable high output thrust [6]. In order to enhance the performance of the LSRM, a special kind of LSRM with double excitation windings are proposed.

1.1 Schematic Structure

The structure diagram of the tubular double excitation winding LSRM is shown in Fig. 1. This motor is a kind of four-phase LSRM, and consists of stator and mover, there are winding phases embedded on the stator and mover. We spilt the stator into multiple parts to make sure that stator windings can be embedded. The stator windings are the motor primary field windings, its energized mode determines the working status of the motor, and the mover windings are secondary field windings. The introduction of mover windings enhances the work performance of this motor. It can produce a continuous linear motion, and has a higher thrust density. In addition, thanks to the tubular structure, the radial force value is substantially zero.

1.2 Operating Principle

The tubular double excitation windings LSRM follow principle of minimum reluctance. The motor is under a state of maximum reluctance when the rotor and stator in the relative position shown in Fig. 2a. The motor is under a state of minimum reluctance when the rotor and stator in the relative position shown in Fig. 2b. When the motor need to provide driving force, the stator windings which associated stator tooth under maximum reluctance to minimum reluctance should be energized, if the mover windings which are in the flux linkage are energized, the

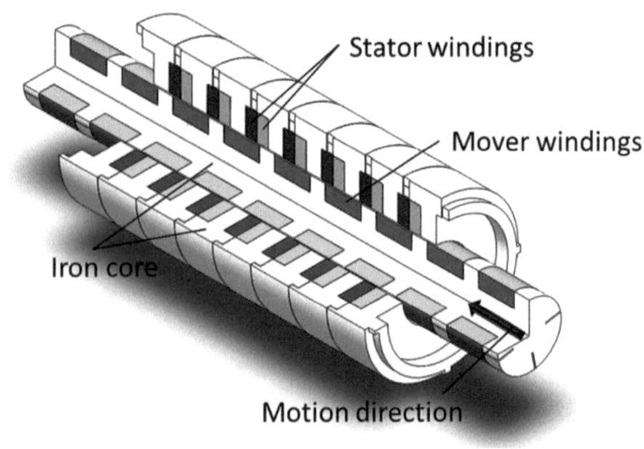

Fig. 1 Schematic structure of the tubular double excitation windings linear switched reluctance motor

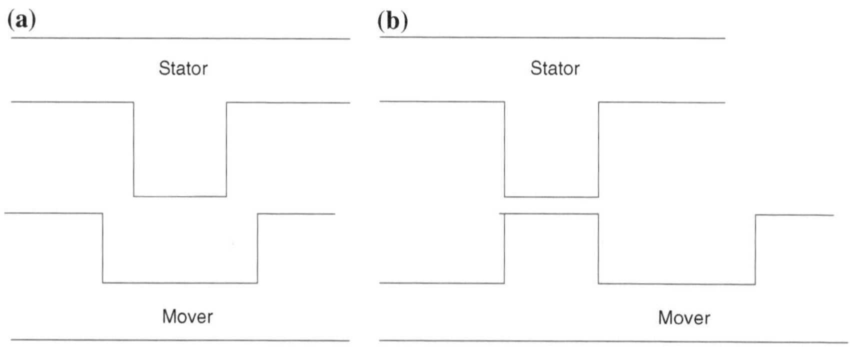

Fig. 2 **a** The relative position of rotor and stator when motor under maximum reluctance. **b** The relative position of rotor and stator when motor under minimum reluctance

output capacity of the motor will be increased. When the motor need to brake, the stator windings energized condition contrary to previous. And the mover windings energized can speed up the motor to brake.

2 Choice of Key Parameters

The motor structure parameters selection has a significant influence of the motor performance, there are some key parameters which include the motor length (L), diameter (D), mover diameter (d), air gap (g), width of stator pole (w_1), width of mover pole (w_2), depth of stator slot (h_1) and depth of mover slot (h_2).

The motor length and motor diameter has the following relationship:

$$\lambda = \frac{L}{D}. \tag{1}$$

where λ has an impact on motor performance and costs. When λ is taken a large value, it's benefit to improve thrust, but will bring cooling difficulties. In general, the motor has the best performance under $\lambda = 0.5-5.0$. In this paper, λ is equal to 4.5.

In order to get more thrust, difference is as large as possible between maximum reluctance and minimum reluctance, so air gap should be minimized. But Constrained by the process and operating conditions, air gap g is taken equal to 1 mm.

Stator tooth pole and mover tooth pole is major work parts of this motor, so value of w_1 and w_2 is very important to motor performance. Ignore the leakage flux of the motor, the area of the stator pole can be expressed as follows:

$$A_s = \frac{D}{2} L k_s. \tag{2}$$

where k_s is correction factor, in general, $k_s = 0.85-1.0$, then we can get the stator pole width w_1:

$$w_1 = \frac{As}{L} = \frac{D}{2L} L k_s = \frac{D}{2} k_s. \tag{3}$$

Similarly, the mover pole width w_2 can be expressed:

$$w_2 = \frac{Am}{L} = \frac{D}{2L} L k_m = \frac{D}{2} k_m. \tag{4}$$

where $k_m = 0.3-1.0$

The depth of stator slot is related to motor diameter and stator yoke height, stator yoke has a minimum height restriction, to ensure that stator yoke is not saturated when motor is under maximum flux density. So the depth of stator slot is:

$$h_1 = \frac{D}{2} - \frac{d}{2} - (1.2-1.4)\frac{w_1}{2}. \tag{5}$$

Similarly, the depth of mover slot can be expressed:

$$h_2 = \frac{d}{2} - (1.2-1.6)\frac{w_2}{2}. \tag{6}$$

The above parameters were taken step 0.5 mm to co-simulation by FEA, and the result is shown in Fig. 3. The motor can achieve maximum output force, it is because that magnetic saturation emerged in LSRM. We can get the motor structure parameters when the motor has the best output performance. The values of those parameters are shown in Table 1, So the motor design in accordance with the major structural parameters shown in Table 1.

3 Characteristic Analysis

Figure 4a show the flux linkage of mover and stator when the machine is running, we can find the mover winding flux linkage curve changes smoothly, and the stators' curve violent change. Figure 4b show the induced voltage of stator and mover windings when the machine running at a speed of 1 m/s. The induced voltage of mover winding curve is more smooth than stators' induced voltage, so the mover windings can be used for sensorless technology.

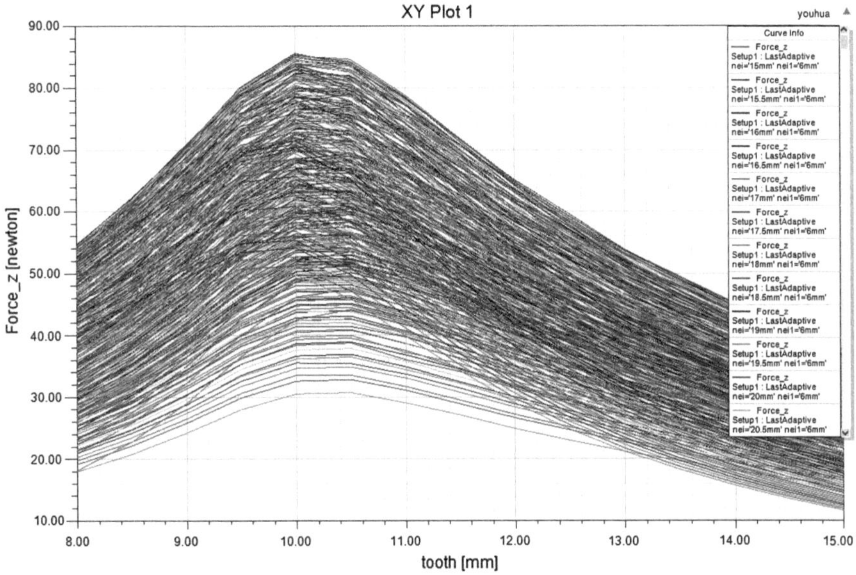

Fig. 3 Output performance of different parameters about LSRM by FEA

Table 1 Key parameters of the proposed motor

Parameters	Value
Phase number	4
Motor length (L)	360 mm
Motor diameter (D)	80 mm
Tooth width (W)	10 mm
Mover slot depth (τm)	8 mm
Stator slot depth (τs)	16 mm
Mover diameter (d)	16 mm

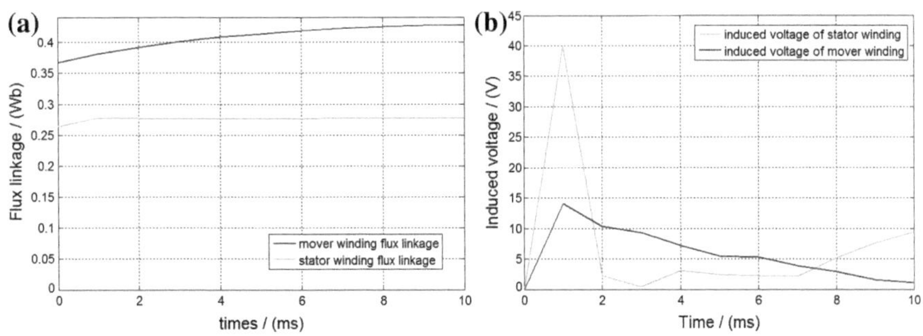

Fig. 4 **a** The flux linkage of mover and stator when the machine is running. **b** The induced voltage of stator and mover windings when the machine running at a speed of 1 m/s

4 Conclusions

In this paper, we analyzed and studied the impact of a novel tubular double excitation windings LSRM's key parameters on motor performance, as well as some of the characteristics of the motor. Laid the theoretical foundation for the motor design and processing, and then to verify the correctness of the theory by experiment.

Acknowledgments The authors acknowledge the financial support from the National Natural Science Foundation of China (NSFC) under grant 51175012, the National Key Basic Research Program of China under grant 2014CB046406, NSFC 51235002, the Program for New Century Excellent Talents in University of China under grant NCET-12-0032, the Fundamental Research Funds for the Central Universities, and the Technology on Aircraft Control Laboratory.

References

1. Pan, J.F., Cheung, N.C., Cao, G., L.-ming, L.: Design and analysis of a DSP-based linear switched reluctance motor. In: 3rd International Conference on Power Electronics Systems and Applications PESA, pp. 1–4, 20–22 May 2009
2. Hirayama, T., Uwada, K., Kawabata, S.: Static characteristics of a double-sided linear switched reluctance motor with high-temperature superconducting excitation winding. In: 15th International Conference on Electrical Machines and Systems (ICEMS), pp. 1–4, 21–24 Oct 2012
3. LEE, B.-S., et al.: Design of a linear switched reluctance machine. IEEE Trans. Ind. Appl. vol. 36, pp. 1571–1580 (2000)
4. Pan, J.F., Zou, Y., Cao, G.: An asymmetric linear switched reluctance motor. IEEE Trans. Energy Convers. vol. 28(2), pp. 444–451 (2013)
5. Siadatan, A., Najmi, V., Afjei, E.: Modeling, simulation and analysis of a novel two layer 8/6 hybrid switched reluctance motor/field-assisted generator. In: 20th Iranian Conference on Electrical Engineering (ICEE), pp. 495−500, 15–17 May 2012
6. Inderka, R.B., Menne, M., De Doncker, R.: Control of switched reluctance drives for electric vehicle applications. Ind. Electron. IEEE Trans. **49**(1), 48–53 (2002)

Optimizing Tracking Performance of XY Repositioning System with ILC

Sigurd Villumsen and Casper Schou

Abstract Controlling complex mechanical systems is often a difficult task, requiring a skilled developer with experience in control engineering. In practice however, the theoretical difficulties of designing a good controller is only a first step as the implementation itself on the various pieces of equipment is also often challenging. This paper investigates if iterative learning control (ILC) can be used as an alternative to tuning existing controllers for improving system performance. This is evaluated by a case study on a high speed XY-positioning system used for laser cutting. An ILC algorithm is implemented by using a server client structure from Matlab. After tuning the parameters an implementation is found which is able to increase the tracking accuracy significantly for cutting speeds up to 0.5 m/s. This is done only by implementing code on the master control unit and thus without changing subsystem controllers.

Keywords Iterative learning control · Machine learning · XY table · Laser cutting

1 Introduction

When designing control algorithms for robots and other mechanical systems it is of high importance to choose a good control strategy to ensure the performance of the system. Much research has been done within the field of control engineering which again has led to many significant contributions [1–3]. The complexity of the control algorithm for a given mechanism is however often limited by the skill of the developer doing the implementation and how much time he can spend optimizing it. This entails that the feedback control loops in one-of-a-kind systems are often composed of simple PID controllers. This is due to the fact that PID controllers are

S. Villumsen (✉) · C. Schou
Department of Mechanical and Manufacturing Engineering, Aalborg University, Aalborg, Denmark
e-mail: sv@m-tech.aau.dk

S. Bai and M. Ceccarelli (eds.), *Recent Advances in Mechanism Design for Robotics*, Mechanisms and Machine Science 33,
DOI 10.1007/978-3-319-18126-4_20

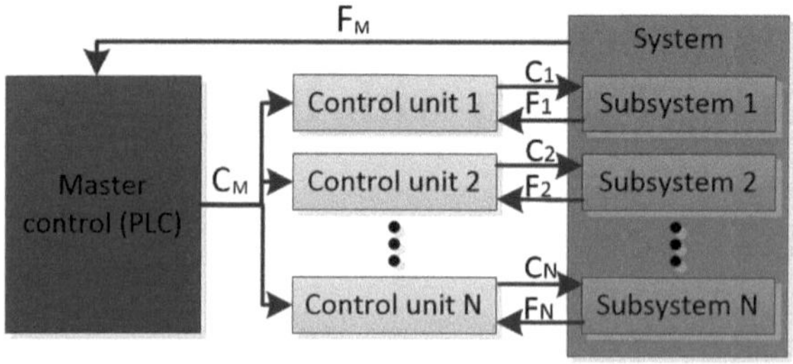

Fig. 1 A generic control system which is composed of several controllers controlling subsystems and a master controller controlling the entire system behavior

generally considered to be easy to implement and can be generated by step-by-step procedures as e.g. Ziegler-Nichols PID tuning rules [4, 5]. Many complex mechanical systems, such as what is seen within the robotics community,are however often composed of several subsystems each with a separate controller. These subsystems are in motion control scenarios often in the form of servo drives connected to a control unit e.g. a PLC or industrial PC which controls the overall system behavior. These servo drives often have internal controllers that controls position, speed, torque or motor current. Such a set-up is depicted on Fig. 1. In systems with this type of topology, overall system optimality is not necessarily obtained by optimizing the individual sub system controllers e.g. with Ziegler Nichols type PID controllers. To achieve good system performance, a developer also needs to look at the interaction of all subcomponents and how they change the dynamic properties of the complete system. This process is further complicated by the fact that most pieces of industrial control equipment can only be programmed by using the manufacturers proprietary programming software. This entails that an engineer essentially needs to be an expert not only in control engineering but also in the software used to program all sub components of the system. To overcome these problems it is proposed that the concept of Iterative Learning Control (ILC) could be used as an approach to optimize performance in systems with similar control architectures. This enables the developers to optimize performance at system level only by programming the main control unit. To determine if ILC can be used as a way of optimizing these kinds of control architecture, it has been chosen to do a case study. The chosen case to investigate is a laser cutting repositioning system found in the laser cutting laboratory at Aalborg University.

In Sect. 2 the ILC algorithm is briefly described. Hereafter Sect. 3 describes the XY repositioning system and it's application. Section 4 presents how the implementation was conducted on the XY repositioning system. Section 5 presents the results of the implementation. Finally, a conclusion of our work is presented in Sect. 6.

2 Iterative Learning Control

The background of ILC is based on the observation that human beings can gain skills from repetitive practice. With ILC this behavior is adapted to machines doing repetitive work in such a way that the system gradually performs better and better [6]. ILC is not based on creating an optimal control law for a given dynamic system, instead it tries to learn a refined reference trajectory in such a way that the executed trajectory converges towards the original desired Ref. [7]. The structure of an ILC controller is shown on Fig. 2.

The input signal to the system u_k and the output y_k are sampled and stored in memory. The subscript k denotes the current trial. By using these stored signals and comparing them to the true reference y_d, a new input signal u_{k+1} is generated. This new input signal is also stored in memory and executed on the next iteration. By repeating this numerous times, the input signal u_k is constantly being refined. This iterative process will then reduce the tracking error for each iteration and thus the system output y_k should converge towards the desired output y_d. Multiple algorithms and approaches for updating u_k (i.e. calculating u_{k+1}) exists but as even the simplest correction models have shown to have good performance [9] it has been chosen implement an ILC algorithm based on the initial research presented by Arimoto in 1984 [10]. The governing equation of this type of algorithm can be seen from Eq. 1.

$$u_{k+1}(t) = u_k(t) + \Gamma e_k(t+1) \tag{1}$$

where Γ is a matrix containing the learning gains, $u_k(t)$ is the control signal, $u_{k+1}(t)$ is the updated control signal and e_k is the error between the reference trajectory and the executed trajectory. A time shift d is often modelled into the equation to compensate for any inherent time delays [11]. This was however found to have a negative effect on the system; thus, d was set to zero. When considering convergence of

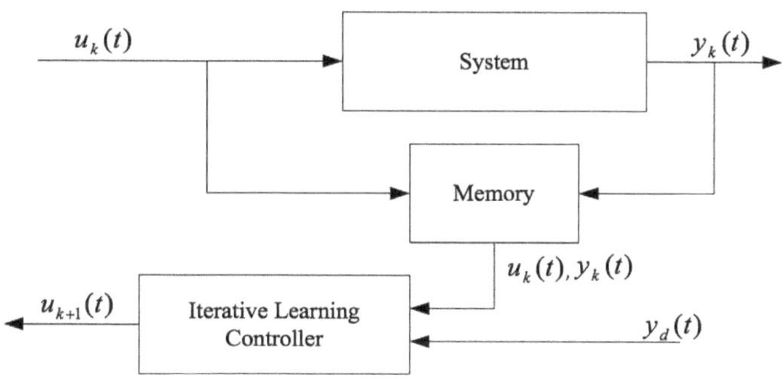

Fig. 2 General ILC architecture [8]

the ILC algorithm it has been proven that $e_k(t) \to 0$ monotonically as $k \to \infty$ if the following necessary and sufficient condition is satisfied.

$$||I - H\Gamma||_i < 1 \quad (2)$$

where H is a lower-triangular Toeplitz matrix whose elements are the Markov parameters of the system to be controlled. Γ is again the learning gain and i is the norm under which this is evaluated, e.g. ∞. Other convergence criteria has been expressed giving conditions for non monotonic convergence. The correct procedure for designing ILC controllers would be to evaluate Eq. 2. However, for one of a kind systems a developer rarely has the time to develop system models to check for the possibility of convergence. Instead a trial and error approach will be used as it is often the case in the industry. This is a feasible approach as long as the process is not inherently dangerous and the iterations don't take too long to complete. Still, as a precaution, the learning gain should be kept at a relatively low level. This might lead to slower convergence, but should ensure that the error in fact is minimized by the iterations. If the gain causes an increase in error, smaller gains will also ensure that the reference trajectory changes at a slower rate, which again enables the developer to stop the process safely before severe instabilities are introduced in the reference signal. The drawback to this approach is of course that it can be difficult to predict if convergence indeed can be ensured.

For ILC to be a feasible solution it is required that the system is initially stable and that it is not affected by heavy stochastic variations. In the following section the case will be introduced and these prerequisites will then be verified by conducting a series of tests on the repositioning system.

3 Case Study—Repositioning System for Laser Cutting

Laser cutting of steel has been done since 1967 where the first oxygen assisted laser cut was conducted [12]. In traditional metal laser cutting a laser beam is focused on the work piece by means of lenses, mirrors and other optical components embedded in a unit called a cutting head. When the beam hits the metal surface, the intensity of the laser heats up the metal and it melts, evaporates and sublimates. Molten material is then removed by cutting gas which is applied through a gas nozzle. If the work piece or the cutting head is moved, a cut is obtained [13, 14]. If the executed path of the repositioning system deviates from the reference trajectory the resulting cut on the work piece will show the same deviation. A conceptual system for repositioning a work piece can be seen on Fig. 3. From this figure it can be seen that the cutting head is mounted on an industrial robot, this robot will however be treated as a stationary mount for the cutting head. This entails that all repositioning will be conducted by the XY-table. The position control structure of the repositioning system can be seen on Fig. 4.

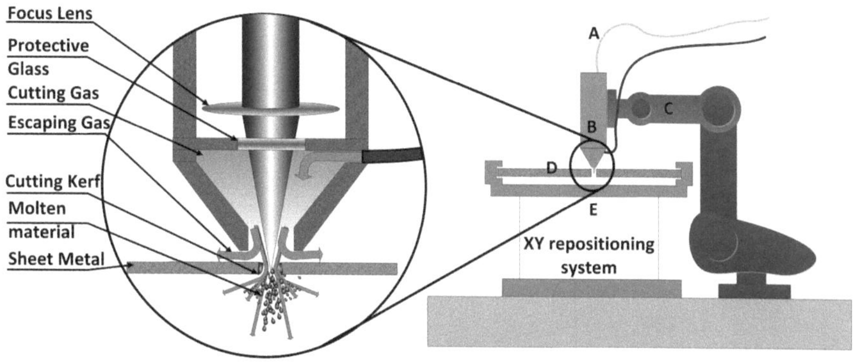

Fig. 3 The lab setup when conducting the angular stability experiments. On the figure the following elements are shown: A laser fiber (*A*), a cutting head (*B*), an industrial robot used as a static mount (*C*), a work piece and fixture (*D*) and finally an XY repositioning system (*E*)

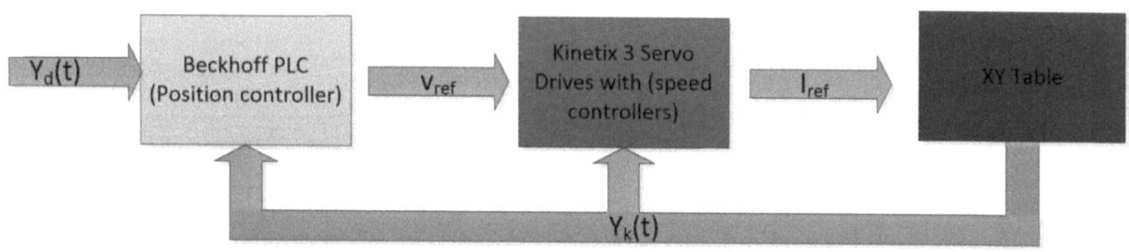

Fig. 4 The control structure of the XY repositioning system. The *red block* is executed on a Beckhoff PLC, the *green* on the servo drive and the *blue* depicts the physical XY table itself

From this figure it is seen that the control of the system is implemented partly on a PLC and partly on servo drives. In addition, several different interacting pieces of software implements the control algorithm. The classical solution to improving the tracking performance is to redesign the speed controllers of the servo drives and then the position controller on the PLC. However, the division between several hardware and software instances makes it difficult to create and tune a new controller. It is therefore proposed to use ILC as a means of correcting for a suboptimal trajectory tracking controller. To determine if the system suffers from tracking error, three runs where a square reference is tracked at three different speeds can be seen on Fig. 5.

From this figure it can be seen that in the vicinity of the corners, large trajectory deviations occur due to the dynamics of the repositioning system. The internal trajectory controller is able to track the sides, but as soon as a new corner is reached the deviation reoccurs. However, some deviations are to be expected as the trajectory dictates a constant speed around the sharp corners, which inherently constitutes a problem as this requires an infinite acceleration. From the figure it is also seen that the overshoot only happens in the X direction and not in the Y direction. This difference is due to the fact that the Y axis is mounted on top of the X axis,

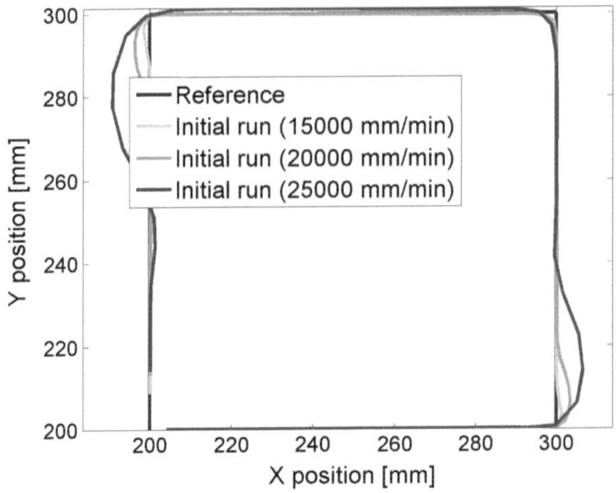

Fig. 5 Three trajectories executed on the XY table with different speeds

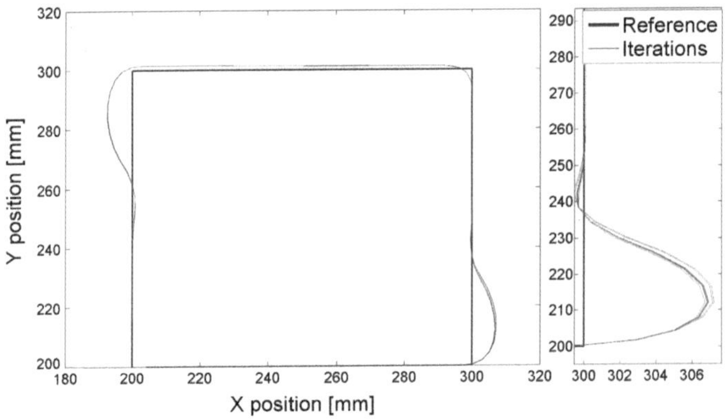

Fig. 6 Ten consecutive runs of the square trajectory with a speed of 25,000 mm/min

which makes the load on the X axis larger than the load on the Y axis; yet they still have the same motor power. The Y axis actually undershoots the reference instead. This is due to the fact that the Y axis hasn't been able to keep up with the reference signal. Which entails that it starts to move in the X direction before it reaches the referenced Y level. The final thing that is seen from the figure is that the system in fact is capable of tracking a reference, which entails that the controller is stable, which is one of the prerequisites for ILC described in Sect. 2.

The second condition described in Sect. 2 is that the tracking errors shouldn't be stochastic in nature, but be caused by the dynamics of the system. If the errors are of a stochastic nature the corrections made by ILC and thus the reference signal will also be stochastic. This is tested by repeating the square run seen on Fig. 5 ten times with a speed of 25,000 mm/min and logging the executed trajectory. The results from this test can be seen on Fig. 6.

This figure shows that the maximum deviation between the resulting trajectories of the ten repetitions is approximately 1 mm. This is significantly less than the overshoot that amounts to approximately 8 mm. This makes it likely that ILC will be able to reduce the deviations and increase the performance of the system.

4 Implementational Considerations

To ease the implementation of the ILC algorithm on the XY repositioning system it has been chosen to modify the control structure seen on Fig. 4 with a server/client structure. This can be seen from Fig. 7. The PLC and XY table acts as a server and a PC with MATLAB works as a client. The server on the PLC only provides the service to execute a given trajectory $u_k(t)$ and return the measured error $e_k(t)$. As a result trajectory generation, data interpretation and ILC are implemented on the

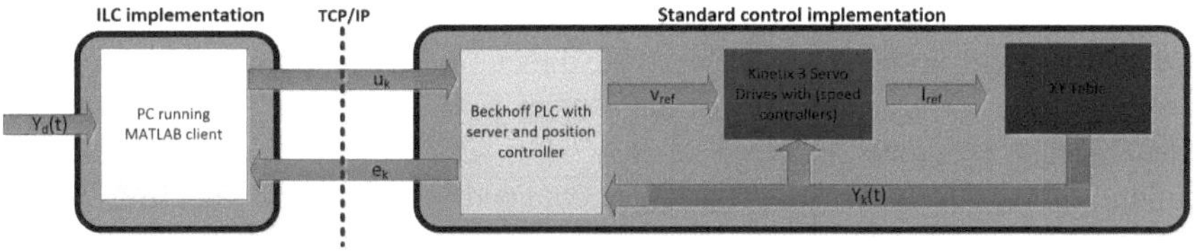

Fig. 7 A diagram showing the server/client structure of the program. The communication between client PC and server is done via TCP/IP communication

Table 1 Overview of the ILC server and client programs program

Client program	Server program
Generate initial reference signal	**Wait** for connection
Establish connection to XY table server	**While** ILC client is connected
Send reference signal to XY table server	**WAIT** for new reference trajectory
WAIT for execution done	Start logging position and velocity
Receive feedback from XY table server	Execute trajectory
Calculate error vector	Send logged data to connected client
FOR 1 to number of iterations	Send execution done
Update reference signal based on error vector	
Send reference signal to XY table server	
WAIT for execution	
Receive feedback from XY table server	
Calculate error vector	

client in MATLAB. A pseudo code representation of the implemented code can be seen in Table 1.

The next implementational consideration is the learning factor Γ. The learning factor was found experimentally as discussed in Sect. 2 to be 0.1 in both the X and Y direction based on the progress of the learning on different shapes. This value proved to be robust with all reference shapes and yielded an adequate level of convergence. Finally, the reference for the system needs to be constructed. It should consist of a series of x, y positions and a corresponding time vector. The number of points generated is given by $\frac{processtime}{sampletime}$ while the velocity is determined from the path length and the process time, $v = \frac{pathlength}{processtime}$. Using this simple sampling the velocity along the path becomes constant which is desirable in laser cutting.

5 Results

In the following section the results obtained from the implementation of the ILC algorithm will be presented. The first results were obtained by using a square reference signal. This should yield a 100 × 100 mm^2 with origin in x = 200 mm y = 200 mm cut in the counter clockwise direction. This is challenging due to the limited acceleration capabilities of the system as described in 4. By utilizing the implemented program described in Sect. 4 a series of 20 ILC iterations were conducted with the square reference. The obtained results can be seen from Fig. 8a where it is seen that the trajectory moves closer to the real reference with more iterations. The final executed trajectory lies close to the reference trajectory, but

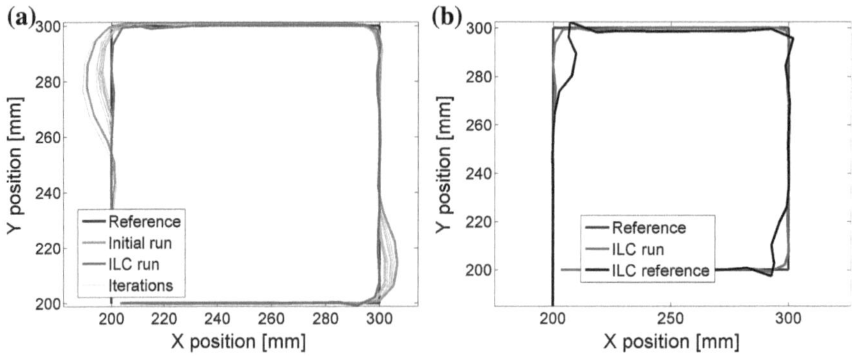

Fig. 8 The results from an ILC run with 20 iterations on a 100 mm^2 at 25,000 mm/min. It is seen that the tracking performance is significantly improved by the altered trajectory reference. **a** Shows how the executed trajectory converges after 20 ILC iterations. **b** Shows how the resulting reference signal after 20 ILC iterations

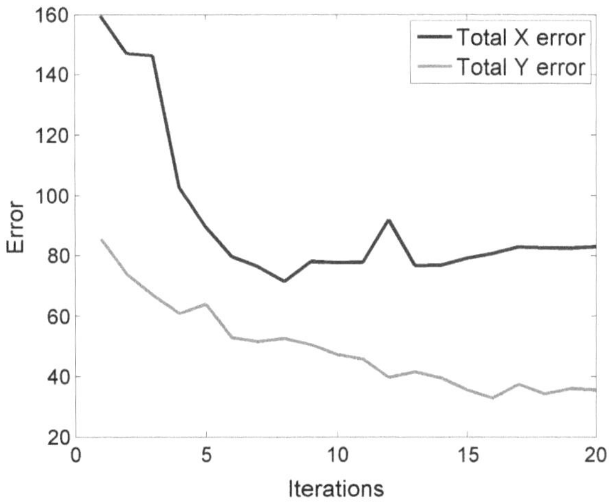

Fig. 9 The summed error in the X and Y direction. It can be seen that the error falls in the beginning and starts to stagnate after a number of iterations

with the corners somewhat cut off. This enables the system to conserve the system velocity and change direction more gradually. From Fig. 9 the summed positional error can be seen. From this figure it is seen that the error is minimized and reaches a minimum after 8 iterations in the X direction and 16 in the Y direction. The figure also shows that the positional error in the X direction drops from approximately 160 to 80 and the Y error drops from approximately 80 to 35. It should however be noted that the system shows signs of over training in the X direction as the error increases from iteration 8 to iteration 20. This originates from the differences in the dynamic capabilities of the X and Y axes. This difference in convergence point could be compensated for by using different learning factors for the X and Y directions. Figure 8b shows the final input reference u_k. It is seen that the reference is moved away from the corners of the square to compensate for the overshoot of the initial run.

The second set of results comes from tracking a circular reference. If the dynamic properties of the X axis are different from the Y axis the executed circle will be warped. In Sect. 1 it was in fact concluded that the two axes are different as the X axis needs to move the weight of the Y axis. A test was conducted on a circle to prove the concept and the result can be seen on Fig. 10a.

From this figure it is seen that the initial executed trajectory is not perfectly circular. By using ILC these deviations are reduced significantly. This can also be seen from Fig. 10b where the euclidian error is plotted versus the angle around the circle. Note that the cumulative error has been reduced from 408 to 88.

The experiments presented in this section have shown, that the implementation of the ILC algorithm from Eq. 1 can be used to increase the tracking accuracy of the XY positioning system.

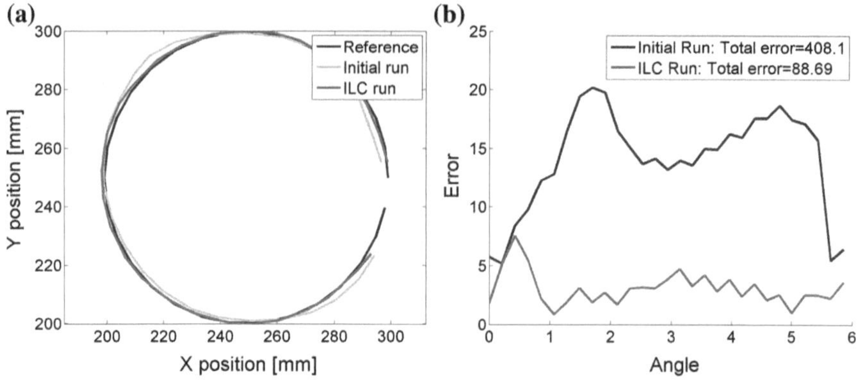

Fig. 10 The trajectory of a circle executed at a speed of 600 mm/s. Note that the circle is askew in the initial run due to the dynamics of the XY table. **a** Shows how the executed trajectory of a circle. **b** Shows the deviation of the executed trajectory as a function of the angle

6 Summary/Conclusion

In the introduction it was proposed that ILC could be a good tool for enhancing performance of mechanical systems with complex or inaccessible control architectures. This was investigated by a case study involving an implementation of an ILC algorithm on a high speed XY table used for laser cutting with a control architecture involving several independent controllers. The results in Sect. 5 have shown that when tracking a square reference the positional error in the X and Y direction were approximately halved after 20 iterations of ILC. After 20 iterations the system showed small signs of over-training in the X direction, but almost none in the Y direction. From the data presented in Sect. 5 it can be concluded that the ILC algorithm improves the dynamic properties of the system and thus makes it capable of improving the accuracy of laser cuts. These results have indicated that ILC is a viable way to increase the tracking performance of mechanical systems with a non standard control structure. The improved performance was not based on prior mathematical modelling and only required implementation on the master control unit. This entails that a developer can increase system performance without changing sub system controllers. This also makes the implementation independent of special software required to program sub system controllers, such as servo drives. The main drawback with the chosen type of ILC algorithm is however that it only optimizes known trajectories which is only applicable for systems doing repetitive work.

References

1. Åström, K.J., Hägglund, T.: The future of pid control. Control Eng. Pract. **9**(11), 1163–1175 (2001)
2. Bertsekas, D.P.: Dynamic Programming and Optimal Control, vol. 1. Athena Scientific Belmont, Belmont (1995)
3. Zhou, K., Khargonekar, P.P., Stoustrup, J., Niemann, H.H.: Robust performance of systems with structured uncertainties in state space. Automatica **31**(2), 249–255 (1995)
4. Ziegler, J.G., Nichols, N.B.: Optimum Settings for Automatic Controllers, *trans. ASME*, vol. 64, no. 11, pp. 759–768 (1942)
5. Åström, K., Hägglund, T.: Revisiting the ziegler—nichols step response method for pid control. J. Process Control **14**(6), 635–650 (2004)
6. Moore, K.: Iterative learning control for deterministic systems. In: Advances in industrial control, Springer (1993)
7. Moore, K.L., Chen, Y., Ahn, H.-S.: Iterative learning control: a tutorial and big picture view. In: Proceedings of the IEEE Conference on Decision and Control (2006)
8. Ahn, H.-S., Chen, Y., Moore, K.L.: Iterative Learning Control: Robustness and Monotonic Convergence for Interval Systems. Springer, New York (2007)
9. Xu, J.-X.: A survey on iterative learning control for nonlinear systems. Int. J. Control **84**(7), 1275–1294 (2011)
10. Arimoto, S., Kawamura, S., Miyazaki, F.: Bettering operation of robots by learning. J. Robot. Syst. **1**(2), 123–140 (1984)
11. Chen, C.-K., Hwang, J.: Iterative learning control for position tracking of a pneumatic actuated x-y table. In: Proceedings of the IEEE International Conference on Control Applications, vol. 1, pp. 388–393 (2004)
12. Sullivan, A.B.D., Houldcroft, P.: Metal cutting by oxygen: assisted laser. Opt. Technol, vol. 1, pp. 37–44 (1967)
13. Schulz, W., Simon, G., Urbassek, H.M., Decker, I.: On laser fusion cutting of metals. J. Phys. D Appl. Phys. **20**(4), 481 (1987)
14. Villumsen, S., Joergensen, S., Kristiansen, M.: Flexible laser metal cutting: an Introduction to the robocut laser cutting technique, In: Proceedings of the 7th World Conference on Mass Customization, Personalization, and Co-Creation (MCPC 2014), pp. 217−228 (2014)

Part IV
Mechanism Design

Shift Strategy for Railway Vehicle Transmissions

Xiaodong Tan, Siyu Bo and Yanlei Lei

Abstract Nowadays, with the construction of the railway has been more and more consummate and the environment requirements of the place where locomotive drives are also increasing, so the demand for a series of rail engineering vehicles have increased. In order to adapt for the various working condition of the rail engineering vehicles, so the rail engineering vehicles must have a good dynamic performance, and there need a hydraulic transmission device which used the hydraulic torque converter and the hydraulic coupler to control speed. In the transmission system, the hydraulic coupler and the hydraulic torque converter can switch flexibly. And this makes the rail engineering vehicles can deal with all kinds of different working conditions, and thus it improve the efficiency of the vehicle. This paper introduces the characteristics of the hydraulic torque converter and the hydraulic coupler, the selection of the shift gear parameters, strategy formulation of shift, and the principle of the shift strategy with optimal traction, thus we formed the final shift strategy.

Keywords Hydraulic coupler · Hydraulic torque converter · Rail engineering vehicles · Shift strategy

1 Introduction

Railway vehicles drive in the poor work environment and its working condition is complicated. In order to improve the ability to adapt to changeable load, especially the adaptability of the working condition which it has low speed and heavy loading,

X. Tan (✉) · S. Bo · Y. Lei
School of Mechanical Engineering, Dalian Jiaotong University, Dalian, China
e-mail: txd_f@sina.com

S. Bo
e-mail: 494738034@qq.com

Y. Lei
e-mail: leiyanlei0501@126.com

S. Bai and M. Ceccarelli (eds.), *Recent Advances in Mechanism Design for Robotics*, Mechanisms and Machine Science 33,
DOI 10.1007/978-3-319-18126-4_21

so we adopt the mode of hydraulic automatic transmission. With the high-speed rail technology has been increasingly perfect and popularity, and due to the high speed rail vehicle is always maintaining high speed during it is running, and the demand for external driving environment is relatively higher, so we need the railway vehicles with high performance and high efficiency to maintain the trajectory. So the requirements of the rail engineering vehicles' performance should be improved. In order to improve the work efficiency, the rail engineering vehicles should have good dynamic performance which can deal with all kinds of conditions, get enough traction under overloading condition and have a relative higher speed under the non-working condition [1].

The transmission device should meet the following requirements: (1) As the locomotive starts, it need to provide big traction. (2) Under the condition of low speed and heavy-load, it should provide enough traction. (3) Under the light load non-working conditions it required to provide a relatively high speed. The transmission includes two main gears, GearI: The hydraulic torque converter works. GearII: The hydraulic coupler works. When the locomotive is operating, it should be keep high efficiency, so the locomotive should frequently shift its gears to maintain this. If we want to ensure the working efficiency of the locomotives, we must ensure that the performance of the transmission device, and the transmission performance is good or bad that depends on the merits of the shift strategy, so the merits of the shift strategy determines the efficiency of the railway vehicles.

2 The Introduction of Hydraulic Transmission Device

The transmission device uses engine to transmit the input power to the input shaft, and then it uses the hydraulic torque converter and the hydraulic coupler to transmit the power. So we control the hydraulic torque converter and the hydraulic coupler's oil charge or extraction to control the output power.

2.1 The Original Characteristics of Hydraulic Torque Converter

Hydraulic torque converter plays an important role in this transmission system. Because the converter can provide large torque, so under the conditions of locomotive starting up, heavy-load and low-speed, we need the locomotive get large traction, and this is the reason why we deign the transmission which used hydraulic torque converter.

The original characteristic curve is always used as the common indicator which can reflect the performance of hydraulic torque converter. The original characteristic curve can intuitively and correctly reflect the basic properties of the hydraulic torque converter [2].

Hydraulic torque converter's impeller torque calculation formula is as follows

$$M_B = \lambda_B \rho g n_B^2 D^5 \tag{1}$$

M_B The torque of hydraulic torque converter pump wheel.
λ_B Pump rotation torque coefficient.
ρ The torque converter working medium's density under the working oil temperature.
n_B The speed of Hydraulic torque converter pump wheel.
g The weight of acceleration.
D The diameter of hydraulic torque converter guide roller.

Once the type of torque converter has been selected in this transmission devise, the only variable coefficient is the pump wheel speed, other coefficients are fixed. λ_B has been measured by the formula calculation 1 on the basis of a great deal of data, and $\lambda_B = \mathrm{f(i)}$ is one of the important curve which can reflect the performance of the torque converter.

The relationship between hydraulic torque converter's input torque and output torque is as follows

$$-M_T = KM_B \tag{2}$$

The original characteristic curves of hydraulic torque converter are shown in Fig. 1.

2.2 The Original Characteristics of Hydraulic Coupler

The hydraulic coupler is also very important in this transmission devise. One of the characteristics of the hydraulic coupler is that its efficiency and transmission ratio are equal, therefore, we usually use it to ensure efficient operation while the speed is high. So in this transmission system, in order to provide a relatively high speed and the good performance for rail vehicle, we make the hydraulic coupler worked under the condition of the non-working. And it also improved the efficiency of the vehicles during it's working.

The original characteristic curve is always used as the common indicator which can reflect the performance of hydraulic coupler. As the same as the original characteristic of the torque converter, it also can directly reflect the performance of the hydraulic coupler [3].

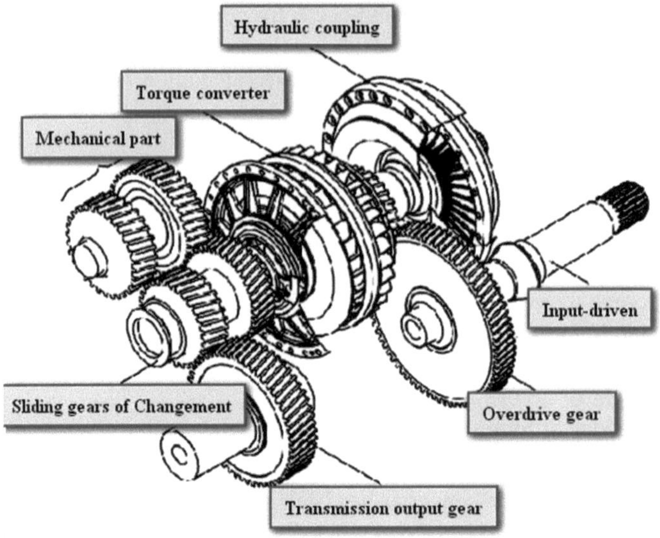

Fig. 1 The geartrain design

The hydraulic coupler's impeller torque calculation formula is as the same as the torque converter's impeller torque calculation formula (1). The relationship between hydraulic coupler's input torque and output torque is as follows

$$-M_T = M_B \tag{3}$$

M_T Hydraulic coupler's output torque.
M_B Hydraulic coupler's input torque.

The relationship between the pump torque coefficient and transmission ratio is described as $\lambda_B = \mathrm{f(i)}$ [4]. And it is measured with the Eq. (1) on the basis of large data, then we drawn curve by the calculated point. As it shows in Fig. 2, it is the hydraulic coupler's original characteristic curve.

3 The Shift Strategy of Railway Vehicles

Shift schedule, i.e. shift strategy, refers that the automatic shifting timing between the adjacent gears change with the shift parameter. And it is the core of the research and development of vehicle automatic transmission control system [5].

There are two classification principles of the shift strategy of railway vehicles that mainly includes shift parameters and the optimization goal. According to the

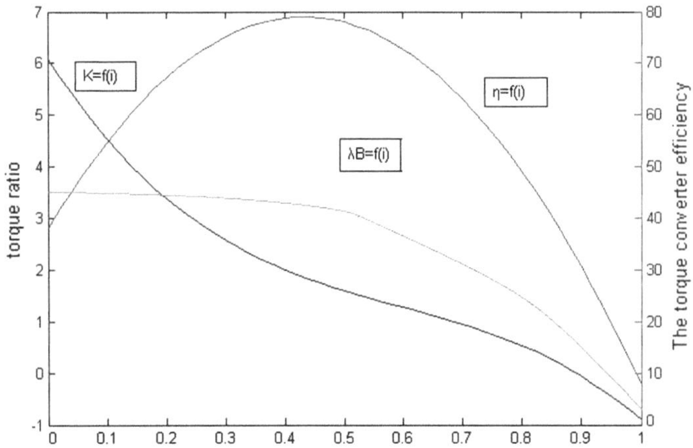

Fig. 2 The original characteristic curves of hydraulic torque converter

different number of shift parameters, we usually divided into two parameters, three and four parameters of automatic shift schedule. And according to the optimization goal is different, we generally divided into dynamic shift strategy and economy shift strategy.

3.1 The Selection of Shift Parameters

The correct selection of shift parameter relates to that whether the vehicle automatic shift theory is reasonable, and it has important effects on shift control effect. Because of the cost of the transmission device and the restrictions on the size of it, we only selected two speed sensors as the data acquisition of the shift parameter. They respectively are the torque converter's and the coupler's input and output rotating speed.

First, torque converter and the coupler turbine's rotating speed are easily measured, and it converts the vehicle speed through the Eq. (4). So they are important parameter which describe the vehicle driving status. Second, drivers can also very intuitively felt the change of speed. So we choose the torque converter, coupler turbine's rotating speed as one of the shift parameters.

$$\mathrm{v} = \mathrm{a}\frac{n_t r}{i_s} \tag{4}$$

The torque converter and the coupler turbine's output torque can converse through the rotating speed, and then the traction can converse through the torque by Eq. (5). And the traction directly affects the running status while the locomotive runs under the conditions of heavy-load operation and the other. Because the physical quantity can't get directly, so we adopt the indirect way of conversion to get the parameters. It can be seen that the output torque and the traction force has a positive relationship, that is to say, we can used the traction to replace the turbine output torque in the shift strategy.

$$F_k = \frac{M_T i_s \eta_s}{r} \quad (5)$$

The torque converter and the coupler's pump rotating speed can be also easily got as the same as the torque converter and the coupler turbine's rotating speed. The engine output shaft is always directly connected to the torque converter and the coupler input shaft, so it can real-time response state of engine. So we choose the torque converter and coupler's pump rotating speed as one of the shift parameter.

3.2 The Analysis of the Shift Strategy of Railway Vehicles

Dynamic shift strategy is divided into the best traction shift strategy and dynamic state shift strategy. Best traction shift strategy is that compare the traction before and after shifting. And its standard is that whether they are equal. So it uses this measure to ensure that the vehicle has good traction performance before and after the shifting. Dynamic shift strategy compare the acceleration before and after shifting and as its standard. Because of rail engineering vehicles has a high requirement on the power performance, so the transmission device adopts the dynamic shift strategy [6]. Engineering vehicle is used to build or maintain the railways construction, therefore it has a high requirement on the performance of the power output. So the device adopts the best traction shift strategy, in order to ensure the capability of the locomotive, we start from the basic requirements of rail engineering vehicles.

You saw the best traction shift strategy, we will specify the best traction shift strategy in the follows. As it is showed in Fig. 3, there are the two working curves of the torque converter and the coupler under the condition of the pump impeller rotational speed are constant. As shown in figure, point A is the intersection of two curves. When the output speed up to the speed which is at point A, in order to obtain larger output torque, transmission device should be converted to coupler working (GearII). Instead if the output speed down to the speed which is at point A, in order to obtain larger output torque, transmission device should be converted to torque converter working (GearI). So point A is the shift point at this output speed.

If we want to obtains the different shift point, we only need measured the output speed under the different input speed, and figure out the relationship between the output speed and output torque through formula which has given in the front, and then connect these point, we will get the shift curve. In Fig. 4, there are two output speed-output torque fitting curves which in the input speed of 1800 r/min and 2100 r/min, and A and B points are the corresponding shift points.

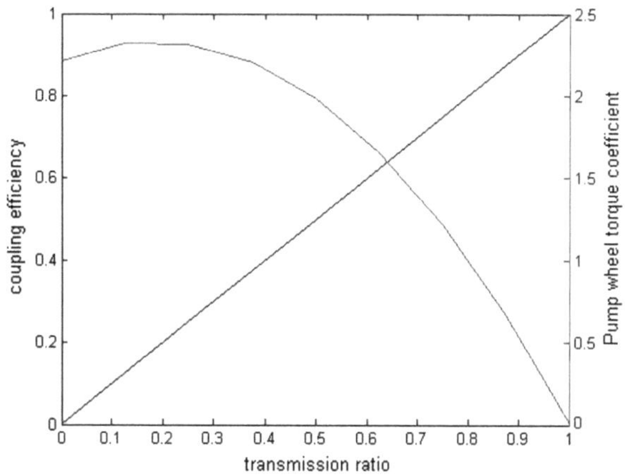

Fig. 3 The original characteristic curves of hydraulic coupler

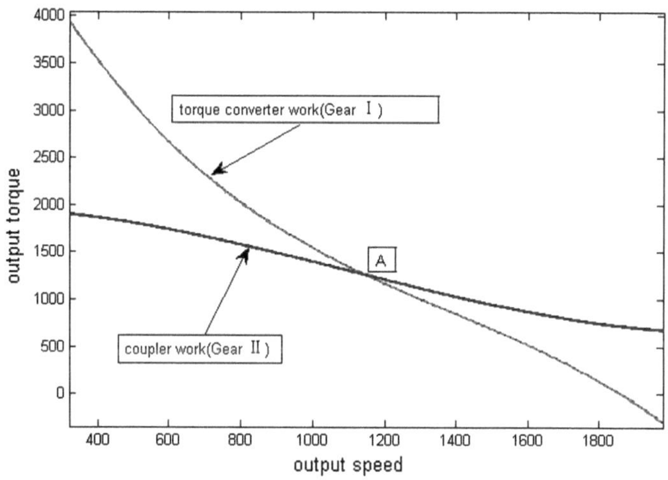

Fig. 4 The curve of the best traction shift

As it can be seen from the figure, there are different shift points A and B under different input speed, so we just need measured enough data and use this principle, we can fitted out the ideal shift point curve.

As shown in Fig. 5, it is the fitting curve of the gear shift point which obtained by the above method and combining with the existing data (Fig. 6).

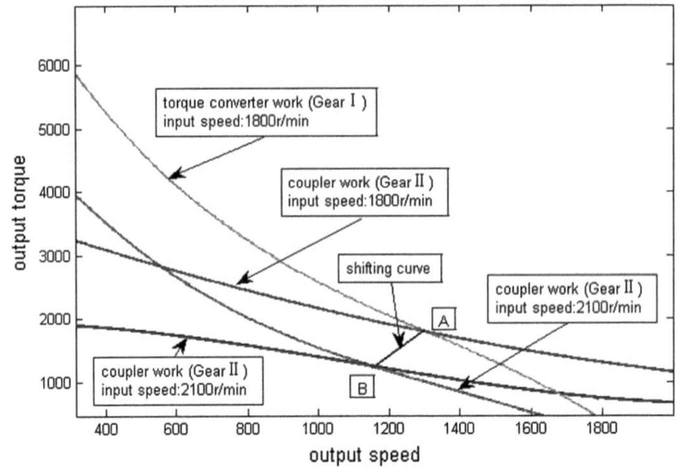

Fig. 5 The curves of different input speed

Fig. 6 The curves of shift with the existing data

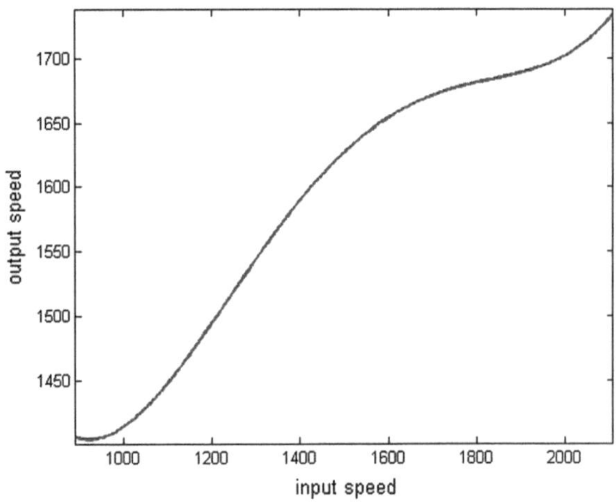

4 Conclusions

This article mainly analyzed shift strategy of the transmission device which consists of the hydraulic torque converter and the hydraulic coupler, and the work characteristics of the hydraulic coupler and the hydraulic torque converter, and the study of shift strategy through the understanding on this basis. The research of shift strategy is mainly aimed at the selection of shift parameters, and the study of best traction shift strategy. In the future during the continual study, we should add more corresponding appropriate data under the different input speed, doing so can get better shift curve, after all, this study is established under the condition of the large amounts of data supported.

References

1. Kim, D., Peng, H., Bai, S., Maguire, J.M.: Control of integrated powertrain with electronic throttle and automatic transmission. IEEE Trans.Contr.Syst.Technol. **15**, 474–482 (2007)
2. Ejiri, E., Kubo, M.: Performance analysis of automotive torque converter elements. ASME J. Fluids Eng. **121**, 266–275 (1999)
3. Zou, Z., Liu, H.: Work principle and applications of hydraulic coupler. Appl. Energy Technol. **119** (11), 11–12 (2007)
4. Zheng, Z., Wang, Y.:A study of the application of hydraulic couplers in engineering. J. Kaifeng Univ. **20**(2), 87–93 (2006)
5. Hongyan, Z., Dingxuan, Z., Xingxing, T.: Four-parameter automatic transmission technology for construction vehicle based on Elman recursive neural network. Chin. J. Mech. Eng. **21**(1), 20–24 (2008)
6. Thoma, J.: Hydrostatic power split drive. Hydraul. Pneumatic Power **16**(188), 464–469 (1970)

Research and Analysis on Transmission Error of RV Reducer Used in Robot

Weidong He and Lijun Shan

Abstract RV transmission device is a new type of few teeth difference planetary transmission, which is widely used in robot joints, elbow and other precision transmission parts due to the high driving accuracy. Taking the RV-40E reducer as an example, transmission error dynamical model and mathematical equations of RV reducer used in robot were established using dynamic sub-structure method. The max transmission error of RV-40E reducer was calculated by theoretical calculation and experiment respectively. The max transmission error calculated from two methods changes periodically within 1′. But the experiment measured transmission error is greater than the theoretical transmission error, due to the bearing clearance and friction in the experimental process were not considered.

Keywords RV transmission · Transmission error · Equivalence model · Kinetic equations

1 Introduction

RV transmission device is a new type of two stages closed and few teeth difference planetary transmission developed on the basis of Cycloidal Gearing. It has a series of advantages as large range of transmission ratio, small volume, light quality, high precision, small retrace tolerance, higher efficiency, driving steady and higher load bearing capacity. It is widely used in robot joints, elbow and other precision transmission parts due to these advantages. A lot of research work has been done since the occurrence of RV transmission, the stress analysis, the tooth profile optimization, the static transmission precision and a lot of studies have reached a certain theoretical depth, but the system dynamic performance study is less, especially the

W. He (✉) · L. Shan (✉)
School of Mechanical Engineering, Dalian Jiaotong University, 116028 Dalian, China
e-mail: hwd5870@163.com

L. Shan
e-mail: slj@djtu.edu.cn

S. Bai and M. Ceccarelli (eds.), *Recent Advances in Mechanism Design for Robotics*, Mechanisms and Machine Science 33,
DOI 10.1007/978-3-319-18126-4_22

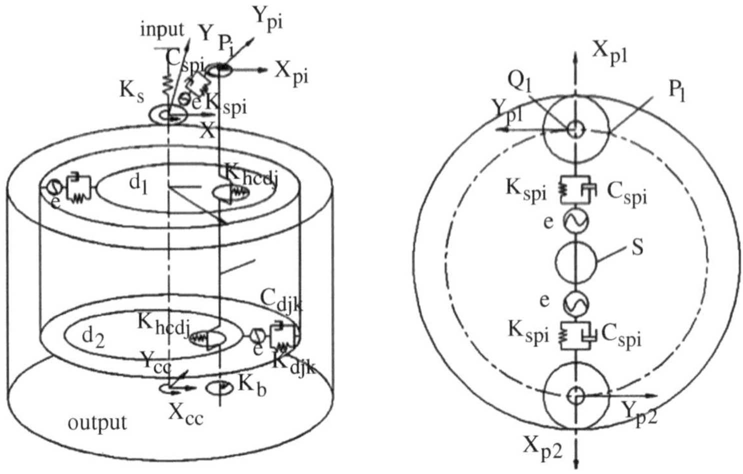

Fig. 1 Dynamics equivalent model of RV reducer

dynamic transmission error of RV transmission. The dynamic transmission error is main factors determining the precision of robot localization. In order to improve the precision of RV transmission, it is necessary to analysis and research on transmission error of the RV reducer system dynamic [1, 2].

2 The Establishment of RV Transmission Error Dynamic Model

According to the RV reducer structure and transmission principle in Ref. [1], RV reducer used in robot was divided into input shaft, the sun wheel, involute planet gear (including crankshaft), cycloid gear and planet carrier as the output side for dynamics analysis using the dynamic substructure method. Dynamic model of the RV reducer transmission error is shown in Fig. 1.

3 The Establishment and Solution of RV Transmission Error Mathematical Models

Dynamic substructure method [3, 4] was used to solve the dynamic transmission error of system, small displacement caused by elastic deformation of substructure component should be considered in addition to the various components processing error and assembly error, which is tiny variation caused by each component location relative to theory position of structure (including rotation angle variation). For the dynamic equilibrium equation, the factors influencing the transmission error should be equivalent to its meshing line of equivalent model element.

(1) Motion differential equation of the input shaft

In actual structure of the RV reducer, transmission centers of involute gear and input shaft were made into a gear shaft, based on dynamic substructure method, it was divided into two parts, here only consider the input shaft torsionally rigidity, ignoring the displacement of the input shaft floating amount in the center, motion differential equation of input shaft movement is:

$$J_1 \ddot{\theta}_1 - k_1(\theta_s - \theta_1) - c\left(\dot{\theta}_s - \dot{\theta}_1\right) - J_s \ddot{\theta}_s = 0 \tag{1}$$

Among them, JI is moment of inertia of input-shaft, kI is torsional stiffness between the input shaft and the sun wheel, CI is torsional vibration damping between the input shaft and the sun wheel.

(2) Motion differential equations of the center wheel

$$m_s \ddot{X}_s - \sum_{i=1}^{2} \left(Q_{spi} + C_{spi}\right) \cos \gamma_i = 0$$

$$m_s \ddot{Y}_s + \sum_{i=1}^{2} \left(Q_{spi} + C_{spi}\right) \sin \gamma_i = 0$$

$$J_s \ddot{\theta}_s - k_1(\theta_s - \theta_1) - c\left(\dot{\theta}_s - \dot{\theta}_1\right) - r_{sb} \sum_{i=1}^{2} \left(Q_{spi} + C_{spi}\right) = 0 \tag{2}$$

where: m_s is the quality of the sun wheel, J_s is the rotational inertia of the sun wheel, r_{sb} is the base circle radius of sun wheel.

(3) Motion differential equation of planetary wheel (including crankshaft)

According to the stress of the planet wheel (including crankshaft), got movement differential equations of the part as follows:

$$m_p \left[\ddot{X}_{pi} - a_{sp}\left(\omega_c^2 \cos P_i + \ddot{\theta}_{cc} \sin P_i\right)\right] + \left(Q_{spi} + C_{spi}\right) \cos \gamma_i$$
$$+ \sum_{j=1}^{2} \left(Q_{djix} + C_{djix}\right) + Q_{cix} = 0$$
$$m_p \left[\ddot{Y}_{pi} - a_{sp}\left(\omega_c^2 \sin P_i + \ddot{\theta}_{cc} \cos P_i\right)\right] - \left(Q_{spi} + C_{spi}\right) \sin \gamma_i$$
$$+ \sum_{j=1}^{2} \left(Q_{djiy} + C_{djiy}\right) + Q_{ciy} = 0$$
$$J_p \ddot{\theta}_{pi} - a_{sp}\left(Q_{spi} + C_{spi}\right) - \sum_{j=1}^{2} \left[\left(Q_{djix} + C_{djix}\right) \sin B_j + \left(Q_{djix} + C_{djix}\right) \cos B_j\right]$$
$$= 0 (i = 1, 2) \tag{3}$$

where: r_{sb} is base circle radius of the sun wheel, m_p is quality of a single involute planet gear (including crankshaft), J_p is mass moment of inertia of a single involute planet gear (including crankshaft) to the center.

(4) The differential equation of cycloid gear

$$m_b\left[\ddot{\eta}_{dj}\sin B_j - e\left(\omega_p^2\sin B_j - \ddot{\theta}_{doj}\cos B_j\right)\right]$$
$$+\sum_{i=1}^{2}\left(Q_{djiy}+C_{djiy}\right)+\sum_{k=m}^{n}\left(Q_{djk}+C_{djk}\right)\sin\left(a_k+B_j\right)=0$$
$$J_b\ddot{\theta}_{dj} - r_d\sum_{k=m}^{n}\left(Q_{djk}+C_{djk}\right)\sin a_k$$
$$+r_{dc}\sum_{i=1}^{2}\left[\left(Q_{djix}+C_{djix}\right)\sin P_i-\left(Q_{djiy}+C_{djiy}\right)\cos P_i\right]=0(j=1,2) \quad (4)$$

where: m_b is the quality of cycloid gear, J_b is rotational inertia of cycloid gear around its own axis.

(5) The motion differential equations of output planet carrier

$$m_c\ddot{X}_{cc}-\sum_{i=1}^{2}Q_{cix}=0$$
$$m_c\ddot{Y}_{cc}-\sum_{i=1}^{2}Q_{ciy}=0$$
$$J_c\ddot{\theta}_{cc}+a_{sp}\sum_{i=1}^{2}\left(Q_{cix}\sin P_i-Q_{ciy}\cos P_i\right)+T=0 \quad (5)$$

where: m_c is the quality of output planet carrier, J_c is rotational inertia of the planet carrier around the axis.

The Eqs. (1)–(5) totally 19 dynamic equilibrium equations are written in matrix form.

$$M\ddot{X}+KX+C\dot{X}=F \quad (6)$$

where: X is the displacement of a column for the equivalent model (19 × 1), M is mass matrix of the equivalent model (19 × 19), K is stiffness matrix of the equivalent model of (19 × 19), C is damping matrix of the equivalent model (19 × 19), F is generalized force vector of the equivalent model (19 × 1).

4 Calculation and Analysis of System Transmission Error

Nonlinear Wilson method [5, 6] was used to solve the dynamic equilibrium equation of RV transmission error. Taking RV-40 E reducer as the research object, the related parameters is: central circle radius of needle teeth dp = 128.1 mm, diameter of needle tooth DRP = 6 mm, eccentricity e = 1.3, needle wheel teeth Zp = 40, cycloid gear teeth = 39, center wheel gear teeth Z_1 = 12, the planet wheel gear teeth Z_2 = 36, modulus m = 1.5, pressure angle = 200, rated output torque T = 420 Nm, cycloidal gear and needle tooth material are all GCr15, elastic modulus E = 2.06 × 105 Mpa, poisson's ratio mu = 0.3.

The error formula of whole machine transmission is,

$$\Delta\theta = \theta_{cc} - \frac{\theta_s}{i} = \theta_{ca} - \theta_c$$

In the formula, θ_s is the sun wheel theoretical angle and theoretical angle of input shaft, θ_{cc} and θ_c are the actual and theoretical output shaft angle respectively, θ_{cc} and θ_c are the actual angle and the theoretical angle of planet carrier respectively, i is transmission ratio.

The 14th circle transmission error curve of output shaft in RV-40E reducer is shown in Fig. 2.

From Fig. 2, it can be concluded that the biggest transmission error of the whole machine is 32.3″ in the 14th running circle of output shaft. Though the machine has got into the stable operation status, but the transmission error still changes periodically in small range, and hysteresis error and error in advance occur alternately, which showing that different state institutions and different force component of the state lead to different dynamic characteristics, and then lead to the different transmission error amplitude of system in different time. RV reducer transmission error will affect the running and positioning accuracy of robot arm, therefore, further analyzing the main factors influencing on the transmission error is necessary.

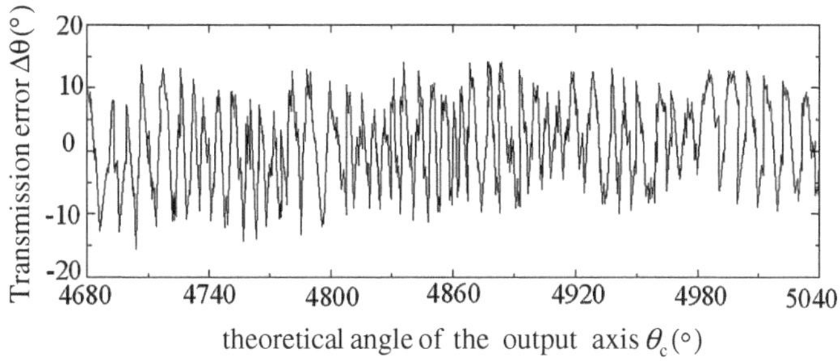

Fig. 2 Error of the whole machine

5 Experimental Study of the Transmission Error

Respectively on the RV-40E reducer input shaft speed for 5, 15, 25 r/min, transmission error is tested in three kinds of working conditions. When the rotational speed is 5, 15, 25 r/min, the max transmission error is 58.28″, 56.33″, 58.85″ as is shown in Figs. 3, 4 and 5. Visibly, it suggests that the speed is not main factors that influence the transmission error. The experiment measured the transmission error and the calculation of transmission error are all within 1′. But the experiment measured the transmission error is greater than the calculation transmission error, due to the bearing clearance and friction in the experimental process were not considered.

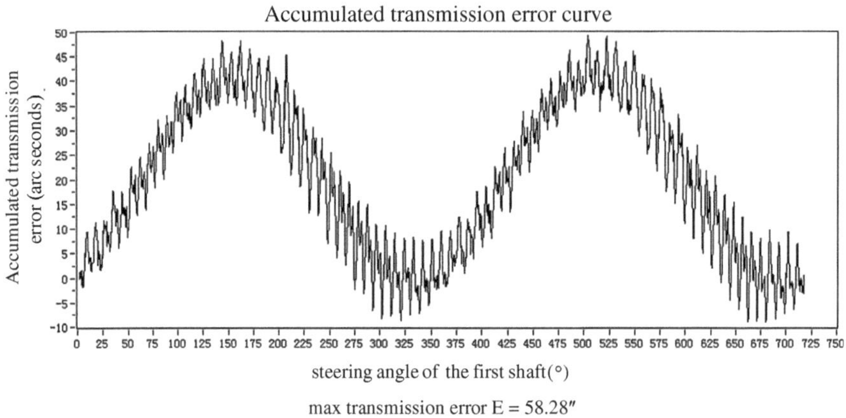

Fig. 3 Accumulated transmission error curve at 5 r/min speed

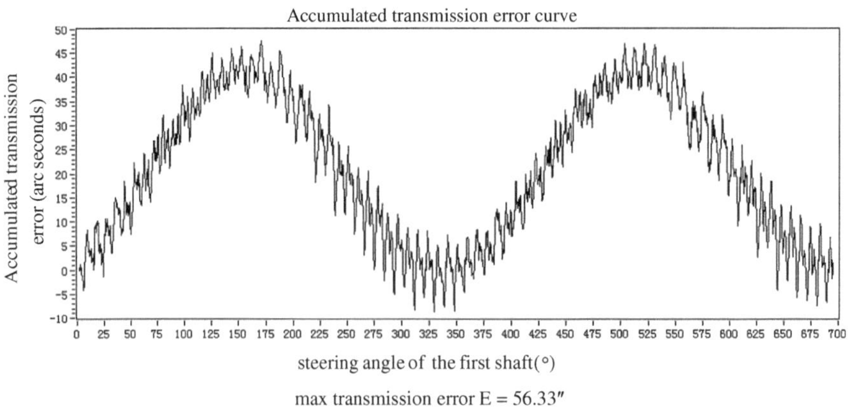

Fig. 4 Accumulated transmission error curve at 15 r/min

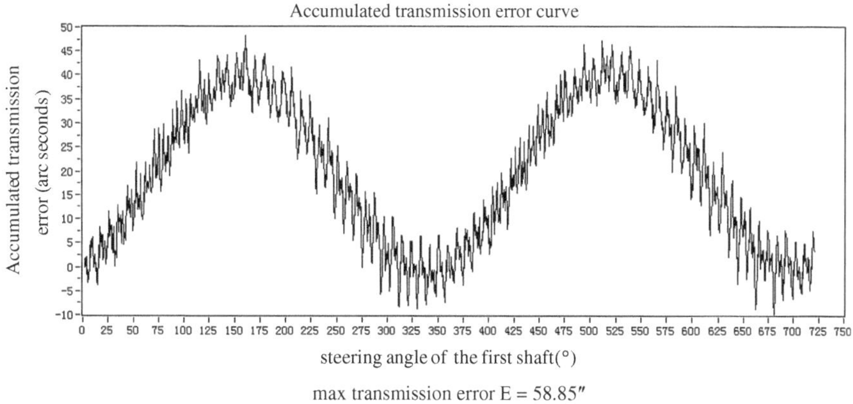

Fig. 5 Accumulated transmission error curve at 25 r/min

6 Conclusions

(1) Dynamic substructure method was adopted to establish the transmission error dynamic model of the RV reducer used in robot considering various components machining error and assembly error and small displacement caused by elastic deformation of substructure component.

(2) Nonlinear Wilson-method was used to calculate the transmission error mathematical analysis equations and transmission error curve after system getting into stable was got. The results show that maximum transmission error changes periodically after stable operation cycle, and the system maximal transmission error is 32.3″.

(3) Experiment method was adopted to test transmission error of RV-40E. The experimental results show that the speed have little influence on the transmission error, Comparing with the calculation of transmission error, the maximum transmission error gotten from two methods is cyclical fluctuations, range within 1′, but the experiment measured transmission error is greater than the calculated transmission error, due to the bearing clearance and the friction were not considered.

Acknowledgments This research is supported by the National Natural Science Foundation of China (Project No. 51375064). The authors would like to thank for their support of the research and development of the new type reducer presented in this paper.

References

1. Weidong, H., Lixing, L., Xin, L.: New optimized tooth-profile of cycloidal gear of high precision rv reducer used in robot. Chin. J. Mech. Eng. **36**, 51–55 (2000)
2. Duanling, L., Yundi, W., Yangyu, F.: On the behavior of instantaneous center and mechanical characteristics of a RV reducer used in robot. Mech. Sci. Technol. **20**(6), 860–861 (2001)
3. Linshan, H., Yunwen, S.: The correlational research on dynamic transmission accuracy for 2 K-V type drive. J. Mech. Eng. **43**(6), 81–86 (2007)
4. Linshan, H., Yunwen, S., Haijun, D.: Research on the transmission accuracy for cycloid drive based on the nonlinear analysis method. Chin. Mech. Eng. **18**(9), 1039–1043 (2007)
5. Yajun, Z., Jiahao, L.: Basic Structural dynamics. Dalian University of Technology Press, Dalian (2007)
6. Ruqing, Z., Xuegang, Y., Ming, D.: Computational structure dynamics. Chongqing University Press, Chongqing (1987)

Structural Design and Kinematic Analysis of Moving Mechanism of Insulator Inspection Robot

S.J. Li, Q. Yang, M. Geng, H.G. Wang and X.P. Li

Abstract The insulator is a part of high-voltage power transmission lines. The security and reliability of power system is mainly determined by safe working of insulator. Moving mechanism of insulator inspection robot is developed to inspect potential risk and operating status of insulator, which has important application in power engineering. Considering the working condition of insulator and mobile program of existing inspection robot, a novel foot-simulated moving mechanism of insulator inspection robot is proposed. Mobile program and structural configuration are determined, structural parameters are designed and kinematics analysis are completed. Finally, a 3D virtual prototyping of the moving mechanism is built with Solidworks, the moving process on insulator string of the mechanism is simulated with Adams, kinematic performances are calculated by simulating. The simulation results verify structural rationality of the moving mechanism and correctness of kinematic analysis proposed in this paper.

Keywords Inspection robot · Insulator · Structural design · Moving mechanism · Kinematic analysis

S.J. Li · Q. Yang (✉) · M. Geng · X.P. Li
College of Mechanical Engineering & Automation, Northeastern University, Shenyang, China
e-mail: qiangyang@mail.neu.edu.cn

S.J. Li
e-mail: shjunli@mail.neu.edu.cn

M. Geng
e-mail: gengmeng@mail.neu.edu.cn

X.P. Li
e-mail: xpengli@mail.neu.edu.cn

H.G. Wang
State Key Laboratory of Robotics, Shenyang, China
e-mail: hgwang@sia.ac.cn

S. Bai and M. Ceccarelli (eds.), *Recent Advances in Mechanism Design for Robotics*, Mechanisms and Machine Science 33,
DOI 10.1007/978-3-319-18126-4_23

1 Introduction

Insulator is the insulated supporting of live part of high-voltage electrical and high voltage electrical installations, and it makes grounding and conductor mechanically connected and electrically insulated. The insulator is a part of high-voltage power transmission lines. The security and reliability of power system is mainly determined by safe working of insulator. Insulator is a kind of ceramic part after high temperature firing; it will have some flaws due to materials and firing process, etc. Because of the working requirements, the life of insulator will be shortened by outdoor natural conditions such as wind, snow and rain, according to Fig. 1a–c, insulator string works in aerial condition and its shape can be divided into hanging string, level string and V-string with a tilt angle. Inspection is an important way to find fault of insulator and eliminate potential risk. The traditional inspection methods contain direct observation, ultrasonic measurement, the infrared thermal imager test and Laser Doppler. The mentioned method is not only inefficient, but also inaccurate. Intelligent inspection robot is developed to inspect potential risk and operating status of insulator, which has important application value for the reliability of power transmission. Moving mechanism is an important component of the inspection robot system, researchers have done some work on designing of moving mechanism. Zhou invented a charged intelligent detector for insulator string of high-voltage power transmission lines. The detector consists of supporting frame, moving mechanism and detection device. The detector is used for inspection of double shed insulator [1]. In 2008, Shenyang Institute of Automation (Chinese Academy of Sciences) patented a new type of level parallel detection robot [2]. At present, the invention has been practically used, but its drawback is the inability to detect single level string and V-string. In 2012, Shandong Electric Power Research Institute proposed intelligent detection robot system for a level insulator strings. The system has simple structure, steady crawl, safe and reliable operation, but its drawback is unable to detect suspension insulator strings [3]. In 2013, Shandong

Fig. 1 The working condition of insulators

Power Special Robotics Engineering Laboratory and Shandong Luneng Intelligent Technology Co., Ltd. jointly developed an insulator inspection robot. This robot has high detection efficiency and simple structure, but further research is needed to verify the operation for V-type suspension insulator string [4]. Korea Electric Power Research Institute designed a special detection robot for suspension insulator strings. The moving mechanism of this robot achieves crawling operation by rotation of claws. When the claws contact with the insulator surface, the direction of contact force always change, this is likely to cause the jam [5, 6].

2 Structural Design of Moving Mechanism

2.1 Moving Mechanism Design

The moving mechanism designed in this paper consists of the same three mobile units, which is 120° with each other. Each includes claws, a drive motor, a parallelogram mechanism with changeable frame, a cam mechanism, a guide link and the frame; and the three mobile units connect each other with the frame.

As shown in Fig. 2, translational step of the mobile unit is implemented by working together of the parallelogram mechanism with changeable frame and the cam mechanism. Two claws are fixed on each translational link of the parallelogram mechanism; the contact surface of claws coincides with outline of insulator surface. The cam mechanism is mainly used to ensure that one translational link of parallelogram mechanism keeps static as the frame any time during the robot operation. In Fig. 2, radius of the cam profile groove is half the length of side link, follower of cam mechanism (i.e. pin) hinges with side link through rotational joint at the midpoint of side link. The cam mechanism, guide link and detecting equipment are integrally connected through the frame. Side link *AB* rotates counterclockwise driven by a motor, climb operation of parallelogram mechanism with changeable frame is achieved by the claws, and then translational step of detecting equipment is obtained through the cam mechanism driven by side link (details in Sect. 2.2).

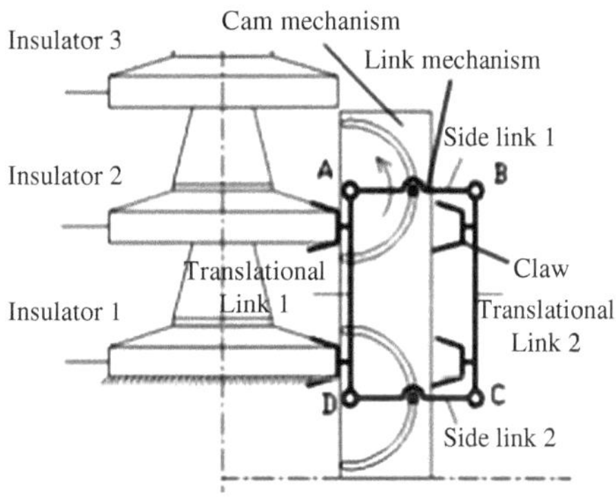

Fig. 2 Main structural diagram of foot-simulated moving mechanism

2.2 Works Principle

Works principle of the moving mechanism during an inspection cycle is shown in Fig. 3. Figure 3a is the initial position when the robot was installed on insulator strings firstly; translational links and side links are perpendicular to each other at this time. Because centers of the cam profile grooves are the hinged positions (point A and point D) of translational link 1, when side links turn first 90°, translational link 1 keeps static as the frame.

After installation of the robot is complete, side link 1 rotates counterclockwise driven by the motor. Figure 3b shows the claws are nearly to touch the insulator strings after translational link 2 turns 90° driven by the motor. Figure 3c shows the four links of parallelogram mechanism are collinear, the claws on translational link 2 contact insulator 2 and insulator 3 completely, and the claws of translational link 1 start going apart from insulator 1 and insulator 2. During the movement from Fig. 3a–c, translational link 1 keeps static as the frame with geometric and force constraint of cam and insulator. Because centers of the two cams profile grooves are the hinged positions (point A and point D) of translational link 1, the cam mechanism keeps static during this movement. Detecting equipment begins to inspect at the position of Fig. 3a, finishes it at Fig. 3c, which is also the position that cam is going to move. Figure 3d shows the moving states after translational link 1 and 2 run alternately. At this time, translational link 2 keeps static as the frame and translational link 1 rotates counterclockwise with the movement of side links. Figure 3e shows the position of which side links have turned 45° and a displacement of the cam occurs in the vertical direction to the original position. In Fig. 3f, translational links and side links are perpendicular to each other again. At this time, because centers of the cam profile grooves are the hinged positions (point B and point C) of translational link 2, the cam stops moving and an inspection cycle finishes. With continued rotation of the drive motor, the mechanisms will return to the position shown as in Fig. 3a, and then is ready to begin the next repeat movement cycle.

Translational step of the moving mechanism can achieve the surface contact of claws and insulators without rolling and sliding wear, so detection efficiency and assessment accuracy are improved. Parallelogram mechanism with changeable frame is shown in Fig. 3a–f, under the geometric and force constraint of the cam mechanism, translational step of the moving mechanism can be achieved. The crawling works of moving mechanism is as the same when it climbs hanging string, level string and V-string.

3 Kinematic Analysis of the Moving Mechanism

The purpose of forward kinematics of the moving mechanism is to solve the displacement, velocity and acceleration of four claws when the uniform angular velocity θ of driven motor is known. Figure 4 is the coordinate arrangement of main

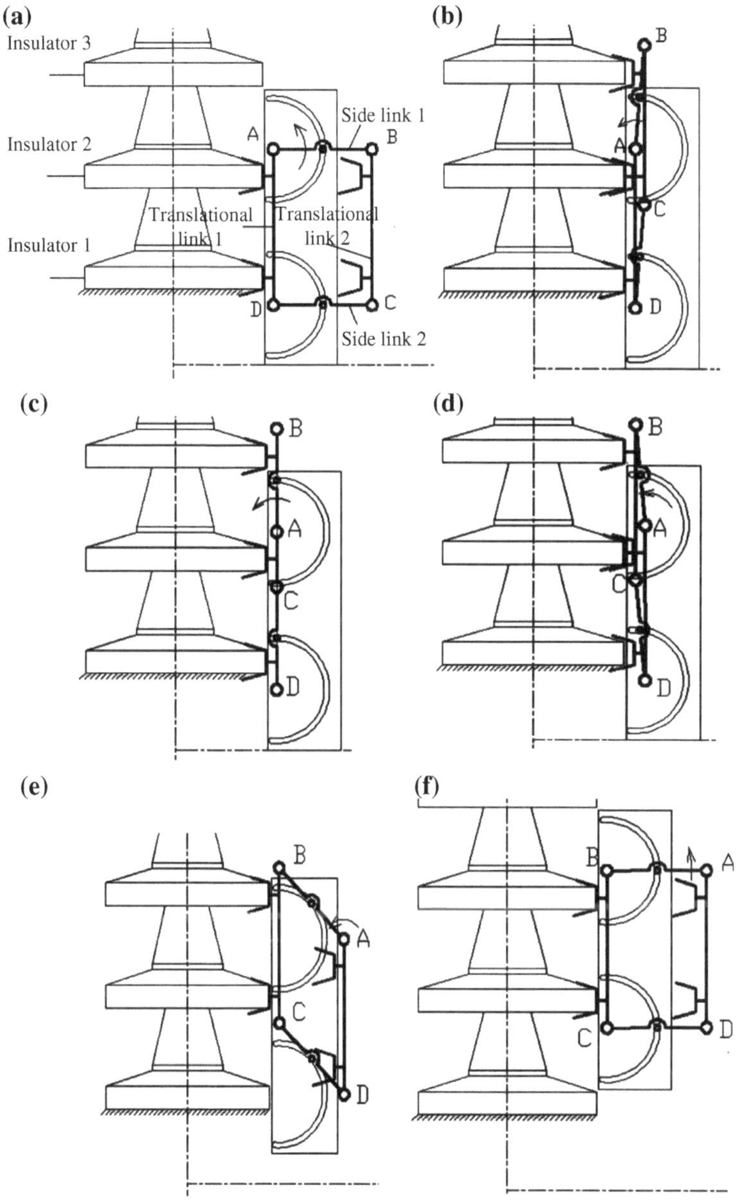

Fig. 3 The schematic diagrams of climbing work of the main moving mechanism

moving mechanism. Kinematic parameters of moving mechanism are designed according to the structural size of XP-210 insulator, $H = 170$ mm and $D = 280$ mm. Structural and Kinematic parameters of the parallelogram mechanism in Fig. 4 are shown in Table 1.

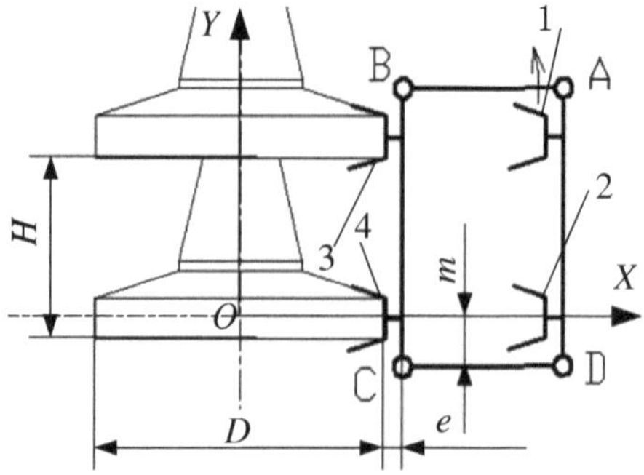

Fig. 4 Coordinate arrangement of the mechanism

Table 1 Kinematic parameters of the mechanism in Fig. 4

Quantity	H	D	L_{AB} (L_{CD})	L_{BC}(L_{AD})	e	m
Value (mm)	170	280	170	230	20	30

As shown in Fig. 4, the axis of insulator string is *Y*-axis direction, the coordinate origin is placed at the center of the insulator contacted with claw 4, and the direction of *X*-axis is to horizontal right. Figure 4 is the initial position when the robot was installed on insulator strings firstly; translational links and side links are perpendicular to each other. Side link rotates counterclockwise driven by the motor, suppose the uniform angular velocity of driven motor is θ, the displacement calculation equations of the claws at any time t can be easily deduced as following:

$$\begin{cases} (X_1, \ Y_1) = (140 + 170\cos\theta \cdot t, 170 + 170\sin\theta \cdot t) \\ (X_2, \ Y_2) = (140 + 170\cos\theta \cdot t, 170\sin\theta \cdot t) \\ (X_3, \ Y_3) = (140, 170) \\ (X_4, \ Y_4) = (140, 0) \end{cases} \quad t \in (0, \frac{\pi}{2\theta}) \tag{1}$$

When $t \in \left(\frac{\pi}{2\theta} + \frac{\pi}{\theta}(k-1), \frac{\pi}{2\theta} + \frac{\pi}{\theta}k\right)$ $k = 1, 3, 5, 7, 9\ldots,$

$$\begin{cases} (X_1, Y_1) = (140, 170(\mathrm{k}+1)) \\ (X_2, Y_2) = (140, 170k) \\ (X_3, Y_3) = (140 - 170\cos\theta \cdot t, 170(k+1) - 170\sin\theta \cdot t) \\ (X_4, Y_4) = (140 - 170\cos\theta \cdot t, 170k - 170\sin\theta \cdot t) \end{cases} \tag{2}$$

When $t \in \left(\frac{\pi}{2\theta} + \frac{\pi}{\theta}k, \frac{\pi}{2\theta} + \frac{\pi}{\theta}(k+1)\right)$ $k = 1, 3, 5, 7, 9\ldots,$

$$\begin{cases} (X_1, Y_1) = (140 + 170\cos\theta \cdot t, 170(\mathrm{k}+2) + 170\sin\theta \cdot t) \\ (X_2, Y_2) = (140 + 170\cos\theta \cdot t, 170(k+1) + 170\sin\theta \cdot t) \\ (X_3, Y_3) = (140, 170(k+2)) \\ (X_4, Y_4) = (140, 170(k+1)) \end{cases} \tag{3}$$

When the robot climbs level insulator strings, forward position can be got by exchanging the values of the horizontal and vertical axis coordinate. When the robot climbs V-string with a tilt angle, there are relationships between the new coordinate values with forward position and the values climbing suspension insulator string, and the new values can be obtained by the coordinate transformation. Velocity and acceleration expressions can be calculated by seeking a first and second order derivative of time t according to claws displacement Eqs. (1)–(3).

4 Simulation

As shown in Fig. 5, a 3D assembly model of the robot moving mechanism is built with a modeling software package, Solidworks 2010.

Set the speed of active side link $60^\circ/s$ and simulation time 5 s in ADAMS. And then displacement curve of claws are got by simulation. Displacement curve of claw 3 in Fig. 4 in the Y-axis direction is shown in Fig. 6. Compared the displacement Eqs. (2) and (3) of claw 3 in the Y-axis direction, they are sinusoidal curve, so the simulation result in Fig. 6 accords with the mathematical analysis.

Figure 7 shows displacement curve of claw 3 in the X-axis direction, from the figure, its movement is cosine curve when claw 3 is apart form the insulator, which accords with Eqs. (2) and (3) completely. Linear region in the figure indicates claw 3's movement when it contacted with the surface of the insulator.

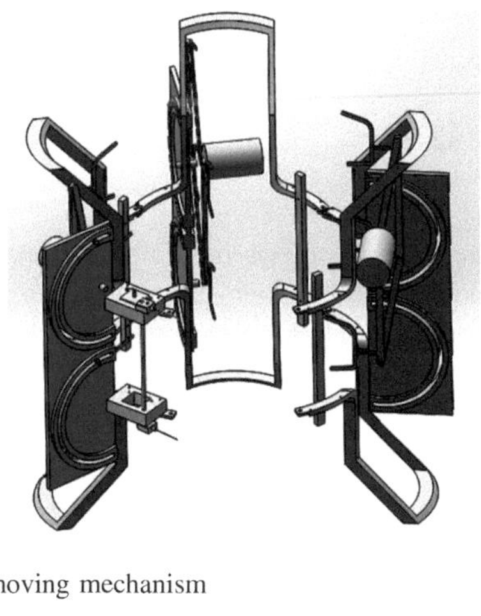

Fig. 5 3-D model of the moving mechanism

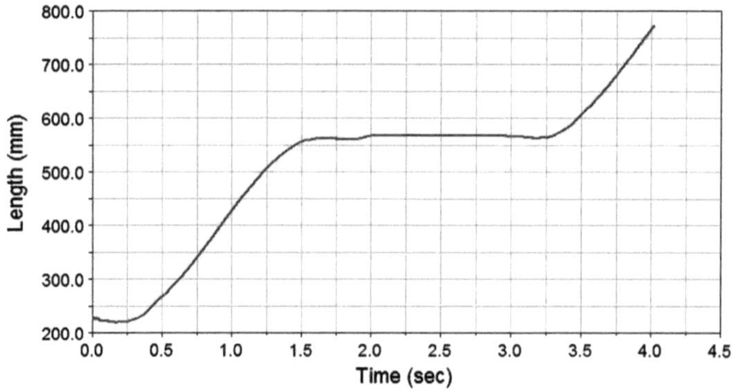

Fig. 6 The displacement curve of claw in *Y* direction

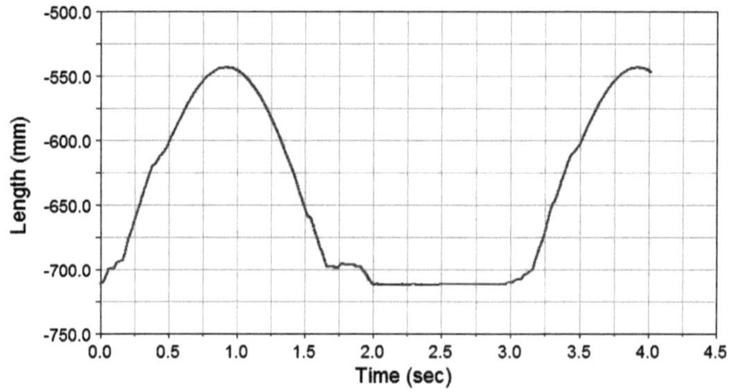

Fig. 7 The displacement curve of claw in *X* direction

5 Conclusions

A novel translational step method of foot-simulated moving mechanism of insulator inspection robot is proposed in this paper, and the moving mechanism is designed which is the combination of a parallelogram mechanism with changeable frame and a kind of cam mechanism. Kinematic analysis of the mechanism is completed; displacement equations are deduced. Based on software package Solidworks and Adams, simulation analysis is finished to verify feasibility of translational step and theoretical analysis results. The moving mechanism proposed in this paper has some advantages: the structure is simple, it can be widely applied for insulator strings (hanging string, level string and V-string), and the surface wear of insulator is light, which will get widely practical applications.

Acknowledgments The authors would like to acknowledge the financial support of the National Natural Science Foundation of China (Grant NO. 51175069, 51205052), the State Key Laboratory of Robotics of China (2012-O16), and the Basic Science and Research Project of Chinese National University under grant No. N140304004

References

1. Zhou, Y.L.: A charged intelligent detector for insulator string of high-voltage power transmission lines. China CN1367391A, 4 Sept 2002
2. Wang, H.G.: An inspection robot for the insulator. China CN201331558Y, 21 Oct 2009
3. Guo, R.: An intelligent detection robot system for level insulator strings. China CN102621430A, 1 Aug 2012
4. Fu, C.: A climbing mechanism of intelligent detection robot for insulator strings. China CN203025284U, 26 June 2013
5. Park, J.Y., Lee, J.K., Cho, B.H.: Development of inspection robot system for live-line suspension insulator strings in 345 KV power transmission lines. In: International Conference on Control, Automation and System, Seoul, pp. 2062–2065 (2008)
6. Park, J.Y., Lee, J.K., Cho, B.H. et al.: Development of advanced insulator inspection robot for 345 KV suspension insulator strings. In: Proceeding of the International MultiConference on Engineers and Computer Scientists, Hong Kong, pp. 17–19 (2010)

Atlas Motion Platform Full-Scale Prototype Design

Z. Copeland, B. Jung, M.J.D. Hayes and R.G. Langlois

Abstract This paper presents an overview of the design of the first full-scale prototype of the Atlas flight simulator motion platform for pilot training. The Atlas concept was introduced in 2005, and is unique in that orienting is decoupled from positioning, and unlimited rotations are possible about any axis of the mechanism. Detail design and manufacturing are complete, and assembly is in progress. The key to the design is three mecanum wheels in an equilateral arrangement, which impart angular displacements to a sphere that houses the cockpit, thereby providing rotational actuation. Since the Atlas sphere rests on these mecanum wheels, there are no joints or levers constraining its motion, allowing full 360° rotation about all axes, yielding an unbounded orientation workspace that is singularity free. In this paper, the current state of the design and assembly regarding actuation, the spherical S-glass shell, and modelling for motion control are discussed.

Keywords Unbounded orientation workspace · Structural design · Actuation · Simulated dynamic response

1 Introduction

The Carleton University Simulator Project (CUSP) is a fourth year capstone design project in mechanical and aerospace engineering. Unlike capstone design projects at most other universities, CUSP is run as a small design office with as many as

Z. Copeland · B. Jung · M.J.D. Hayes (✉) · R.G. Langlois
Department of Mechanical and Aerospace Engineering, Carleton University, Ottawa, Canada
e-mail: John.Hayes@carleton.ca

Z. Copeland
e-mail: Zack.Copeland@carleton.ca

B. Jung
e-mail: Bradley.Jung@carleton.ca

R.G. Langlois
e-mail: Robert.Langlois@carleton.ca

S. Bai and M. Ceccarelli (eds.), *Recent Advances in Mechanism Design for Robotics*, Mechanisms and Machine Science 33,
DOI 10.1007/978-3-319-18126-4_24

30 students led by three or four faculty members. The students experience working in a virtual enterprise environment. Among developing other soft skills, they experience a design project matrix: experience having to prioritise among multiple supervisors and multiple tasks; communicate in a large multidisciplinary team; develop verbal and written communication skills; but also develop the capacity for unbiased listening. Since its inception in 2002, approximately 300 fourth-year students from mechanical, aerospace, and computer systems engineering have incrementally advanced the development of a novel simulator system that overcomes motion limitations associated with industry-standard simulator motion platforms based on hexapods. The full-scale design is called the *Atlas simulator motion platform* and is illustrated in Fig. 1. The unique design of Atlas decouples its three-dimensional translational workspace from its unbounded, singularity-free orientation workspace [5]. In the current configuration, a MOOG MB-EP-6DOF Gough-Stewart platform [4, 11] is used to provide translation while Atlas provides the rotation.

The concept of spherical actuation is not new. Spherical dc induction motors were introduced in 1959 [13] and developed over the next 30 years, see [2, 10], for example. However, due to physical limitations imposed by the stator and commutator, angular displacements are limited. Unbounded rotational motion is achieved by the Eclipse II architecture [6]; however, there is no closed-form algebraic model for its kinematics and the velocity-level kinematics require estimating parameters numerically. The Desdemona motion platform [1] uses a fully-gimballed system to enable rotation about any axis. However, because of the gimbal arrangement, the orientation workspace is not free of singularities because of the potential of gimbal-lock.

Based on the performance of several proof-of-concept small-scale demonstrators, design of the Atlas full-scale prototype began in 2011 and manufacturing

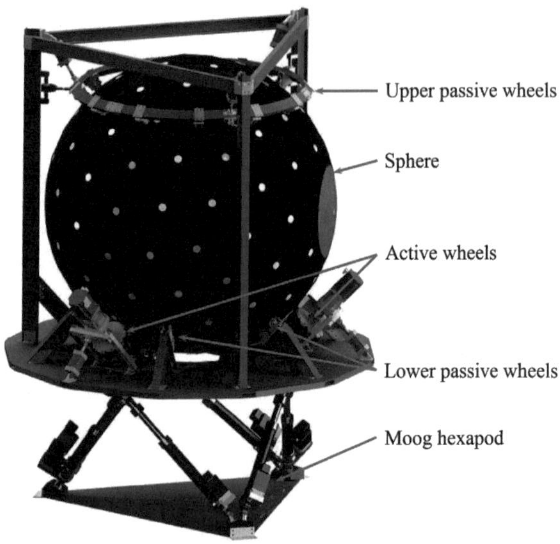

Fig. 1 A 3D rendering of the Atlas simulator

began in earnest in 2013. All of the individual components are currently in assembly. In this paper, the current state of the design and assembly regarding actuation, the spherical S-glass shell, and modelling for motion control are discussed. Please note that the use of dual metric and Imperial dimensioning reflects the reality of design in Canada: the standard is metric; however, many stock components are sized in Imperial units.

2 Structural Components

Structural components within the Atlas motion platform can be broken into three main categories for consideration: external structures; the spherical cockpit; and the internal support structures. Due to weight restrictions, a large number of the Atlas components are machined out of 6061 T6 aluminum. Bolt-together designs are employed wherever possible.

The primary focus of the external structure of the Atlas prototype is to provide support and stability throughout a wide range of simulated conditions. In order to accomplish this task, it must perform several simultaneous functions: it must allow for the spherical cockpit to be constrained under expected operational loading conditions and provide enough space for the actuation components to be mounted and stabilized.

Support is provided by three vertical I-beams, in an equilateral arrangement along the outside of the sphere, see Fig. 1. The sphere is constrained with two sets of passive mecanum wheels, connected to the I-beams, located at the top and bottom of the sphere. The 12 passive mecanum wheels placed along the bottom of the sphere help to distribute its weight and that of the internal structures, while the upper set of 12 passive mecanum wheels provide downward force to ensure sufficient contact force between the sphere and the three active mecanum wheels such that the sphere can be rotated. The downward force is supplied by three pneumatic cylinders connected between the I-beams and the ring of 12 upper wheels.

2.1 Sphere Structure

In order to create unbounded rotation, the cockpit of the Atlas prototype is housed in a 9.5 ft diameter fiberglass sphere, consisting of an internal support structure for increased rigidity as well as two hatches to facilitate entry and egress. Due to mechanical and spacial design constraints, the sphere shell has been designed to comprise four identical quarter spheres. Using epoxy S-glass, the strength required to maintain integrity under the loading from the active mecanum wheels was achieved, with an estimated failure load in excess of 1000 psi.

Each flange between sphere segments is bound together with a series of bolts, while an aluminum reinforcement is applied to both sides of the joining flanges in

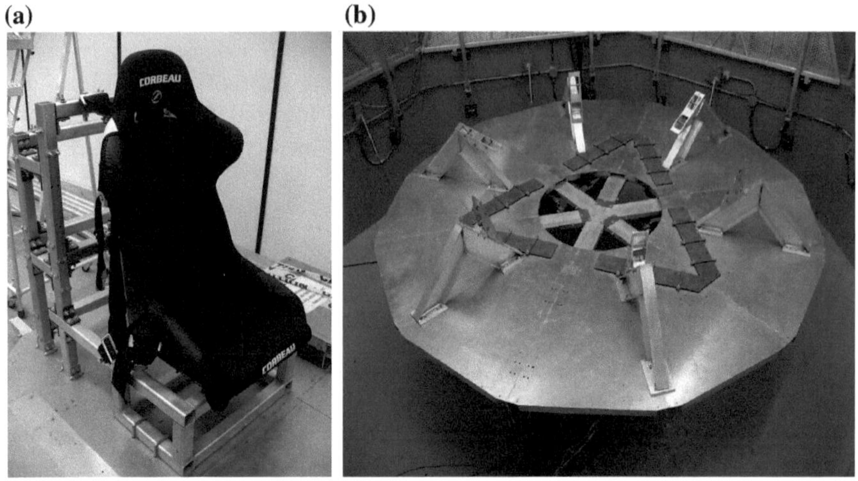

Fig. 2 **a** Floor structure and pilot support system; **b** interface platform connected to the Moog hexapod

order to provide additional stiffness to the sphere, as well as serving as a washer for the bolts to prevent damage to the sphere flanges. In addition to these stiffeners, a series of ribs that serve as a mounting interface for the internal structures is also connected at a 90° angle to the flange washers.

Entry points are included at two antipodal points of the sphere. They comprise two 30 in. diameter hatches that are locked in place with a striker bolt system which connects to the internal support structure of the sphere. The hatches are formed from 3/8 in. thick machined aluminum. Air circulation is provided to the pilot with four fans placed on each hatch, which make use of 151 holes in the hatches to allow for each fan to draw and expel air from and to the surroundings. Attached to the internal support structures are the brackets and flooring that provide a base for the cockpit seen in Fig. 2a. Reconfigurability created the demand for a highly-modular internal structure, so it is bracket and bolt based, allowing for any single component to be reconfigured without requiring disassembly of the sphere. The flooring support structure interfaces with the stiffeners via four brackets which are bolted to them directly.

3 Actuation

The Atlas motion platform possesses 6° of freedom. What distinguishes Atlas from conventional motion platforms is that linear displacements are decoupled from angular displacements. Moreover, the orientation workspace is unbounded and free from singularities [7]. This is accomplished by attaching the sphere orientation system to an interface platform atop a hexapod, see Fig. 2b. A MOOG MB-EP-6DOF

Gough-Stewart platform was selected to be the hexapod motion base. While this platform is capable of full 6° of freedom motion, it's controller enables the use of only its translation capabilities [8], thereby providing linear combinations of surge, sway, and heave.

The orientation degrees of freedom are supplied by the three active mecanum wheels which supply torque to the sphere, and the 12 pneumatically-controlled upper wheels ensure sufficient normal forces at the sphere-active wheel contact patches, while the 12 lower passive mecanum wheels help support the sphere load. Initially, omniwheels were to provide this rotational actuation, as seen in Fig. 3b but mecanum wheels were eventually selected in order to reduce actuation-induced vibrations [12]. The reduction in vibration is similar to that achieved when helical gear pairs are used in place of corresponding spur gear pairs. Figure 3a is an image of one of the active mecanum wheels designed to apply torque to the sphere. While commercial mecanum wheels exist, appropriate sizing and weights were problematic. Hence, the active wheels were designed in several iterations leading to the final versions illustrated in the figures. The passive mecanum wheels are substantially smaller than their active counterparts, as illustrated in Fig. 4.

Analysis revealed that a normal force of 1500 lbf between each active roller and the sphere is required to move the sphere. To ensure an appropriate distribution of this force at each contact point without destroying the S-glass required the forces be distributed over an area of at least 2.5 in^2, while ensuring that the roller material was stiff enough to avoid deflecting to the point that the hubs would abrade the sphere during operation. Urethane was selected because of favourable wear characteristics and strength limits that are adequate for the expected loading of the Atlas prototype, as well as having a large range of durometers (levels of compliance) to choose from. Testing was conducted in an MTS press, shown in Fig. 5a to establish suitable

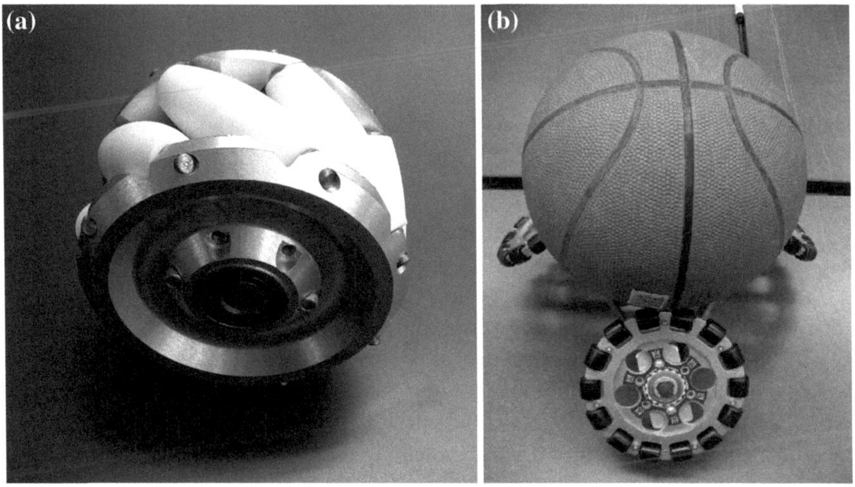

Fig. 3 **a** CUSP mecanum wheel design; **b** omniwheel proof of concept model

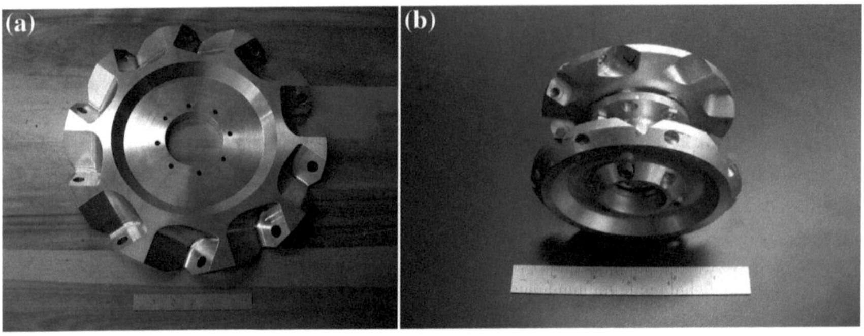

Fig. 4 **a** Atlas active wheel hub. **b** Atlas passive wheel hub assembly

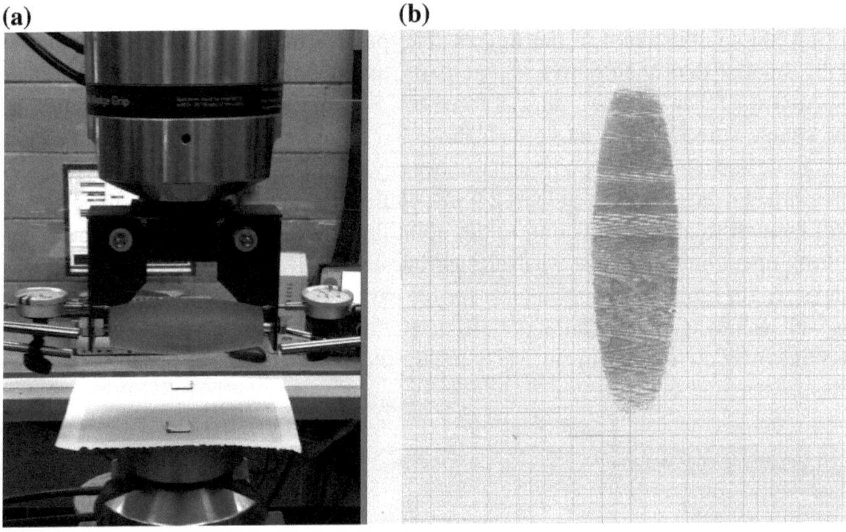

Fig. 5 **a** MTS press test frame with sample roller inserted. **b** Resulting ink blot after compression testing a durometer 55A urethane to 1300 lbf

urethane durometer. Contact patch size was recorded via an ink stamp test where the roller was compressed onto a sheet of paper and analyzed photogrammetrically. A sample ink blot can be seen in Fig. 5b. This led to the determination of the effective pressure exerted on the sphere surface.

The analysis revealed that the two durometers of urethane considered, 55A and 85A, both satisfied the minimum required surface contact patch size of 2.5 $in.^2$. While the durometer 55A urethane had surface areas in excess of 5 $in.^2$, it experienced deflections in excess of 0.5 in., indicating that it would likely jam and abrade the sphere during torque transmission. For this reason, the durometer 85A was selected for the rolling surfaces on both the passive and active mecanum wheels.

Force limits on the sphere S-glass surface mean that the maximum surface pressure is 700 psi. The durometer 85A sample was able to maintain a safety factor of at least 1.5 over the operable pressure range expected.

4 Control, Dynamics, and Simulation

Motion control of the Atlas platform is currently under development. At this point, the controller is being developed using a dynamic model to simulate the sphere motions given pilot flight control inputs. The following gives a high-level description of the feedback motion control [7], motor torque requirements [9], and the simulated dynamic response of the sphere.

4.1 Control

The overall Atlas control system comprises the vehicle simulation which determines the motion of the simulated vehicle in response to pilot input and environmental disturbances, a two-part sphere orientation sensing system which fuses orientation data from a visual orientation system (VOS) that actively tracks circular barcodes affixed to the outer surface of the sphere and an on-board three-axis gyroscope [7]; a washout algorithm that converts the vehicle state information to the desired platform state using classical washout theory that has been extended to accommodate large angular motions that are possible with the Atlas simulator; and an actuator control system that controls the six legs of the MOOG motion base to provide desired translational motions and the three rotation drive motors that control the sphere orientation by applying the required combination of sphere motor torques.

4.2 Torque Jacobian and Simulated Dynamic Response

The three motors that actuate each of the active mecanum wheels are positioned equilaterally around the sphere's vertical axis and 45° below the equator, indicated by $\theta = -45°$ for all three wheels, as illustrated in Fig. 6a. The mecanum wheel castor rollers are at a 45° angle relative to the mecanum wheel axis of rotation, $\gamma = -45°$ for all rollers, as indicated in Fig. 6b.

The torque vector, **M**, generated by the *i*th mecanum wheel acting on the sphere is given by the cross product of the position vector of the centre of the wheel's contact patch relative to the sphere centre with the force injected to the sphere by the wheel, resulting in the equation,

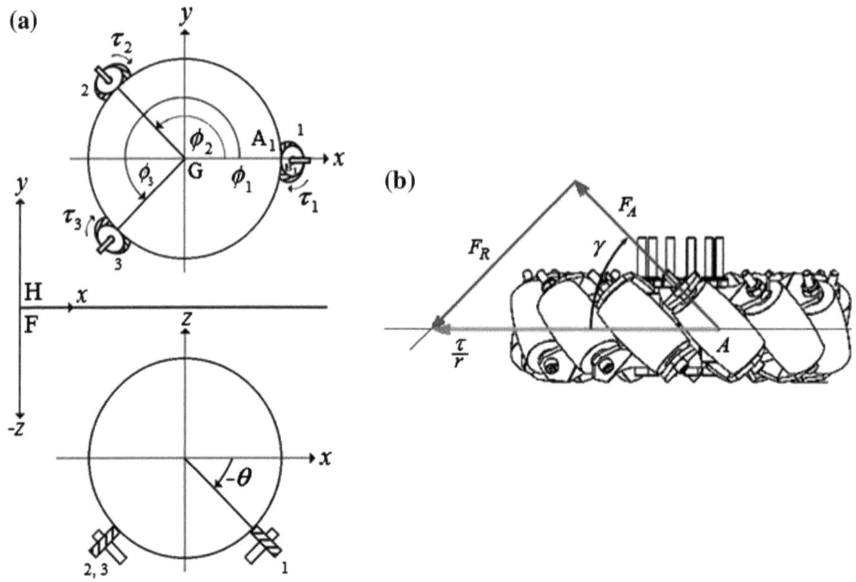

Fig. 6 **a** Mecanum wheel position around sphere; **b** roller angles around Mecanum wheel

$$\begin{aligned}\mathbf{M}_i &= \mathbf{R}_{\mathrm{A_i/G}} \times \mathbf{F}_{\mathrm{A_i}} \\ &= \frac{\tau_i R}{rC\gamma_i}\left[-C\phi_i C\gamma_i S\theta_i + S\phi_i S\gamma_i, \quad -S\phi_i C\gamma_i S\theta_i - C\phi_i S\gamma_i, \quad C\gamma_i C\theta_i\right]^T\end{aligned}$$

where the position of the contact patch around the sphere is specified by angle, ϕ_i, and inclination angle, θ_i; the angle the roller axis makes with the mecanum wheel axis is γ_i; the radius of the sphere is R, the effective radius of the mecanum wheel is r; the individual mecanum wheel torques are τ_i; and C, S, and T are the *cosine*, *sine*, and *tangent* trigonometric functions, respectively. Full details of the derivation of the mecanum-wheel-generated torque can be found in [9].

Combining the torque contributions for each wheel into a single matrix expression results in,

$$\mathbf{M} = \mathbf{J}_\tau \boldsymbol{\tau}, \tag{1}$$

where $\mathbf{M}$ is the total moment applied to the sphere and $\mathbf{J}_\tau$ is the torque Jacobian given by,

$$\mathbf{J}_\tau = \frac{R}{r}\begin{bmatrix} -C\phi_1 S\theta_1 + S\phi_1 T\gamma_1 & -C\phi_2 S\theta_2 + S\phi_2 T\gamma_2 & -C\phi_3 S\theta_3 + S\phi_2 T\gamma_3 \\ -C\phi_1 S\theta_1 - S\phi_1 T\gamma_1 & -C\phi_2 S\theta_2 - S\phi_2 T\gamma_2 & -C\phi_3 S\theta_3 - S\phi_2 T\gamma_3 \\ C\theta_1 & C\theta_2 & C\theta_3 \end{bmatrix}. \tag{2}$$

A sphere emulator program was designed for the purposes of dry testing sphere reactions to motor torques, as well as recording the simulation run for subsequent playback and data analysis. The emulator reads the desired mecanum wheel torque data in real time from the custom Atlas simulation computer network or from a previously-saved file and calculates the incremental orientation of the Atlas sphere. The program takes the sphere's inertial matrix and torque Jacobian matrix into account to provide a dynamic model of the sphere movement. The emulator program was written in MATLAB and includes a graphical user interface displaying the sphere animation, data readouts, and user controls. Rotation calculations were completed using quaternion mathematics to avoid representational singularities.

Quaternion mathematics are used in the emulator because quaternion rotation does not have a hierarchy of rotations. Rather, quaternions represent the rotation of the object in 4-D hyperspace about an arbitrary axis. Once the desired motion is converted into quaternion notation, the sphere will rotate in a single sweeping motion. This makes motion interpolation between large angle rotations possible. Quaternion mathematics and conversions can be found in [3].

The moment of inertia tensor **I** necessary for dynamic modelling was extracted from the Atlas composite CAD model by estimating the approximate densities of each component in the assembly. This is a 3 × 3 symmetric matrix that is variable to accommodate changes to the sphere weight distribution between simulation runs.

At each simulation cycle, the initial motor torque, angular rate, and sphere attitude $(\tau, \omega_0, \theta_0)$ are read by the emulator and then converted to quaternion form denoted as e_0. The angular momentum equation can be rearranged as

$$\alpha_1 = \mathbf{I}^{-1}(\mathbf{M} - \omega_0 \times \mathbf{I}\omega_0), \tag{3}$$

where α is the sphere angular acceleration.

The angular acceleration is numerically integrated resulting in an updated angular velocity:

$$\omega_1 = \omega_0 + \int \alpha_1 \mathrm{d}t. \tag{4}$$

The angular rate in Euler terms is converted to the quaternion rate using [3], yielding

$$\dot{e}_1 = \frac{1}{2} L\omega_1. \tag{5}$$

The quaternion rate is then numerically integrated to obtain an updated orientation:

$$e_1 = e_0 + \int \dot{e}_1 \mathrm{d}t. \tag{6}$$

Finally the quaternion rate is converted back into Euler terms. The calculated angular rate and angle are then fed back into the system for the next iteration, $e_1 \rightarrow \theta_1$. This process is then repeated for the duration of the dynamic simulation.

5 Concluding Remarks

This paper has presented an overview of the current state of development of the Atlas simulator motion base with emphasis on the rotational actuation system, as well as computational tools that have been developed to support the design and operation of the system. As is apparent from the design, the unusual actuation method provides unique motion capabilities for the simulator—most notably unbounded, singularity-free rotation. However, with that come some practical challenges relating to the need for tether-free power, data transfer, and ventilation. Additionally, since the drive system is unconventional and relies on technologies for which exact closed-form solutions do not exist in many cases, and despite the fact that the basic design principles are low risk, it is anticipated that much will be learned through the commissioning and calibration phases of the full-scale prototype.

Three related areas of particular interest will be the contact forces, mecanum wheel tractive forces, and wheel slip. In total, the sphere is held in place by patch contact with 27 mecanum wheels. The system is clearly statically indeterminate and time-varying due to the inertial loads acting on the system as the simulator is in motion and also due to the time-varying tractive forces applied by the active mecanum wheels. Design tools that were developed to approximate the worst-case contact forces necessarily made assumptions—such as the relative rigidity of the external supporting structure, the effective stiffnesses of both the small and large mecanum wheel castor assemblies, and the characteristics of urethane in this application. The maximum tractive effort of the active mecanum wheels will depend both on the prevailing contact forces and the friction characteristics at the interface between the active mecanum wheels and sphere surface. Initial testing was performed to determine the required durometer of urethane and its friction properties; however, some notable differences between test conditions and in situ operation are anticipated. Finally, wheel slip due to tangential compression of the urethane material prior to it entering the active mecanum wheel contact patches is expected to result in longitudinal slip, similar to what occurs with pneumatic tires. The extent of this and its impact on the idealized kinematic equations, based upon which the system has been designed, will have to be assessed and integrated into the Atlas control system.

Development of the Atlas simulator has been a provider of, and continues to provide excellent educational opportunities for both undergraduate and graduate students, and continues to present a series of well-bounded research questions. The results of this work directly benefit the CUSP project but also contribute better understanding to the broader problem of spherical actuation using mecanum wheels.

References

1. Bles, W., Groen, E.: The DESDEMONA motion facility: applications for space research. Microgravity Sci. Technol. **21**(4), 281–286 (2009)
2. Chirikjian, G.S., Stein, D.: Kinematic design and commutation of a spherical stepper motor. IEEE/ASME Trans. Mechatron. **4**(4), 342–353 (1965)
3. Diebel, J.: Representing Attitude: Euler Angles, Unit Quaternions, and Rotation Vectors. Stanford University, Stanford (2006)
4. Gough, V.E.: Discussion in London: automobile stability, control, and tyre performance. In: Proceedings of the Automobile Division, Institution of Mechanical Engineers. pp. 392–394 (1956)
5. Hayes, M.J.D., Langlois, R.G.: Atlas: a novel kinematic architecture for six DOF motion platforms. Trans. Can. Soc. Mech. Eng. **29**(4), 701–709 (2005)
6. Kim, J., Hwang, J.C., Kim, J.S., Iurascu, C., Park, F.C., Cho, Y.M.: Eclipse-11: a new parallel mechanism enabling continuous 360-degree spinning plus three-axis translational motions. IEEE Trans. Robot. Autom. **18**(3), 367–373 (2002)
7. Klumper, K., Morbi, A., Chisholm, K.J., Beranek, R., Ahmadi, M., Langlois, R.G.: orientation control of atlas: a novel motion simulation platform. No. 13-CSME-192, E.I.C. Accession 3650 (Sept 2013)
8. Moog Inc.: Moog motion systems overview (2009). Review 3 Nov 2009
9. Plumpton, J.J., Hayes, M.J.D., Langlois, R.G., Burlton, B.V.: Atlas motion platform mecanum wheel Jacobian in the velocity and static force domains. No. 13-CSME-192, E.I.C. Accession 3650 (Sept 2013)
10. Roth, R.B., Lee, K.M.: Design optimization of a three-degree-of-freedom variable reluctance spherical wrist motor. ASME J. Eng. Industry **117**, 378–388 (1995)
11. Stewart, D.: A platform with six degrees of freedom. Proc. Instn. Mech. Engr. **180**(15), 371–378 (1965)
12. Weiss, A., Langlois, R.G., Hayes, M.J.D.: Dynamics and vibration analysis of the interface between a non-rigid sphere and omnidirectional wheel actuators. Robotica (May 2014)
13. Williams, F., Laithwaite, E.R., Eastham, G.F.: Development and design of spherical induction motors. Proc. IEEE **47**, 471–484 (1959)

Simulation of Kinematic Pairs in the Calculation Mechanisms by Finite Element Method

Y.S. Temirbekov

Abstract Developed an original method, which allows us to apply the finite element method to analyze the stiffness and strength of planar and spatial lever mechanisms with kinematic pairs of arbitrary orientation in space. The idea of the method is that the basic equation of equilibrium of a system of finite elements is made and solved in the local coordinate systems of nodes. This allows to take into account the lack of connections in the kinematic pair in any direction. For the finite element modeling is used also a method of hard knots.

Keywords Mechanism · Kinematic pair · Finite element method

1 Introduction

One of the methods for calculating of constructions lever mechanisms (LM) is the finite element method (FEM) [1–9]. Here is given the original approach, which allows the use of the finite element method for the analysis of stiffness and strength LM with kinematic pairs with arbitrary orientation in space. The idea of the proposed method is that the basic equilibrium equation solved by the method rigid of nodes in local coordinate systems [10]. The basic algorithm FEM is not changed.

2 Types of Rod Constructions Considered in FEM

Flat LM have kinematic pairs with mutually parallel axes, so their calculation on the stiffness and strength of FEM can be applied. For spatial LM—orientation kinematic pairs is arbitrary. Let as consider the problem of accounting for such kinematic pairs

Y.S. Temirbekov (✉)
Institute of Mechanic & Mechanical Engineering, Kazakhstan,
Almaty Technological University, Almaty, Kazakhstan
e-mail: temirbekove@mail.ru

S. Bai and M. Ceccarelli (eds.), *Recent Advances in Mechanism Design for Robotics*, Mechanisms and Machine Science 33,
DOI 10.1007/978-3-319-18126-4_25

in the calculation FEM—given the lack connections between some elements of kinematic pairs. As is well known in the rotational pair—the reaction torque is zero and of the slider—is zero reaction from its direction of motion. To account for the missing component, i.e. in order to equate it to zero, it is necessary to consider the equilibrium equation containing these components of the reaction. Obviously, such equations are the equations of equilibrium in the projections on the axis kinematic pairs. But in the FEM basic system of equations is composed of the equilibrium equations of nodes in projections on the axes of the global coordinate system (GCS). Hence, GCS should be chosen so that its axis is parallel to the kinematic axis. But, GCS can not simultaneously be all parallel to the axes of the kinematic pairs of a spatial mechanism.

Here we propose a method that allows to use the FEM for the analysis of any kinds of spatial LM with arbitrarily oriented kinematic pairs. The basic ideas and scheme of FEM realization in this case practically do not change. To make the missing components equate to zero it is necessary to consider the vectors of reactions to the balance equation of kinematic pairs in projections onto axis containing these components. To get these equations into each kinematic pair introduce local coordinate system (LCS) in such a way that the axis of kinematic pair and LCS axis coincide. Then the equilibrium equation kinematic pair to the LCS this kinematic pair will include zero components of reactions. For example, if the hinge axis or the axis of the slide—is not parallel to the axis of the GCS, the equilibrium equation kinematic pairs in the projection on the axis GCS does not contain zero reaction components. Consequently, the possibility of accounting the absence of these reaction components is lost. Therefore, the FEM can always be used for flat LM since all kinematic pairs parallel to the some axis of GCS.

3 Simulation of Kinematic Pairs. Method of Hard Nodes

For the modeling of kinematic pairs it is used a method that was developed by the author [10]. It is called the method of hard nodes—by analogy with [11]. In general, if the kinematic pair are connected "n" the groups of rods ($n \geq 1$). Each group consists of $k_i \geq 1, \quad i = 1, \ldots, n$; rigidly connected rods. And for each degree of freedom, these groups have their own kinematic and force parameters. Then any kinematic pair is a combination of hard knots in one coordinate point and having "K" common degrees of freedom ("K"—class kinematic pairs). In other words, the kinematic pair is modeled not as a whole, as it is usually modeled in the FEM for rod constructions, but as each element of the kinematic pair.

Let as consider an arbitrary spatial kinematic pair. Total number of degrees of freedom of its constituent units is equal to the number of connections "K" imposed on it. That is the class of kinematic pair. Consider kinematic pairs for which $3 \leq K \leq 5$. Let us find degree of freedom W (in terms of FEM) kinematic pairs consisting of k hard nodes. Obviously, this is the sum of common degrees of freedom and the additional degrees of freedom of nodes:

$$W = K + k(6 - K)$$

Number N of degrees of freedom of the LM model are:

$$N = 6n_{Ж} + \sum_{i=1}^{n_{Ш}} (K_i + k_i(6 - K_i)) - n_{Г}$$

where $n_{Ж}$—the number of hard nodes without kinematic pairs (single nodes); $n_{Ш}$—the number of kinematic pairs; $n_{Г}$—the number of degrees of freedom of the boundary conditions; k_i $(i = 1,\ldots, n_{Ш})$—number of hard nodes included in the ith kinetic pair; K_i $(i = 1,\ldots, n_{Ш})$—i—the grade kinematic pair.

Thus, in compiling the finite element model LM any i—the kinematic pair represents a combination k_i hard knots. It is located in one coordinate point and has K_i common of degrees of freedom (class of kinematic pair).

Method of hard nodes has the following advantages over the known method of accounting in the kinematic pairs FEM [10]:

- Possibility of accounting of complicated hinge connections, for which have can not be applied the traditional methods of modeling. For example, the hinge between two or more base triangles (classification Artobolevsky I.).
- No need to transform the stiffness matrices of elements kinematically connected elements before building a global SMS.
- Introduction to the number of unknowns of the problem is linearly dependent displacement components of the kinematic pair.

4 Simulation of Kinematic Pairs. Types of Kinematic Pairs

Consider the scheme LM in some position. Position of the axis kinematic pairs with respect to GSC OXYZ is known through corners α, β, γ between the kinematic axis and the axes OX, OY, OZ, respectively.

Number of the kinematic axes depends on type of a kinematic pair. Translational, rotational and cylindrical couples have only one axis "S", which coincides with the direction of the translational motion or rotation axis. Spherical hinge with a finger has two axes "S_1" and "S_2". One coincides with the axis of the finger and the other—with the line extending perpendicularly the slit of a finger.

For each kinematic pair LM is necessary to build LCS OXYZ. LSC single (non-paired) nodes and nodes spherical pair can be selected in parallel GSC OXYZ because of an arbitrary orientation of degrees of freedom of these nodes. For pairs class IV and V with one axis "S", the local axis Ox will direct along the axis "S". Axis OY, OZ is obtained from the condition that the triple OXYZ was right. For a spherical pair with a finger we will direct the Ox-axis along the axis of the finger S_1, and the OY-axis along the axis S_2 extending perpendicularly the slit of a finger. Then the direction the axis Oz is uniquely determined.

Obviously, each pair is having a single common LSC. Information about the arrangement LSC of nodes relative to GSC is obtained by using matrix of direction cosines. For kinematic pair also we give degrees of freedom.

That is for a pair you must specify the information—which are the degree of freedom nodes are common, and what—kinematic.

Table 1 shows the types of kinematic pairs used in the LM. There are shown the orientation in space on their LCS, and also are shown the components of the reactions which are absent.

There was a software developed on the basis of a computer program STAP [10]. In this program there was modified on input array ID, which is now formed as follows:

- If for jth degree of freedom (j = 1, …, 6) i-node imposed boundary condition, then ID (i, j) = −1.
- If for the jth degree of freedom of movement is possible, the ID (i, j) = 0.
- If the nodes $i_1, i_2, \ldots, i_k$ creates single kinematic pair, then:

$$\text{is selected } l = \min\{i_1, i_2, \ldots, i_k\}, \tag{1}$$

$$ID(l,j) = 0, \quad \forall j, \tag{2}$$

$ID(i_1,j) = ID(i_2,j) = \ldots = ID(i_k,j) = l$, if j—common degree of freedom of the pair,
$ID(i_1,j) = ID(i_2,j) = \ldots = ID(i_k,j) = 0$, if j—the kinematic degrees of freedom.

That is, total jth degree of freedom of the nodes belonging to the same kinematic pair, is described in the ID array one zero in the jth column and in a row, which corresponds to a node of the pair with the minimum number l. For the rest of nodes kinematic pair in the jth column is stored number l. Number l shows that these nodes constitute a kinematic pair with the l-st node. Each row of the matrix ID is given in the LCS corresponding node. That is, the elements of the ith row describe the degree of freedom of the l-st node in its LCS. Then when counting the number of global degrees of freedom and their the numbering:

- "0" in the ID array successively replaced by the global degrees of freedom;
- "−1" is replaced with "0";
- and each of an integer $l > 0$ in the jth column is replaced by the ID(l,j), that is the global number of jth degree of freedom of ith node already previously determined from (1, 2). This procedure is simple and can be written in the form:

Table 1 Types of kinematic pairs

Kinematic pair	Class 5, rational	Class 5, translational	Class 4, cylindrical	Class 4, spherical with a finger
Unit designation		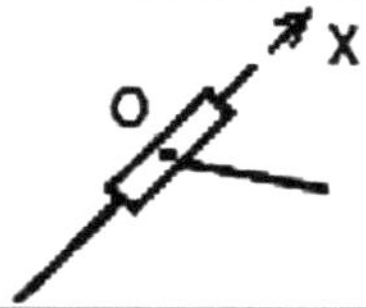	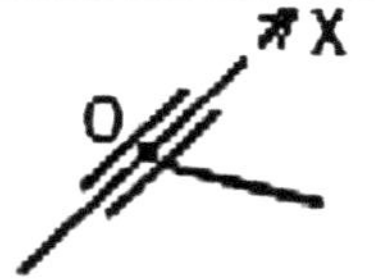	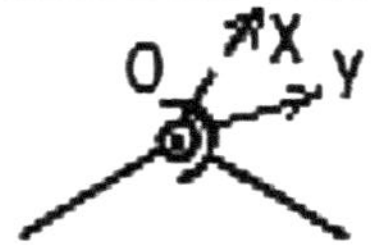
Zero insertion force	M_x	N_x	$N_{x,}$ M_x	$M_{x,}$ M_y
Orientation of local coordinate systems	OX—along the axis of rotation	OX—along the direction of the slide	OX—along the axis of rotation	OX—along the axis of the finger, oy—perpendicular to slit

(1) Assign the N = 1.
(2) In the cycle for i = 1, … n, and j = 1, … 6:

   (1) if ID (i, j) < O, then assign the ID (i, j) = O;
   (2) if ID (i, j) = O, then assign the ID (i, j) = N;
   (3) if ID (i, j) > O, then assign the ID (i, j) = ID (ID (i, j), j);
   (4) assigned to N = N + 1.

Thus, the type and class of a kinematic pair specified using ID. Suppose, for example, nodes *i* and j constitute any kinematic pair and i < j.

Then the *i*th row of the array ID will be: $[0 \quad 0 \quad 0 \quad 0 \quad 0 \quad 0]$.

Row "j" is dependent on the type and class of the kinematic pair. We show a string array ID, corresponding to the jth node.

1. For the rotational pair (Class 5): $[\mathrm{i} \quad \mathrm{i} \quad \mathrm{i} \quad 0 \quad \mathrm{i} \quad \mathrm{i}]$, the 4th degree of freedom, i.e., rotation around the axis Ox LCS pair is the kinematic; and the remaining five degrees of freedom—common to *i*-th node.
2. For translational pair (Class 5): $[0 \quad \mathrm{i} \quad \mathrm{i} \quad \mathrm{i} \quad \mathrm{i} \quad \mathrm{i}]$, 1st degree of freedom, i.e., translational motion along Ox LCS pair is the kinematic; and the remaining five degrees of freedom—common to *i*-th node.
3. For a cylindrical pair (Grade 4): $[0 \quad \mathrm{i} \quad \mathrm{i} \quad 0 \quad \mathrm{i} \quad \mathrm{i}]$, kinematic are 1st degree of freedom, i.e., translational motion along Ox LCS pair and kinematic are 4th degree of freedom, i.e., rotation around the axis Ox LCS pair; while the remaining four degrees of freedom—common to *i*th node.
4. For a spherical pair with a finger (Grade 4): $[\mathrm{i} \quad \mathrm{i} \quad \mathrm{i} \quad 0 \quad 0 \quad \mathrm{i}]$, kinematic are 4th and 5th degrees of freedom, i.e., rotation around the axes Ox and Oy LCS pair; while the remaining four degrees of freedom—common to *i*-node.
5. For a spherical pair (Grade 3): $[\mathrm{i} \quad \mathrm{i} \quad \mathrm{i} \quad 0 \quad 0 \quad 0]$, the kinematic are 4th, 5th and 6th degree of freedom, i.e., rotation around the axes Ox, Oy and Oz LCS pair; remaining 3—common to *i*th node.

5 The Transition to the Local Coordinate Systems

In kinematic pairs will be elastic displacement, coinciding for general degrees of freedom. And also various elastic displacement will appear on kinematic degrees of freedom. External forces are considered in LCS nodes.

Let it be $U_i = \left(u_1^i, \ldots u_6^i\right)^T$, $\tilde{U}_i = \left(\tilde{u}_1^i, \ldots, \tilde{u}_6^i\right)^T$, $F_i = \left(f_1^i, \ldots, f_6^i\right)^T$, $\tilde{F}_I = \left(\tilde{f}_1^i, \ldots, \tilde{f}_6^i\right)^T$, $(i = 1, \ldots, m)$—the displacement vector and the external forces vector of *i*-node, respectively GCS OXYZ and LSC O_ixyz *i*th node, *m*—the total number of nodes. Let it be

$$[T_i^o] = \begin{bmatrix} \cos(X,\tilde{x}_i) & \cos(X,\tilde{y}_i) & \cos(X,\tilde{z}_i) \\ \cos(Y,\tilde{x}_i) & \cos(Y,\tilde{y}_i) & \cos(Y,\tilde{z}_i) \\ \cos(Z,\tilde{x}_i) & \cos(Z,\tilde{y}_i) & \cos(Z,\tilde{z}_i) \end{bmatrix}$$

the matrix of direction cosines O_ixyz *i*th node relative to GCS OXYZ. Then for the 1th node should be the following equations:

$$U_i = [T_i]\tilde{U}_i, \quad F_i = [T_i]\tilde{F}_i, \tilde{U}_i = [T_i]^T U_i, \quad \tilde{F}_i = [T_i]^T F_i, i = 1, \ldots m,$$

where the matrix $[T_i]$—is a transition vector of LSC *i*th node in GCS and looks:

$$[T_i] = \begin{bmatrix} T_i^o & O \\ O & T_i^o \end{bmatrix}, i = 1, \ldots m.$$

$[T_i^0]$—there is rotation matrix, so it is orthogonal: $[T_i^0]^T = [T_i^0]^{-1}$. Consequently, the matrix $[T^i]$ is also orthogonal:

$$[T_i^0]^T \cdot [T_i^0] = [T_i^0] \cdot [T_i^0]^T = [E] \quad \text{or} \quad [T_i]^T = [T_i]^{-1}$$

Let it be U and F—displacement vector and the vector of the external nodal forces in GCS OXYZ:

$$U = (U_1, U_2, \ldots, U_m)^T = (u_1, u_2, \ldots, u_N)^T, F(F_1, F_2, \ldots, F_m)^T = (f_1, f_2, \ldots f_N)^T \tag{3}$$

where N—number of degrees of freedom model. Similarly for LSC nodes:

$$\tilde{U} = (\tilde{U}_1, \tilde{U}_2, \ldots \tilde{U}_m)^T = (\tilde{u}_1, \tilde{u}_2, \ldots, \tilde{u}_N)^T, \tilde{F} = (\tilde{F}_1, \tilde{F}_2, \ldots \tilde{F}_m)^T = (\tilde{f}_1, \tilde{f}_2, \ldots \tilde{f}_N)^T \tag{4}$$

Then clearly, the vectors (3) and (4) are connected by the following equations:

$$U = [T]\tilde{U}, \quad F = [T]\tilde{F}, \quad \tilde{U} = [T]^T U, \quad \tilde{F} = [T]^T F \tag{5}$$

where the transformation matrix $[T]$ is:

$$[T] = \begin{bmatrix} T_1 & O & \ldots & O \\ O & T_2 & \ldots & O \\ \ldots & \ldots & \ldots & \ldots \\ O & O & \ldots & T_m \end{bmatrix}$$

We will show that the matrix $[T]$ is orthogonal:

$$[T]^T[T] = \begin{bmatrix} T_1^T & O & \dots & O \\ O & T_2^T & \dots & O \\ \dots & \dots & \dots & \dots \\ O & O & \dots & T_m^T \end{bmatrix} * \begin{bmatrix} T_1 & O & \dots & O \\ O & T_2 & \dots & O \\ \dots & \dots & \dots & \dots \\ O & O & \dots & T_m \end{bmatrix}$$
$$= \begin{bmatrix} T_1^T T_1 & O & \dots & O \\ O & T_2^T T_2 & \dots & O \\ \dots & \dots & \dots & \dots \\ O & O & \dots & T_m^T T_m \end{bmatrix} = \begin{bmatrix} E & O & \dots & O \\ O & E & \dots & O \\ \dots & \dots & \dots & \dots \\ O & O & \dots & E \end{bmatrix} = [E]$$

Therefore, for $[T]$ performing an orthogonal property:

$$[T]^T \cdot [T] = [T] \cdot [T]^T = [E] \quad \text{or} \quad [T]^T = [T]^{-1}$$

The basic equation of equilibrium is [1–10]:

$$[K] \cdot U = [F] \tag{6}$$

where $[K]$—stiffness matrix models in GCS OXYZ.

Transforming this equation with an orthogonal matrix:

$$\begin{gathered} [T]^T[K]U = [T]^T[F] \text{ or } [T]^T[K][E]U = [T]^T[F] \text{ or} \\ [T]^T[K][T][T]^T U = [T]^T[F] \end{gathered} \tag{7}$$

Using (5), expression (7) can be written as:

$$[T]^T[K][T]\tilde{U} = [\tilde{F}]$$

Denote the $[\tilde{K}] = [T]^T[K][T]$—there is stiffness matrix models in LCS nodes. Thus, the basic equation is equivalent to:

$$[\tilde{K}][\tilde{U}] = [\tilde{F}], \tag{8}$$

Equilibrium equation in LCS nodes, and instead of solving Eq. (6) can seek a solution of Eq. (8). Consequently, the proposed method is correct and is equivalent to the known method, which is based on the solution of the equilibrium equations of the form (6). This means that the basic principles of FEM and its implementation are not changed.

6 Example

As an example, let as consider the scheme the grapple with the following parameters. The model consists of 59 elements connected in 57 nodes (Fig. 1). GCS OXYZ chosen so that the axis OY perpendicular to the plane of the mechanism. According to the model we introduce a matrix ID and the coordinates of nodes. On a design of boundary conditions are imposed—the fixed hinges at the nodes of 1, 2, 44, 45 and fixedly mounted to the node 25. Therefore, for the ID they are form: $[-1 \quad -1 \quad -1 \quad -1 \quad 0 \quad -1]$, here to "0" in a row 5 is the ability to rotate around the site with "Y" and node 25 view $[-1 \quad -1 \quad -1 \quad -1 \quad -1 \quad -1]$.

Finite element model contains 12 pairs of rotational: 3, 4, 14–16, 24, 28, 31, 32, 42, 43 nodes have pivotally connected to the nodes 46–57, respectively. That is, these nodes have a common coordinates and common 5 degrees of freedom in pairs (Fig. 1).

For example, for a pair of nodes 3, 46 ID string have the form: $[0 \quad 0 \quad 0 \quad 0 \quad 0 \quad 0], [3 \quad 3 \quad 3 \quad 3 \quad 0 \quad 3]$.

Here, the number "3" in the 46th row indicates that corresponding degrees of freedom of the 3rd and of 46th nodes are common: for them constitute one equilibrium equation.

"0" in 5th position of the node 46 indicates a difference of rotation angles of the 3rd and of 46th node around the axis "Y".

Finite element model is loaded with the forces at the nodes 7 and 39 to 5 kN, their direction is shown in Fig. 1. Elastic elements of the model with $E = 2 \times 10^6\ \mathrm{H/M}^2$, $\nu = 0.3$ and all have in the cross section—the ring with D = 0.03 and d = 0.02.

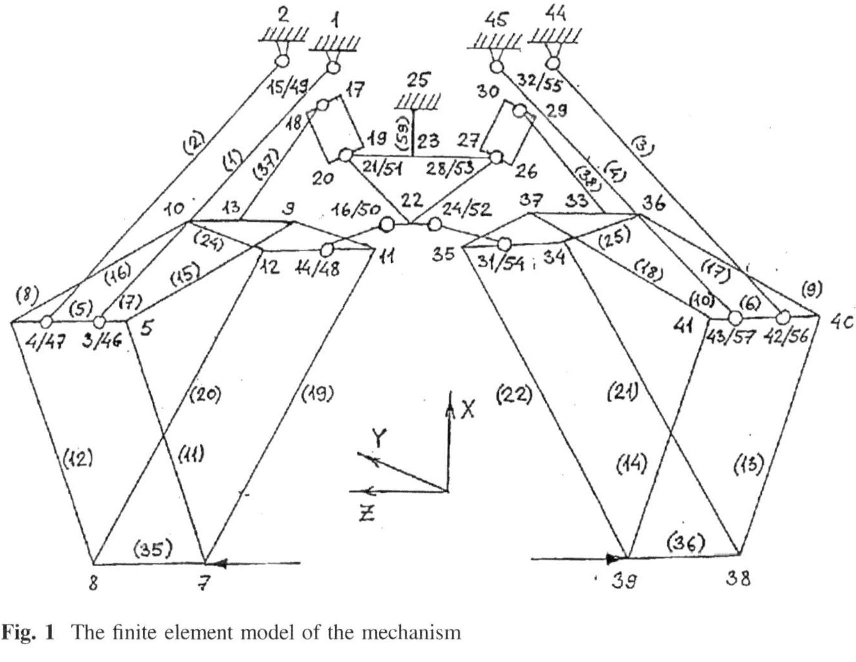

Fig. 1 The finite element model of the mechanism

Figure 2 shows the values of the nodes displacement in the following order: on the abscissa—number of nodes (Fig. 1), the vertical axis—linear displacement along the axes OX, OY, OZ GCS—range 1, range 2, range 3 respectively.

Figure 3 shows the values of displacements of nodes in the following order: on the abscissa—number of nodes (Fig. 1), the vertical axis—angular displacements around the axes OX, OY, OZ GCS—range 1, range 2, range 3 respectively.

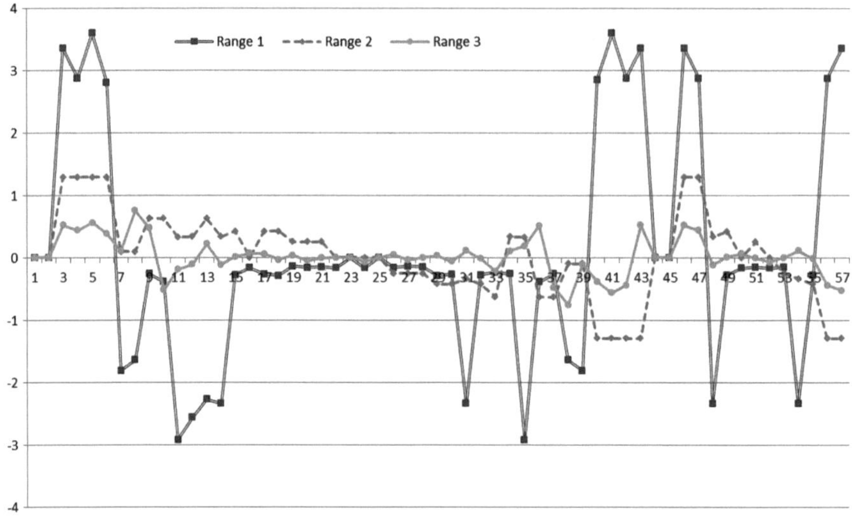

Fig. 2 Linear elastic displacement schemes constructions

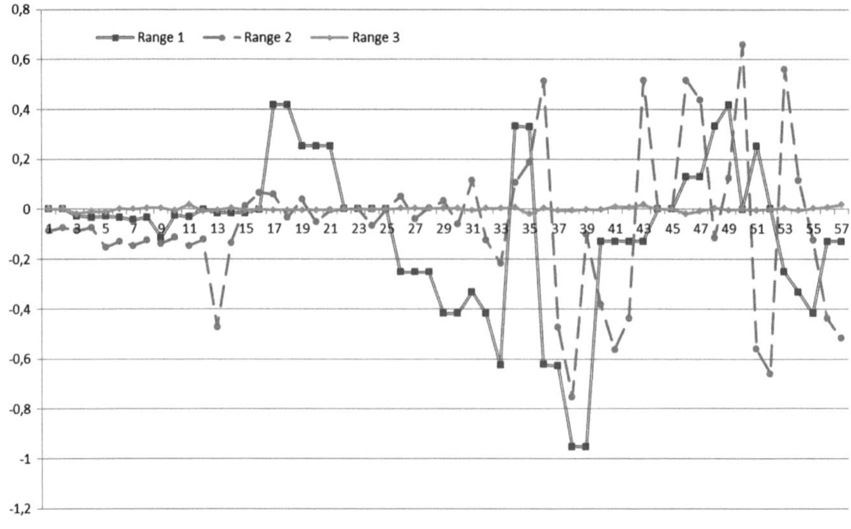

Fig. 3 Angle elastic displacement schemes constructions

Figures 2 and 3 shows that 3, 4, 14–16, 24, 28, 31, 32, 42, 43 nodes and their corresponding nodes 46–57 elastic displacement have on five degrees of freedom of the same value, and only the angular displacement relative to the axis "Y" are different.

7 Conclusions

This approach allows to use FEM for analysis of stiffness and strength mechanisms with kinematic pairs of arbitrary orientation in space. The idea of the proposed method is that the basic equilibrium equation is solved by method of hard nodes in the local coordinate systems. The basic ideas of the FEM are not changed here.

References

1. Kovalev, V.V.: Stiffness elements hinged connections links of in the dynamics caterpillar mover. Dissertation Ph.D. Tekhn. Sciences, Barnaul, p. 140 (2007)
2. Gorobtsov, A.S.: Development of methods for the analysis of spatial kinematics and dynamics of mechanisms and machines with arbitrary structure and nonlinear relationships. Doctor Dissertation. Tekhniki. Sciences, Volgograd, p. 404 (2002)
3. Romlay, F.R.M.: Modeling of a surface contact stress for spur gear mechanism using static and transient finite element method. J. Struct. Durab. Health Monit. (SDHM), **4**(1), 19–27 (2008) (Tech Science Press)
4. Dumitru, N., Cherciu, M., Zuhair, A.: Theoretical and experimental modeling of the dynamic response of the mechanisms with deformable kinematic elements. 12th IFToMM World Congress, Besancon, June 18–21 (2007)
5. Doshi, N.P., Ingole, N.K.: Analysis of connecting rod using analytical and finite element method. Int. J. Mod. Eng. Res. (IJMER) www.ijmer.com **3**(1), 65–68 (2013) (Jan–Feb 2013)
6. Zsolt, M.I., Cristina, M.: Modeling and analyses plan mechanisms with Adams 12. Annals of the faculty of engineering Hunedoara—2004 tome II. Fascicole 3, pp. 101–105 (2004)
7. Zsolt, M.I., Cristina, M.: Up and down car mechanism dynamic simulation. Annals of the faculty of engineering hunedoara—2008 tome VI. Fascicule 3, pp. 240–244 (2008)
8. Dumitru, N., Malciu, R.: Madalina Calbureanu Contributions to the Elastodynamic Analysis of Mobile Mechanical Systems Using Finite Element Method, pp. 116–121. Recent Advances in Robotics, Aeronautical and Mechanical Engineering (2007)
9. Djoldasbekov, U.A., Temirbekov, Y.S.: Some aspects of the analysis and synthesis of mechanisms high classes: Monograph—Astana, Akmolinsky CSTI, p. 299 (2006)
10. Bath, K., Wilson, E.: Numerical methods of analysis and finite element method. M. Mir, p. 486 (1984)
11. Smirnov, A.F., Alexandrov, A.V., Laşcencov, B.J., Shaposhnikov, N.N.: Building mechanics. In: Stroyizdat, M. (ed.) Rod system. p. 587 (1981)

Part V
Mechanics of Robots

Compliance Control and Design for Intelligent Lunar Robot

R.H. Li, Q.L. Fan, X.J. Zhang and Y. Zhang

Abstract In this paper, we focus on the need for the conceptual design of the intelligent lunar robot, and one kind of mechanical structure and compliance control methods are researched. Two-level shock absorber mechanism is adopted to effectively solve the question of the joint between four wheels of the robot and the ground at any road conditions. The control system consists of the on-board *PC*, the lower computer and the position feedback module. Each of them separately use the independent micro computer to process the system tasks in the distributed and parallel way. The experiments are carried on to test the performance of the intelligent lunar robot and the result shows that the whole system has the high precision and the rapid response. Simultaneously, this control system has strong computation ability and secondary development potential.

Keywords Compliance design · Lunar robot · Movement performance

1 Introduction

The research on the mobile robots is one of the current important studies and very useful in many fields. It plays a major role in the lunar exploration, industrial, medical and other fields [1]. At present, this research mainly concentrates on the multi-sensor information fusion, navigation and localization, path planning [2, 3]. Terry [4] proposed *TDE* algorithm to estimate the approximate time delay, which could ensure the correct collection to the discrete sensor data. Portasa [5] analyzed the linear control model for the multi-sensor fusion. Li [6] proposed a general object

R.H. Li (✉) · X.J. Zhang · Y. Zhang
School of Mechanical Engineering, Dalian Jiaotong University, Dalian, China
e-mail: lironghua705@163.com

R.H. Li · Q.L. Fan
Aerospace System Engineering Shanghai, Shanghai, China
e-mail: 67946456@qq.com

S. Bai and M. Ceccarelli (eds.), *Recent Advances in Mechanism Design for Robotics*, Mechanisms and Machine Science 33,
DOI 10.1007/978-3-319-18126-4_26

recognition method and the experiments had shown that the mobile robot navigation technology based on this method is robust and effective. Liu [7] proposed a heuristic search on the active localization method to solve the problem of the calculation of greater complexity and poor real-time.

In this paper, an architecture and principle of the intelligent robot platform is introduced, which is designed and developed aimed at the request of the research on the lunar simulation. Firstly, a practical and novel mobile robot mechanical system is developed on the basis of the virtual prototype technology. Secondly, the control system is designed as follows: the on-board *PC* is used for the path planning and management; the lower computer and the position feedback module are used to control movement directly. Thirdly, the movement analysis is carried on and the result provides the curve of the actual dynamic path data on the human-machine interface. The movement performance indicated that the robot has the high precision, the strong human-machine interactive ability and the rapid system response.

2 System Composition and Working Principle

The platform design uses the modular method, including the control part, the mechanical part and sensing part. The system is highly integrated and reliable. Two-level shock absorber mechanism is adopted to effectively solve the question of the joint between four wheels of the mobile robot and the ground at any road conditions. The control system consists of the on-board *PC*, the lower computer and the position feedback module. Each of them separately uses the independent micro computer to process the system tasks in the distributed and parallel way. The tasks of path planning and control movement can be carried out at the same time, so the system response is rapid. Meanwhile, the system has open interfaces and can be compatible with a variety of sensor interfaces. In this way, it is helpful for the system upgrading. The design of the system meets the needs to simulate the lunar research. The overall scheme of the system design is shown as Fig. 1.

2.1 Mechanical Part

The mechanical system is designed by the *Solidworks* software on the basis of the modular method. 3D model of the robot is shown as Fig. 2.

The robot adopts four wheel drive mechanism. Although the stabilization of this mechanism is good on flat road, it is difficult to ensure the stable joint between the wheels and the ground at any road conditions, such as climbing or on the rough pavement. In addition, the robot tends to vibrate in the process of movement to do harm to the internal components of the lunar robot and lead to various problems.

Some solutions are put forward to solve the problems. The follower wheel is designed to enable the rotational displacement under certain circumstances.

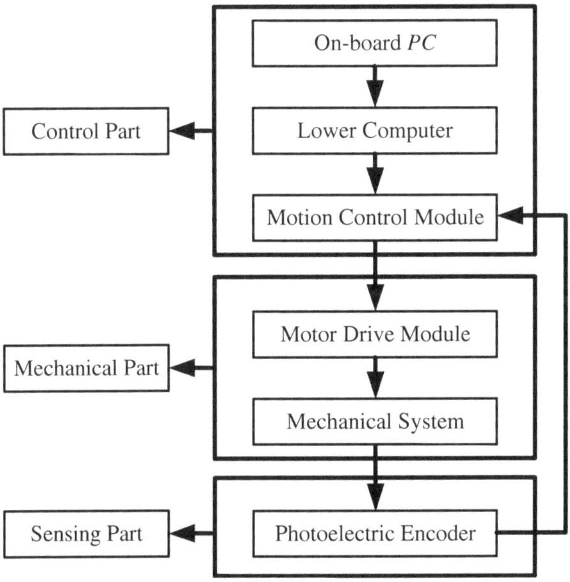

Fig. 1 Diagram of overall project design

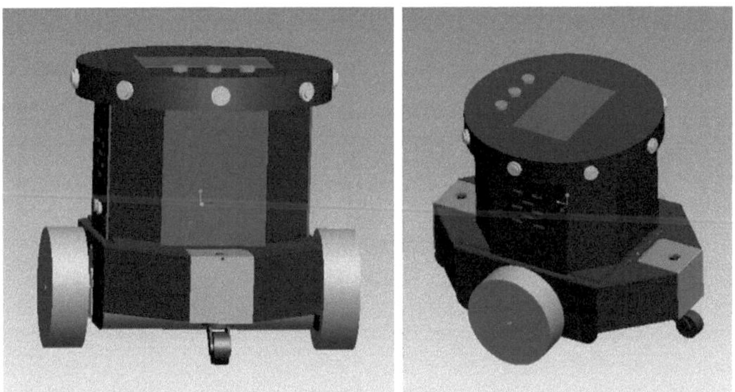

Fig. 2 3D model of the robot

As shown in Fig. 3, the system can be a good solution to the problem of pavement fitting when climbing. However, the vehicle stability is very poor when the driving wheel is elevated. Hence, four damping springs are added to the mechanism as shown in Fig. 3. This system effectively solves the problem of the pavement fitting when the driving wheel is elevated.

The damping system designed in this paper can guarantee the stable joint between the wheels and the ground when climbing as shown in Fig. 4a. At the same

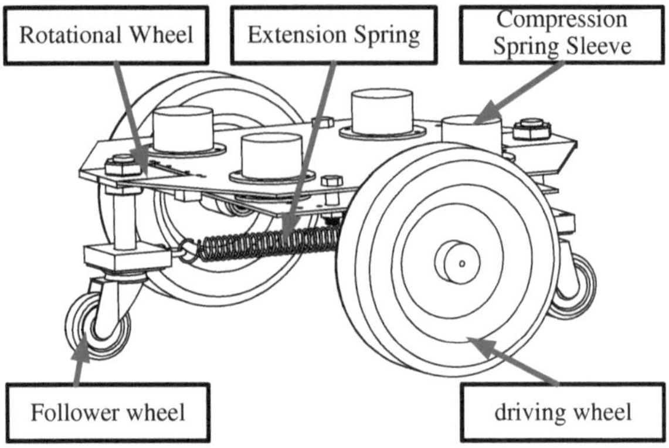

Fig. 3 Mechanical structure

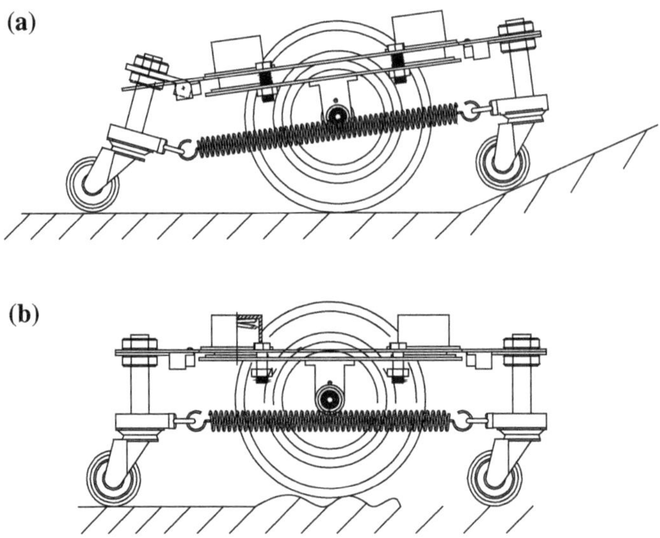

Fig. 4 Movement effects. **a** Status when climbing. **b** Status when the driving wheel is elevated

time, it can ensure that the four wheel pavement fitting when the driving wheel is elevated as shown in Fig. 4b.

The drive subsystem uses two full-bridge driver chip *LMD18200*, which drives two *DC* motors and provides power for the robot mechanical system. The chip integrates four *DMOS* tubes to form a standard *H*-drive. It provides bipolar drive mode and unipolar drive mode. This design adopts unipolar drive mode. Due to the adoption of this integrated driver chip, the whole circuit has fewer components with the small space.

2.2 Control Part

The control system is stratified. The on-board *PC* is for planning and management and the lower computer and the position feedback module are for the direct motion control.

The embedded industrial screen is selected as the on-board *PC*. It provides the powerful computing capabilities for mobile robots and handles more complex computation of the path planning. Meanwhile, the human-machine interface is designed as shown in Fig. 5. The upper left part is the robot coordinate. The upper right part is the visual information collected by the robot. The below part is parameter setting space.

The lower computer selects a high stable performance chip, *STC89C516RD+*, to achieve the simple path planning algorithm and read the switch sensor information. The lower computer communicates with the position feedback module by *I/O* port. The position feedback module mainly consists of *LM629* and photoelectric encoder. *LM629* is a powerful chip to be used for the closed loop motion control by generating the *PWM* code, the velocity and acceleration diagram, encoder pulse counting function, digital *PID* control and a number of collections. The motor position information is feedback to *LM629* by the photoelectric encoder. The lower computer and position feedback module are shown as Fig. 6.

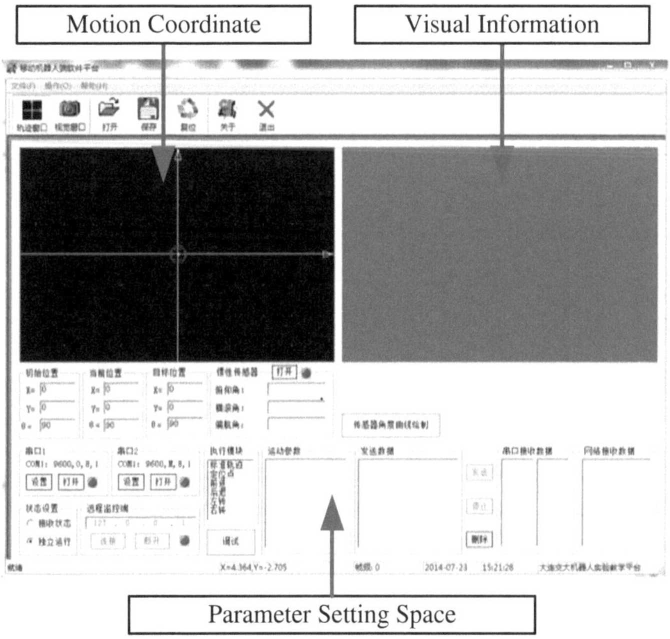

Fig. 5 Human-machine interface

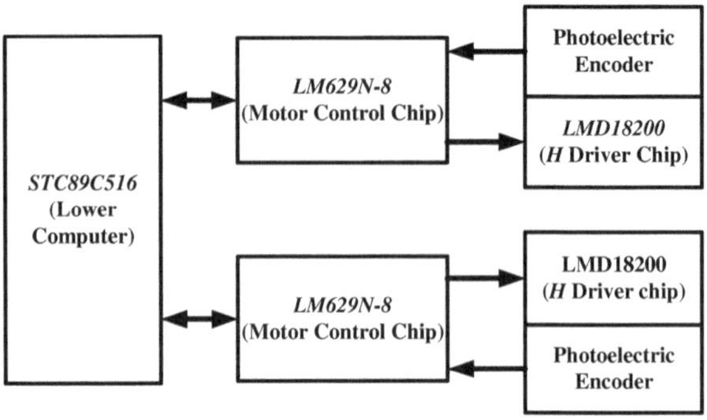

Fig. 6 lower computer and position feedback module

2.3 Motion Analysis

At time $t = 0$, the origin is located at the centre point of the robot C. The x axis is paralleling the wheel axis and the y axis is the direction that the robot moves at. The coordinate system XOY is established as shown in Fig. 7. The $\alpha(0 \le \alpha < 2\pi)$ represents the angle of velocity vector v_c. At time $t = 0$, the $\alpha(0) = \pi/2$. The motion of the robot consists of three basic actions, turning left, turning right and going straight.

As shown in Fig. 7, l is the distance between two wheel center O_1, O_2. θ is the angle of the robot turning from time t_0 to time $(t_0 + \Delta t)$. ω indicates the direction of rotation. R_c is the rotation radius of mass C. S_1 and S_2 respectively represent the travel distances of O_1, O_2 in time Δt.

Analyzing the geometric relationship, we get the robot motion equations.

$$x = x_0 + \frac{l(S_1 + S_2)}{2(S_1 - S_2)}\left[\sin\alpha_0 - \sin\left(\alpha_0 - \frac{S_1 - S_2}{l}\right)\right] \tag{1}$$

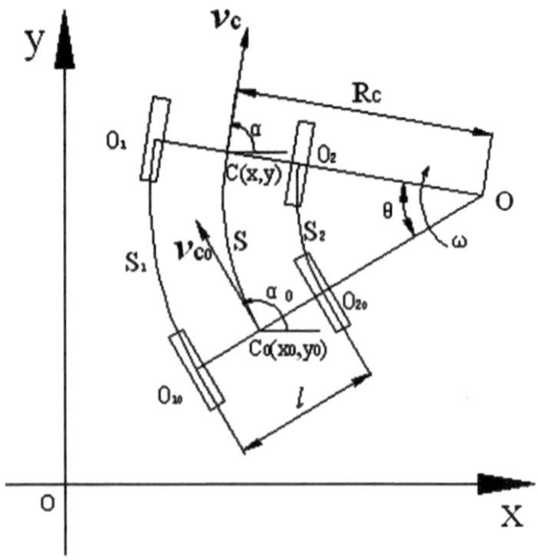

Fig. 7 Diagram of movement analysis

$$y = y_0 + \frac{l(S_1 + S_2)}{2(S_1 - S_2)}\left[\cos\left(\alpha_0 - \frac{S_1 - S_2}{l}\right) - \cos\alpha_0\right] \tag{2}$$

$$\alpha = \alpha_0 - \frac{S_1 - S_2}{l} \tag{3}$$

When it moves straightly, $S_1 = S_2$. The linear equations of motion can be determined by getting the limit of Eqs. (1) and (2).

This equation is important for robot localization. In practical application, firstly S_1 and S_2 can be got by the angle information of the robot wheel shaft. Then, the direction information of the robot can be got through the equation applied. It provides the data for the actual visual trajectory and increases the ability of the human-machine interaction.

3 Performance Test

The intelligent lunar robot platform is shown in Fig. 8. This section provides an experiment to confirm the performance of the platform.

An ideal trajectory of the robot is set by on-board *PC*. The actual trajectory is displayed on the human-machine interface. By comparison with the coincidence degree between the actual trajectory and the ideal trajectory, we can observe the position accuracy of the robot.

The square and sinusoidal wave motions are selected for the performance test. Both of them include two cycles. Each length of square wave is 1000 mm. The amplitude of sinusoidal wave is 1000 mm. The ideal trajectory and actual trajectory curve are shown as Fig. 9. The four pictures qualitatively reflect the coincidence degree between the ideal trajectory and the actual trajectory. The whole movement

Fig. 8 Robot platform

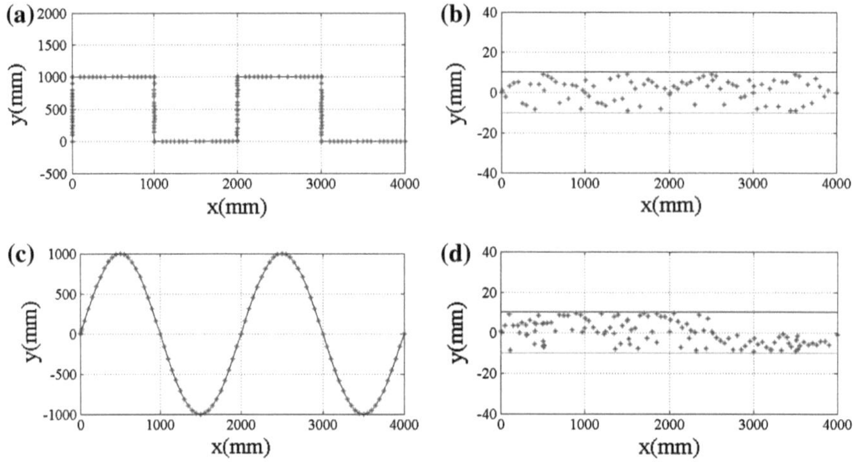

Fig. 9 Ideal trajectory and actual trajectory. **a** Square wave trajectory. **b** Square wave trajectory error. **c** Sinusoidal wave trajectory. **d** Sinusoidal wave trajectory error

process collects a large amount of data to analyze the position error. The test results show that the robot has a higher accuracy (the error is in the range of 10 mm) and rapid response. The motion states are flexible to meet the requirement.

4 Conclusions

The intelligent mobile robot system is introduced, which is developed aimed at the request of the research on the lunar exploration. Two-level shock absorber mechanism is adopted to effectively solve the question of the joint between four wheels of the mobile robot and the ground at any road conditions. The control system consists of the on-board *PC*, the lower computer and the position feedback module. Each of them separately uses the independent micro computer to process the system tasks in the distributed and parallel way. The experiments are carried on to test the performance of the intelligent lunar robot and the result shows that the whole system has the higher precision and the rapid response. Simultaneously, this control system has strong computation ability and secondary development potential.

In the future work, the system will continue to be upgraded and innovated according to the demand.

Acknowledgments This work is supported by Scientific Research Project of the Education Department in Liaoning Province (No. L2013198) and National Nature Science Foundation of China (No. 51305055).

References

1. Stephan, K.-D., et al.: Social implications of technology: the past, the present, and the future. Proc. IEEE **100**, 1752–1781 (2012)
2. Al-Araji, A.-S., et al.: Applying posture identifier in designing an adaptive nonlinear predictive controller for nonholonomic mobile robot. Neurocomputing **99**, 543–554 (2013)
3. Mosadeghzad, M., et al.: Dynamic modeling and stability optimization of a redundant mobile robot using a genetic algorithm. Robot. **30**, 505–514 (2012)
4. Terry, A.-J., et al.: Sensor fusion by a novel algorithm for time delay estimation. Digit. Signal Proc. **22**, 439–452 (2012)
5. Portasa, E., et al.: Multi-sensor fusion for linear control systems with asynchronous. Out-Of-Sequence and erroneous data, Automatica **47**(7), 1399–1408 (2011)
6. Li, X.-D., et al.: A visual navigation method for robot based on a GOR and GPU algorithm. Robot **34**(4), 465–475 (2012)
7. Liu, Y.-L., et al.: Heuristic search assisted active localization for mobile robot. Robot **34**(5), 590–595 (2012)

Design Analysis and Dynamic Modeling of a High-Speed 3T1R Pick-and-Place Parallel Robot

Guanglei Wu, Shaoping Bai and Preben Hjørnet

Abstract This paper introduces a four degree-of-freedom parallel robot producing three translation and one rotation (Schönflies motion). This robot can generate a rectangular workspace that is close to the applicable work envelope and suitable for pick-and-place operations. The kinematics of the robot is studied to analyze the workspace and the isocontours of the local dexterity over the representative regular workspace are visualized. The simplified dynamics is modeled and compared with Adams model to show its effectiveness.

Keywords Pick-and-place robots · Schönflies motion · Parallel manipulators · Dynamics

1 Introduction

The 3T1R (three translations and one rotation) robots have been widely used in material handling for pick-and-place operations. Compared to the serial SCARA robot, the parallel typed robots have the advantages in terms of high speed and light weight to improve productivity and reduce cost.

G. Wu (✉) · S. Bai
Department of Mechanical and Manufacturing Engineering,
Aalborg University, Aalborg, Denmark
e-mail: gwu@m-tech.aau.dk

S. Bai
e-mail: shb@m-tech.aau.dk

P. Hjørnet
Blue WorkForce A/S, Frederikshavn, Denmark
e-mail: ph@blueworkforce.com

S. Bai and M. Ceccarelli (eds.), *Recent Advances in Mechanism Design for Robotics*, Mechanisms and Machine Science 33,
DOI 10.1007/978-3-319-18126-4_27

Up to date, a number of parallel robots which can generate Schönflies motion have been presented. The successful Adept Quattro robot created by Pierrot and Company [1, 2], which has four symmetrical limbs, hit the market in 2007, is the fastest industrial robot available. It can accelerate at 200 m/s^2 with a 2 kg payload, being capable of moving 240 cycles per minute (cpm). By making use of the base structure of the Quattro robot, various robots with different mobile platforms have been introduced [3–5]. These robots usually have a cylindrical workspace. Besides, Angeles et al. proposed a two-limb 4-dof parallel manipulator to achieve high speed and a large workspace [6]. Recently, Angeles introduced the 'pepper mill' [7] robot with more simple structure.

This paper introduces a 3T1R parallel robot, which has a modified structure based on the Quattro robot. The design of this robot determines a rectangular workspace [8, 9] to allow more robots deployed above the conveyor, which is much more suitable for the pick-and-place application than the existing robots. The kinematics of the robot is developed and the workspace and dexterity are analyzed. Moreover, the dynamic model is established and illustrated with the standard testing trajectory, being validated by an Adams model.

2 Manipulator Architecture

The conceptual design of the new robot is shown in Fig. 1, whose motors are mounted at different orientations on the base frame, of which the axes of rotation are not coplanar. Each limb is composed of an inner arm and an outer arm (parallelogram, a.k.a 'Π joint') connected to the base and mobile platforms.

The global coordinate frame xyz is built in Fig. 1b, whose origin is located at the geometric center of the base frame. The moving coordinate frame is attached to the mobile platform and the origin is at the geometric center. The geometric parameters of the first leg is displayed in the figure.

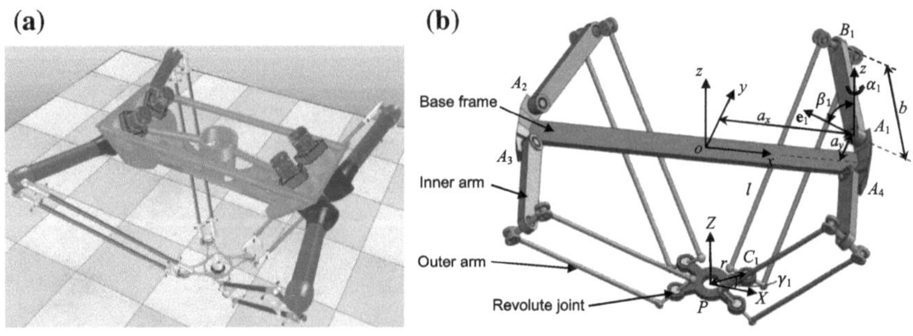

Fig. 1 The conceptual design of the robot: **a** CAD model, **b** schematic view

3 Mobility Analysis

The degree-of-freedom of the manipulator under study is derived by the Group Theory. The robot is composed of four *RRΠRR* typed limbs as shown in Fig. 2 and the bond $\mathcal{L}_i$ of the *i*th limb is the product of the following five bonds:

- The rotation subgroup $\mathcal{R}(\mathcal{A}_i)$ passing through A_i and parallel to $\mathbf{e}_i$.
- The rotation subgroup $\mathcal{R}(\mathcal{B}_i)$ passing through B_i and parallel to $\mathcal{A}_i$.
- The translation subgroup $\mathcal{T}(\mathbf{n}_i)$ corresponding to the Π-joint lying in a plane normal to $\mathbf{n}_i$.
- The rotation subgroup $\mathcal{R}(\mathcal{C}_i)$ passing through C_i and parallel to $\mathcal{B}_i$.
- The rotation subgroup $\mathcal{R}(\mathcal{C}'_i)$ passing through C_i and parallel to $\mathbf{k}$.

and $\mathbf{k}$ is the unit vector of z-axis. Thus, the kinematic bonds of the *i*th limb is

$$\mathcal{L}_i = \mathcal{R}(\mathcal{A}_i) \cdot \mathcal{R}(\mathcal{B}_i) \cdot \mathcal{T}(\mathbf{n}_i) \cdot \mathcal{R}(\mathcal{C}_i) \cdot \mathcal{R}(\mathcal{C}'_i) \quad (1)$$

where the product of the first four bonds leads to the Schönflies group, namely,

$$\mathcal{L}_i = \mathcal{X}(\mathbf{e}_i) \cdot \mathcal{R}(\mathcal{C}'_i) \equiv \mathcal{X}(\mathbf{k}) \cdot \mathcal{R}(\mathcal{C}_i) \quad (2)$$

Therefore, the intersection of the four limbs yields

$$\mathcal{L}_1 \cap \mathcal{L}_2 \cap \mathcal{L}_3 \cap \mathcal{L}_4 = \mathcal{X}(\mathbf{k}) \quad (3)$$

Henceforth, the intersection of all subgroups being a Schönflies subgroup $\mathcal{X}(\mathbf{k})$, the robot generates the Schönflies motion.

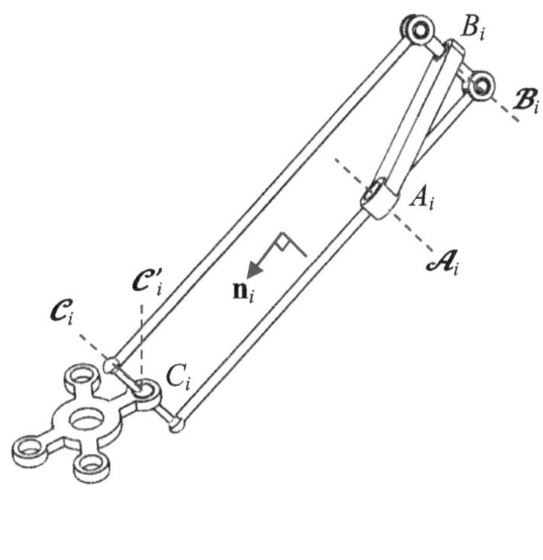

Fig. 2 The joints of the *i*th limb with rotation input

4 Kinematic Analysis of the Robot

Under the coordinate system depicted in Fig. 1b, the axis of rotation of the ith actuated joint is described by

$$\mathbf{e}_i = \mathbf{R}_i\mathbf{k}; \quad \mathbf{R}_i = \mathbf{R}(z, \alpha_i)\mathbf{R}(y, \beta_i)\mathbf{R}(z, \theta_i) \tag{4}$$

where $\alpha_{1(3)} = -\alpha_{2(4)} = \alpha$ and $-\beta_{1(4)} = \beta_{2(3)} = \pi/4$. Moreover, their position vectors in frame xyz are denoted by

$$\mathbf{a}_1 = -\mathbf{a}_3 = \left[a_x \; a_y \; 0\right]^T; \quad \mathbf{a}_2 = -\mathbf{a}_4 = \left[-a_x \; a_y \; 0\right]^T \tag{5}$$

then, the position vector of point B_i is derived as

$$\mathbf{b}_i = \mathbf{R}_i[b \; 0 \; 0]^T + \mathbf{a}_i \tag{6}$$

Let the mobile platform (MP) pose be denoted by $\mathbf{x} = \left[\mathbf{p}^T \; \phi\right]^T$, $\mathbf{p} = [x \; y \; z]^T$, the position vector of point C_i in frame xyz is expressed as

$$\mathbf{c}_i = \mathbf{Q}\mathbf{c}'_i + \mathbf{p} \tag{7}$$

where $\mathbf{Q} = \mathbf{R}(z, \phi)$ is the rotation matrix of the mobile platform and $\mathbf{c}'_i$ is the position vector of C_i in the frame XYZ:

$$\mathbf{c}'_i = r[\cos\gamma_i \; \sin\gamma_i \; 0]^T; \quad \gamma_1 = -\gamma_4 = \gamma, \; \gamma_{2(3)} = \pi \mp \gamma \tag{8}$$

4.1 Inverse Geometric Problem

The inverse geometry problem is solved from the following kinematic constraints:

$$\|\mathbf{c}_i - \mathbf{b}_i\|^2 = l^2, \quad i = 1, \ldots, 4 \tag{9}$$

Expanding and simplifying the above equation leads to

$$I_i \sin\theta_i + J_i \cos\theta_i + K_i = 0 \tag{10}$$

with

$$I_i = 2b(\mathbf{c}_i - \mathbf{a}_i)^T\mathbf{u}_i; \quad \mathbf{u}_i = [\sin\alpha_i \; \cos\alpha_i \; 0]^T \tag{11a}$$

$$J_i = -2b(\mathbf{c}_i - \mathbf{a}_i)^T\mathbf{v}_i; \quad \mathbf{v}_i = [\cos\alpha_i \cos\beta_i \; \sin\alpha_i \cos\beta_i \; - \sin\beta_i]^T \tag{11b}$$

$$K_i = \|\mathbf{c}_i - \mathbf{a}_i\|^2 + b^2 - l^2 \tag{11c}$$

Consequently, the inverse geometry problem is solved as

$$\theta_i = 2\tan^{-1}\frac{-I_i \pm \sqrt{I_i^2 + J_i^2 - K_i^2}}{K_i - J_i} \tag{12}$$

From Eq. (12), it is seen that each limb has two solutions corresponding to two working modes. Here, the "$- + -+$" mode is selected as the working mode.

4.2 Kinematic Jacobian Matrix

Differentiating the four equations in Eq. (9) with respect to time yields

$$\mathbf{A}\underline{\mathbf{x}} = \mathbf{B}\dot{\theta} \tag{13}$$

with

$$\mathbf{A} = [\mathbf{j}_1\ \mathbf{j}_2\ \mathbf{j}_3\ \mathbf{j}_4]^T; \quad \dot{\mathbf{x}} = \left[\dot{x}\ \dot{y}\ \dot{z}\ \dot{\phi}\right]^T \tag{14a}$$

$$\mathbf{B} = \text{diag}[h_1\ h_2\ h_3\ h_4]; \quad \dot{\theta} = \left[\dot{\theta}_1\ \dot{\theta}_2\ \dot{\theta}_3\ \dot{\theta}_4\right]^T \tag{14b}$$

where $\mathbf{A}$ and $\mathbf{B}$ are the forward and backward Jacobians, respectively, and

$$\mathbf{j}_i = \left[(\mathbf{c}_i - \mathbf{b}_i)^T\ (\mathbf{k} \times (\mathbf{c}_i - \mathbf{p}))^T(\mathbf{c}_i - \mathbf{b}_i)\right]^T \tag{15a}$$

$$h_i = (\mathbf{e}_i \times (\mathbf{b}_i - \mathbf{a}_i))^T(\mathbf{c}_i - \mathbf{b}_i) \tag{15b}$$

The kinematic Jacobian matrix is obtained as

$$\mathbf{J} = \mathbf{A}^{-1}\mathbf{B} \tag{16}$$

4.3 Workspace and Dexterity

The workspace (WS) can be found through the geometrical approach reported in [5] and visualized as shown in Fig. 3a, which admits a cuboid workspace. The corresponding geometric parameters are listed in Table 1. By changing the motor position and orientation, the robot topology can formulate workspace of different shapes, such as cylinder/trapezoid displayed in Fig. 3b, c.

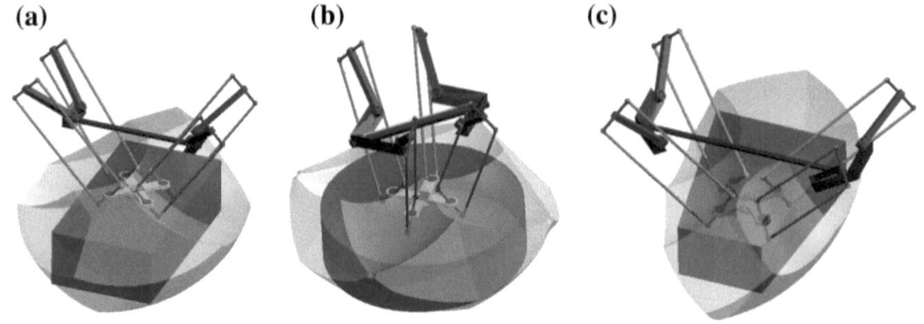

Fig. 3 Robot topologies with different shaped workspace

Table 1 Geometrical parameters of the robot

$[a_x, a_y]$ (mm)	α	b (mm)	l (mm)	r (mm)	γ
[200, 55]	15°	315	596	112	30°

Here, the cuboid WS in Fig. 3a will be investigated. As the the top view of the reachable WS has the cross-section close to superellipse, a superellipsoid is adopted to approximate the WS. Figure 4 shows the maximum superellipsoid workspace.

Dexterity is an utmost important concern, which is usually evaluated by the condition number of the kinematic Jacobian matrix. The entries of the forward Jacobian matrix of the robot are not homogeneous as they contain mixed rotation and translation terms, for which the condition number is limited to indicate the kinematic performance. Here, the characteristic length is introduced to normalize the forward Jacobian matrix below [5]:

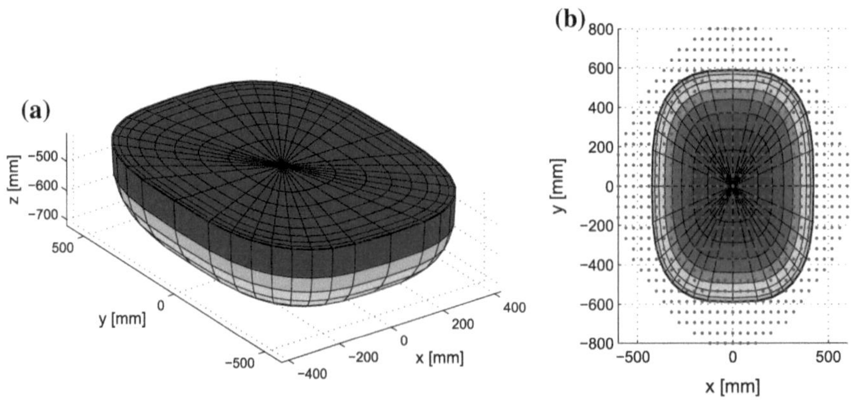

Fig. 4 The superellipsoid workspace with constant orientation $\phi = 0$: **a** 3D view, **b** top view

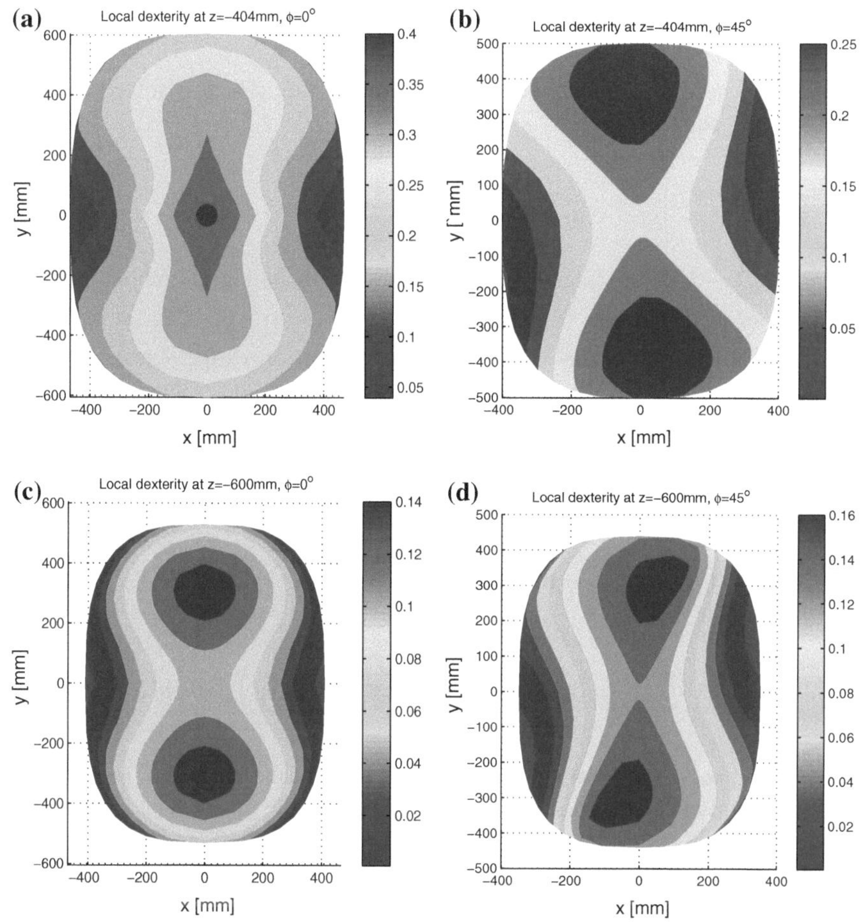

Fig. 5 Local dexterity distributions on *upper* and *lower* cross-sections of the workspace

$$L = \sqrt{\frac{3\mathbf{J}_{\omega}^{T}\mathbf{J}_{\omega}}{\mathrm{tr}(\mathbf{J}_{v}^{T}\mathbf{J}_{v})}} \tag{17}$$

where $\mathbf{J}_v$ is the first three columns and $\mathbf{J}_\omega$ is the last one in $\mathbf{A}$, respectively, and then the forward Jacobian is normalized to $\mathbf{A}' = [\mathbf{J}_v/L\,\mathbf{J}_\omega]$. The local dexterity isocontours are displayed in Fig. 5, where the condition number is based on 2-norm. It is observed that the dexterous workspace nearly forms a rectangular volume that is suitable for pick-and-place operation.

5 Dynamic Modeling

Dynamics is used to compute the actuator torque/power, which is solved by using the Lagrange equations below

$$\frac{\mathrm{d}}{\mathrm{d}t}\left(\frac{\partial L}{\partial \dot{\mathbf{q}}}\right) - \frac{\partial L}{\partial \mathbf{q}} + \mathbf{C}_q^T \boldsymbol{\lambda} = \mathbf{Q}_{ex} \tag{18}$$

where $L \equiv T - V$ is the Lagrangian of the system, including the mobile platform and the four legs, and $\mathbf{q} = \left[\theta^T \ \mathbf{x}^T\right]^T$. Moreover, $\mathbf{Q}_{ex} = \left[\boldsymbol{\tau}^T, \mathbf{0}\right]^T \in \mathbb{R}^8$ is the vector of external forces and vector $\boldsymbol{\tau} = \left[\tau_1 \ \tau_2 \ \tau_3 \ \tau_4\right]^T$ characterizes the actuator torques. Matrix $\mathbf{C}_q = [\mathbf{B} \ -\mathbf{A}]$ is the system's constraint Jacobian. Moreover, $\boldsymbol{\lambda} = \left[\lambda_1 \ \lambda_2 \ \lambda_3 \ \lambda_4\right]^T$ is a vector of Lagrange multipliers.

In this modeling, the rotation of the outer arm is not considered for simplification, thus, the kinetic and potential energies are calculated below:

$$T = \sum_{i=1}^{4}\left(\frac{1}{2} I_b \dot{\theta}_i^2 + \frac{1}{8} m_l (\dot{\mathbf{b}}_i + \dot{\mathbf{c}}_i)^T (\dot{\mathbf{b}}_i + \dot{\mathbf{c}}_i) + \frac{1}{2} m_j \dot{\mathbf{c}}_i^T \dot{\mathbf{c}}_i\right) + \frac{1}{2} \dot{\mathbf{x}}^T \mathbf{M}_p \dot{\mathbf{x}} \tag{19a}$$

$$V = \sum_{i=1}^{4}\left(\frac{1}{2} m_b \mathbf{b}_i^T \mathbf{g} + \frac{1}{2} m_l (\mathbf{b}_i + \mathbf{c}_i)^T \mathbf{g} + m_j \mathbf{c}_i^T \mathbf{g}\right) + m_p \mathbf{p}^T \mathbf{g} \tag{19b}$$

where I_b and m_b are the moment of inertia and mass of the inner arm, respectively, and m_l, m_p and m_j are the masses of the outer arm, mobile platform and the joint on the latter. Moreover, $\mathbf{M}_p = \mathrm{diag}\left[m_p \ m_p \ m_p \ I_p\right]$ is the mass matrix of the mobile platform. Terms $\dot{\mathbf{b}}_i$ and $\dot{\mathbf{c}}_i$ stand for the velocities of points B_i and C_i, respectively, which can be calculated with known $\dot{\theta}_i$ and $\dot{\mathbf{x}}$. In the above formulation, it is assumed that the centers of mass of the outer arms is coincident to their geometric centers.

Substituting Eq. (19) into Eq. (18), the terms in the equation of motion for this system can be derived. With payload $\mathbf{f}$, the actuator torques are expressed as:

$$\tau_a = \tau - \mathbf{J}^T \mathbf{f} \tag{20}$$

In order to evaluate the dynamic equation, a CAD model with geometrical parameters in Table 1 and mass properties in Table 2 was built in MSC Adams for comparison. The dynamic simulation is conducted with a pick-and-place trajectory of 25 mm × 305 mm × 25 mm, where the mobile platform positions in the y and z directions are shown in Fig. 6, and $x = 0$, $\phi = 0$, $\mathbf{f} = 0$.

The simulated velocity and acceleration profiles are shown in Fig. 7, where the solid and dashed lines stand for the Adams model and analytical model, respectively.

Table 2 The mass and moment of inertia of the robot for dynamic simulation

m_b (kg)	I_b (kg m^2)	m_l (kg)	m_j (kg)	m_p (kg)	I_p (kg m^2)
1.8	0.0595	0.42	0.2	2.2	0.026

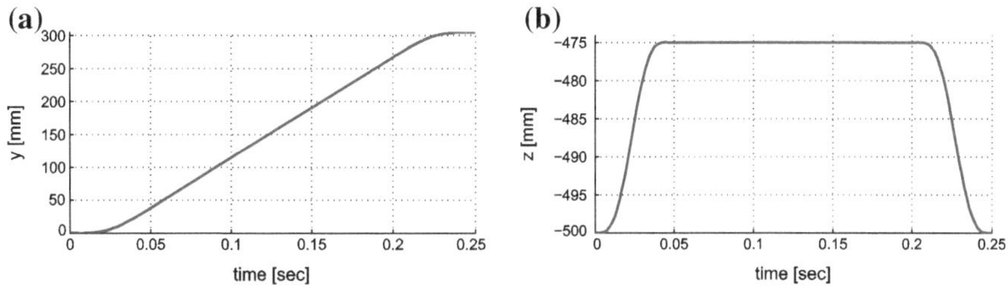

Fig. 6 The mobile platform positions with respect to time

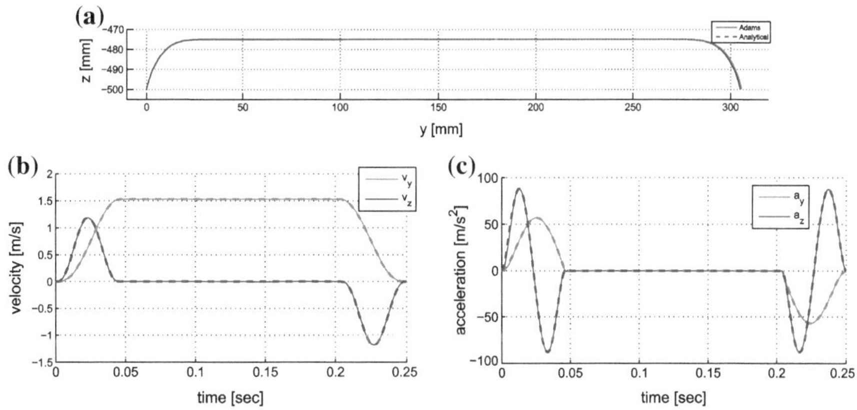

Fig. 7 Motion profiles of the testing trajectory for the Adams and analytical models

As a result, the motor torques and powers are illustrated with Fig. 8. It is noted that a gearbox with reduction ratio $\rho = 30$ is mounted on the output shaft in each limb. By comparison, it can be observed that there is a relatively good correlation between these two models.

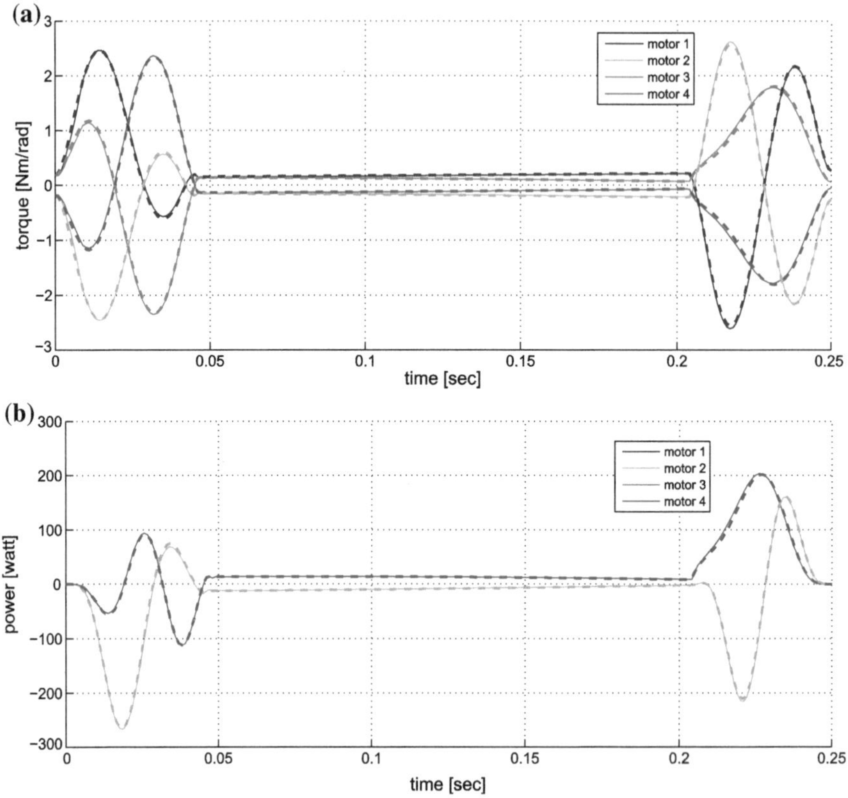

Fig. 8 The motor torques and powers along the testing trajectory

6 Conclusions

This paper introduces a 3T1R parallel robot, which can generate a rectangular workspace that is suitable for pick-and-place operations. The isocontours of the local dexterity distributions show that the robot can have a workspace with better kinematic performance to implement the pick-and-place operations. The simplified dynamic equations are modeled for the computation of motor torques and powers. Moreover, the dynamic model is validated with an Adams model.

Acknowledgments The authors would like to acknowledge Danish Innovationsfonden for the support under the Grant No. 137-2014-5.

References

1. Pierrot, F., et al.: H4: a new family of 4-dof parallel robots. In: IEEE/ASME International Conference on Advanced Intelligent Mechatronics, pp. 508–513 (1999)
2. Pierrot, F., et al.: Optimal design of a 4-DOF parallel manipulator: from academia to industry. IEEE Trans. Robot. **25**(2), 213–224 (2009)
3. Kong, X., et al.: Type synthesis of 3T1R 4-DOF parallel manipulators based on screw theory. IEEE Trans. Robot. Autom. **20**(2), 181–190 (2004)
4. Liu, S., et al.: Optimal design of a 4-DOF SCARA type parallel robot using dynamic performance indices and angular constraints. J. Mech. Robot. **4**(3), 031005 (2012)
5. Altuzarra, O., et al.: A symmetric parallel Schönflies-motion manipulator for pick-and-place operations. Robotica **29**(6), 853–862 (2011)
6. Cammarata, A., et al.: The dynamics of parallel schönflies motion generators: the case of a two-limb system, Proceedings of the IMechE, Part I: J. Syst. Control Eng. 223(11):29–52 (2009)
7. Harada, T., et al.: Kinematics and singularity analysis of a CRRHHRRC parallel Schönflies motion generator. CSME Trans. **38**(2), 173–183 (2014)
8. Bai, S., et al.: Design of a parallel robot with optimized workspace shape for fast and flexible pick-and-place operation. In: Third IFToMM Asian-MMS. Tianjin, China (2014)
9. Bai, S., et al.: Stiffness modeling of a 4-dof parallel robot for Schönflies motion generation. In: ASME ESDA 2014. Copenhagen, Denmark, ESDA2014-20094 (2014)

Dynamic Characteristics of Two Cylinders' Joint Surfaces Based on Fractal Theory

Xiaopeng Li, Xue Wang, Jiasheng Li, Muyan Li and Shujun Li

Abstract In order to reveal the deformation state of elastomer actually, the fractal model of the M-B is modified, and deformation properties of elastic stage, elastic-plastic stage and plastic stage of elastomer are analyzed. From the combination of macro and micro perspective, the fractal model of contact stiffness between two cylinders' joint surfaces is established considering the influence of friction, which is proofed to be feasible by numerical simulation. Moreover, the Fixed curved joint is taken as a research object, the dynamic model is established by the method of spring element, and the first 6 natural frequencies is obtained by the finite element analysis method. Finally, the natural frequency and model analysis obtained from theory and experiment are comparative analyzed. The results show that the established stiffness model is well suitable for the reality of joint surface. Then a new approach for the treatment of joint surface, which is largely present in robotics and NC machine tool etc. are completely provided.

Keywords Joint surface · Friction factor · Contact stiffness · Fractal model · Mechanical dynamics

X. Li (✉) · X. Wang · J. Li · M. Li · S. Li
School of Mechanical Engineering & Automation, Northeastern University, Shenyang, China
e-mail: xpli@me.neu.edu.cn

X. Wang
e-mail: 13940224160@163.com

J. Li
e-mail: 18842577323@163.com

M. Li
e-mail: 15840539775@163.com

S. Li
e-mail: shjunli@me.neu.edu.cn

S. Bai and M. Ceccarelli (eds.), *Recent Advances in Mechanism Design for Robotics*, Mechanisms and Machine Science 33,
DOI 10.1007/978-3-319-18126-4_28

1 Introduction

The joint surface exists in assembling machinery parts such as robotics, NC machine tool etc. It plays an important role to transfer movement, load and energy in the normal operation of mechanical system. The results show that joint surface's dynamic characteristics affect the stiffness, damping, machining accuracy and stability in robotics, NC machine tool etc. So, studies on dynamic characteristics of joint surface have important significance in theory and practical applications [1].

Many scholars have carried out lots of studies on joint surface [2–4]. In Macroscopic aspect, researches of joint surface's dynamic characteristics are largely depends on Hertz theory [5], which is mainly study macroscopic properties of joint surface, such as geometric parameters, material parameters and boundary conditions. But the microscopic characteristics are merely considered in it. In microcosmic aspect, stiffness and damping models [6, 7] of joint surfaces are obtained depends on the M-B fractal model [8]. But M-B fractal model is mainly suitable for the contact analysis of two infinite planes, not for finite surfaces. And macroscopic properties are merely considered. A fractal model of two cylinders' normal contact stiffness is built up by Zhao [9], but the effect of friction is not considered. Actually, friction and gap between joint surfaces has nonlinear vibration characteristics, which can change the topology mechanisms of dynamic systems and cause problems of vibration and noise [10].

Considering the friction and elastic-plastic deformation of joint surface, fractal models of contact stiffness of two cylinders' joint surfaces are deduced based on Hertz theory and modified M-B fractal model in this paper.

2 Fractal Prediction Model of Contact Stiffness

2.1 Deformation Property Analysis of Elastomer

Based on the research of rough-surface contact model by Wang and Komvopoulos [11, 12], the deformation mode of elastomers includes elastic, elastic-plastic and plastic, and the transformation from elastic stage to plastic stage is not mutational. However, the transfer deformation stage, that is to say the elastic-plastic stage, is ignored in M-B fractal model. Moreover, the influence of friction and gap between joint surfaces should be fully considered.

The most basic theory of the contact surface is Hertz theory, and based on the research of elastic deformation and plastic deformation, the critical deformation of micro-bulge from elastic deformation to plastic deformation is Eq. (1).

$$\delta c = (\frac{\pi K \sigma y}{2E})^2 R \tag{1}$$

where, K is the correlation coefficient of hardness H and yield strength σy, and three of them meet the relation $H = K\sigma y$. E is the integrated elastic modulus got from Hertz analysis.

Based on the M-B fractal model [13], the mathematical model of the surface outline of the micro-bulge can be expressed as Eq. (2), and when $x = 0$, it can be expressed as Eq. (3), a is the actual contact area.

$$z(x) = G^{D-1}a^{1-0.5D}\cos\left(\frac{\pi x}{a^{0.5}}\right), \quad -0.5a^{0.5} < x < 0.5a^{0.5} \tag{2}$$

$$\delta = G^{D-1}a^{1-0.5D} \tag{3}$$

Then, curvature radius of elastomers can be gotten as shown in Eq. (4).

$$R = a^{0.5D}G^{1-D}/\pi \tag{4}$$

From the Ref. [14], when the relative sliding friction produced in the joint interfaces, the critical average contact pressure of micro-bulge is:

$$p_{\mu} = 1.1k_{\mu}\sigma_y \tag{5}$$

where, k_{μ} is the friction coefficient correction factor, $k_{\mu} = 1-0.228\,\mu$, $0 \leq \mu \leq 0.3$, $k_{\mu} = 0.932e^{-1.58(\mu-0.3)}$, $0.3 \leq \mu \leq 0.9$.

Based on the Eqs. (1) and (5), while the influence of friction is considered, the critical deformation between the elastic and plastic state of micro-bulge can be expressed as Eq. (6). So the critical area between the elastic and plastic state of micro-bulge can be obtained, and expressed as Eq. (7).

$$\delta_{\mu c} = \left(\frac{3\pi p_{\mu}}{4E}\right)^2 R = \left(\frac{3.3\pi k_{\mu}\sigma_y}{4E}\right)^2 R \tag{6}$$

$$a_{\mu c} = \left(\frac{3.3\pi^{0.5}k_{\mu}\sigma_y}{4E}\right)^{2/(1-D)} G^2 \tag{7}$$

The elastic deformation will happen when the contact area is larger than the critical elastic area ($a_{\mu c}$), while the influence of the friction is considered. Then we introduce the concept of critical plastic area (a_{pt}), and the plastic deformation will happen when the contact area is smaller than the critical plastic area. The state is elastic-plastic deformation when the contact area is larger than the critical plastic area and smaller than the critical elastic area.

From the Ref. [15], Eqs. (3) and (4), the critical plastic area can be expressed as follows:

$$a_{pt} = G^2 \left(\frac{E\beta}{\sigma_y} \right)^{2/(D-1)} \quad (8)$$

where, $\beta = \frac{\sqrt{\pi}\left(\frac{2m+1}{2m}\right)^{2(m-1)}}{(30 \times 0.2^{1/m})^{m/(m-1)}}$, m is the strain hardening index.

2.2 Model of Normal Contact Stiffness

From the Ref. [16] and other references, the relation of the biggest contact area (a_l) and the distribution of the contact points ($n'(a)$) is:

$$n'(a) = -\frac{\mathrm{d}N'(A > a)}{\mathrm{d}a} = \lambda \frac{D}{2} \psi^{(2-D)/2} a_l^{D/2} a^{-(D+2)/2} \quad (9)$$

$$\lambda = \left(\frac{C_1 \left(\frac{\Omega(B,E)}{X_h} \right)^{C_2}}{\sum S} \right)^{X_h} F^{C_2 X_h} \quad (10)$$

where, ΣS is the surface area of two surfaces. X_h is the synthetical curvature radius. B is the width of the contact body. E is the equivalent elastic modulus. $\Omega(B, E)$ is the function of B and E. C_1, C_2 are the coefficient related to the geometry of surfaces.

With Hertz theory, the normal contact stiffness of single micro-bulge is:

$$K_n = \int k_n \cdot n'(a)\mathrm{d}a \quad (11)$$

Considering the elastic deformation, elastic-plastic deformation and plastic deformation, the fractal model of the normal contact stiffness of the whole joint surfaces can be deviated as follows:

$$\begin{aligned} K_n &= \int_{a_{\mu c}}^{a_l} k_n \cdot n'(a)\mathrm{d}a + \int_{a_{pt}}^{a_{\mu c}} k_n \cdot n'(a)\mathrm{d}a \\ &= \frac{2 \cdot \lambda \cdot E \cdot D \cdot \psi^{1-0.5D} \cdot a_l^{\frac{D}{2}}}{\sqrt{\pi}(1-D)} \cdot \left(a_l^{\frac{1-D}{2}} - a_{\mu c}^{\frac{1-D}{2}}\right) + \frac{2 \cdot \lambda \cdot E' \cdot D \cdot \psi^{1-0.5D} \cdot a_l^{\frac{D}{2}}}{\sqrt{\pi}(1-D)} \cdot \left(a_{\mu c}^{\frac{1-D}{2}} - a_{pt}^{\frac{1-D}{2}}\right) \end{aligned} \quad (12)$$

where, E' is the corresponding elastic modulus.

According to the $E = \sigma/\varepsilon = Pl/A\Delta l$, the stress decreases with the increase of contact area in the stage of elastic-plastic deformation, and then the deformation amount becomes bigger. So, the corresponding elastic modulus of elastic-plastic stage is smaller than that of elastic stage. The relation of them we make in this paper is $E' = 0.9E$.

From Eq. (12), it shows that the stiffness of the joint surfaces is combined with the critical plastic area and critical elastic area in friction state. When the fractal characteristic parameters of joint surfaces, such as D, G and E, and the geometric parameters of two cylinders, such as R_1, R_2, F, and B, are obtained, the normal contact stiffness of two cylinders' joint surfaces can be estimated. The accuracy of estimating the actual contact rigidity will be improved by this method.

2.3 Model of Tangential Contact Stiffness

The relation between tangential force and tangential deformation can be expressed as Eq. (13) when the normal force (P) and tangential force (T) is applied to the contact point of equivalent sphere and rigid plane.

$$\mathrm{d}t = \frac{3(2-\upsilon)}{16\overline{G}r}\mu P\left[1-\left(1-\frac{T}{\mu P}\right)^{2/3}\right] \tag{13}$$

where, $\overline{G} = E/2(1+\upsilon)$ is the equivalent shearing elastic modulus. υ is the poisson ratio. μ is the friction coefficient. $a = \pi r^2$, the tangential contact stiffness is:

$$k_t = \frac{8\overline{G}\sqrt{a}}{(2-\upsilon)\sqrt{\pi}}\left(1-\frac{1}{\mu}\cdot\frac{T}{P}\right)^{1/3} \tag{14}$$

According to Eqs. (7)–(9) and (14), the tangential contact stiffness can be expressed as follows:

$$\begin{aligned} K_t &= \int_{a_{\mu c}}^{a_l} k_t \cdot n'(a)\mathrm{d}a + \int_{a_{pt}}^{a_{\mu c}} k_t \cdot n'(a)\mathrm{d}a \\ &= \frac{8\lambda\overline{G}D \cdot a_l^{\frac{D}{2}}}{\sqrt{\pi}(2-\upsilon)(1-D)} \cdot \left(1-\frac{1}{\mu}\cdot\frac{T}{P}\right)^{\frac{1}{3}} \cdot \left(a_l^{\frac{1-D}{2}} - a_{\mu c}^{\frac{1-D}{2}}\right) \\ &\quad + \frac{8\lambda\overline{G}'D \cdot a_l^{\frac{D}{2}}}{\sqrt{\pi}(2-\upsilon)(1-D)} \cdot \left(1-\frac{1}{\mu}\cdot\frac{T}{P}\right)^{\frac{1}{3}} \cdot \left(a_{\mu c}^{\frac{1-D}{2}} - a_{pt}^{\frac{1-D}{2}}\right) \end{aligned} \tag{15}$$

3 Feasibility Analysis of Fractal Prediction Model

3.1 Analysis of Normal Contact Stiffness Model

The relation of the dimensionless actual contacting area and single-point maximal contacting area is:

$$A_r = \frac{D}{2-D}\psi^{\frac{2-D}{2}}a_l \tag{16}$$

When $\sigma_y/E = 0.001$, $\mu = 0.5$, $D = 1.45$, $G = 1.36 \times 10^{-9}$, $R_1 = 100$ mm, $R_2 = 50$ mm, $F = 1000$ N, $B = 50$ mm, $E = 155{,}000$ Mpa, compared with the research of Ref. [16], the feasibility analysis of the normal contact stiffness model established above considering the influence of friction and elastic-plastic deformation can be studied, the results are shown in Fig. 1.

From Fig. 1, we can find that when the effect of friction is taken into consideration, the normal contact stiffness is low. This is mainly because of the existence of friction, the critical elastic area of elastomers will increase, and the proportion of contact point whose contact area is smaller than the critical elastic area will increase. In this condition, the proportion of plastic deformation will increase, while the proportion of elastic deformation will decrease. When both the effect of friction and elastic-plastic deformation is taken into consideration, the normal contact stiffness is high. This is because the proportion of plastic deformation decreases with the consideration of elastic-plastic deformation. Then, the damping of joint surfaces will decrease and the stiffness will increase. The results all proof the normal contact stiffness model we established is feasible.

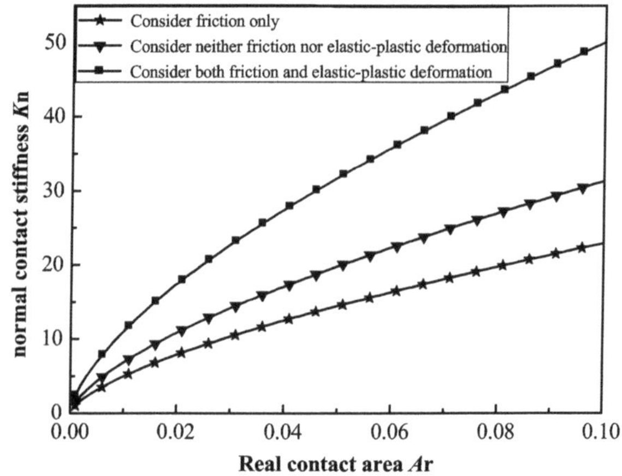

Fig. 1 Contrast curves of normal contact stiffness in different conditions

3.2 Analysis of Tangential Contact Stiffness Model

When $\nu = 0.3$, $\mu = 0.5$, $T/P = 0.02$, $D = 1.50$, $G = 1.36 \times 10^{-9}$, compared with research of Ref. [17], the feasibility analysis of the tangential contact stiffness model established above considering the influence of friction and elastic-plastic deformation can be studied, the results are shown in Fig. 2. The law of curves in Fig. 2 is similar to that in Fig. 1; the results proof the tangential contact stiffness model we established is feasible.

4 Experimental Analysis of Fixed Curved Joint

The fixed curved joint is taken as a research object, shown in Fig. 3. The size of stent is 82 mm × 74 mm × 100 mm, the radius of optical axis is 50 mm, and the length is 150 mm. According to the Ref. [18], the fractal dimension (D) of its profilogram is 1.426 and the fractal scale parameters (G) are $2.34e^{-5}$ with the method of experimental determination and data fitting.

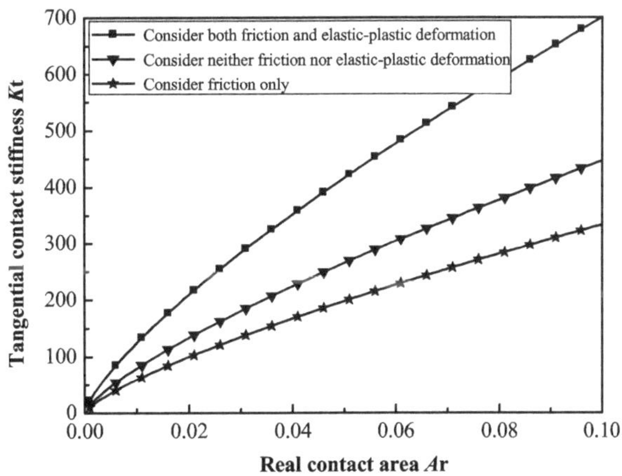

Fig. 2 Contrast curves of tangential contact stiffness in different conditions

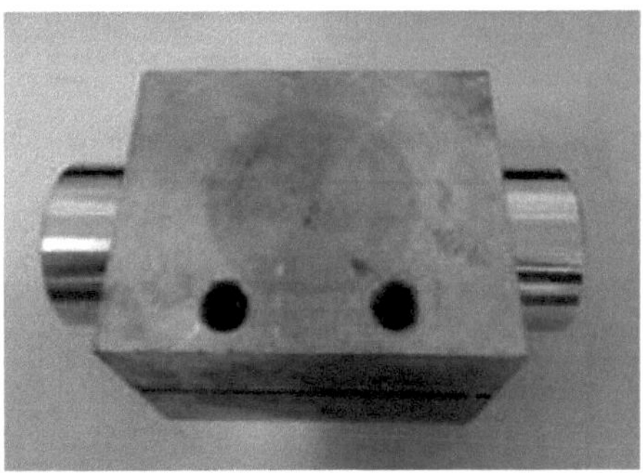

Fig. 3 Fixed curved joints for experiment

Table 1 Material constants and fractal parameters

μ	E (Pa)	v	ρ(kg/m^3)	D	G
0.19	1.1e11	0.3	7850	1.426	2.34e-5

By substituting the fractal parameters of steel plate (D), (G), elastic modulus (E), poisson's ratio (v), density (ρ), and friction coefficient (μ) into Eqs. (12) and (15), the normal contact stiffness (K_n) and tangential contact stiffness (K_t) of joint surfaces are obtained. The parameters are summarized in Table 1. To sake comparison, the normal contact stiffness (K_n) and tangential contact stiffness (K_t) in Ref. [16, 17] are calculated, shown in Table 2.

Then, the equivalent dynamic model of fixed curved joint is established by the method of spring element. Based on the vibration characteristics of the system, cycles of viscoelastic element are established on the location L_1, L_2, L_3, 4 groups of spring elements are arranged in each cycle, and each group is consist of 3 spring elements separately arranged in the X, Y, Z direction. That is to say, the fixed curved joint is equivalent to a three-degrees-of-freedom model through 36 spring elements. The model has two degrees of freedom in the radial direction and single degree of freedom in the axial direction. Each freedom is a parallel connection of 12 spring elements, and its contact stiffness value is $K_n/12$ or $K_t/12$.

According to the structure of the fixed curved joint, the finite element model is built. In the model, the spring element is simulated by combin14, and each layer material is divided by the Solid-92 element with the simulation soft. Each model shape and natural frequency are obtained through the method of Block Lanczos of modal analysis, the result is shown in Table 3. And the comparison of the results of finite element analysis and modal test are listed in Table 3.

From Table 3, each order natural frequency obtained by finite element analysis is agreement with the results of modal test. The error is lower than 10 %, which proofs the contact stiffness fractal model of two cylinders is correct. Compared with the

Table 2 Stiffness of fixed curved joint's joint surface under different calculation models

	Model in this paper	Traditional model
Normal contact stiffness K_n (N/m)	7.27e9	5.79e9
Tangential contact stiffness K_t (N/m)	6.11e9	2.5e9

Table 3 Comparisons of the results of finite element analysis and modal test

Model order	1	2	3	4
Natural frequency of new model (Hz)	4085.4	5623.8	8595.4	11,954
Natural frequency of traditional model (Hz)	3500.7	5208.1	8449.4	11,883
Natural frequency of modal test (Hz)	3820	5470	8720	12,110
Error of new model (%)	6.9	2.8	1.4	1.2
Error of traditional model (%)	8.3	4.7	3.1	1.9

traditional model, the error of the model considering the effect of friction and elastic-plastic deformation is lower. The results all proves the superiority of the model we establish.

5 Conclusions

Based on Hertz theory and modified M-B fractal model, fractal model of contact stiffness between two cylinders' joint surfaces is established considering the influence of friction and elastic-plastic deformation. The model has geometric characteristics and scale independence, and it has been proofed by numerical simulation feasible for the reality of the joint surface. The works all lay certain theoretical basis to solve surface contact problems of cylinders and gears in robotics, NC machine tool, etc.

Substituting the fractal parameters, the normal contact stiffness and tangential contact stiffness of joint surfaces are obtained. And the equivalent dynamic model of the joint surface of fixed curved joint is established by the method of spring element. The natural frequencies are gained with finite element analysis, which are agreement with the results of modal test. Compared with the traditional model, the error of the model considering the effect of friction and elastic-plastic deformation is lower. The results all proves the superiority of the model we establish.

Acknowledgments This paper is supported by National Natural Science Foundation (51275079), Program for New Century Excellent Talents in University (NCET-10-0301) and Fundamental Research Funds for the Central Universities (N110403009).

References

1. Zhang, X., et al.: Review and prospect of the research on the static and dynamic characteristics of machine joint surfaces. J. Taiyuan Heavy Mach. Inst. **23**(3), 276–281 (2002)
2. Jaspreet, S.-D., et al.: Effect of a nonlinear joint on the dynamic performance of a machine tool. J. Manuf. Sci. Eng. **129**(5), 943–950 (2007)
3. Shir, Y.-O., et al.: Parameter estimation and investigation of a bolted joint model. J. Sound Vib. **307**(3/5), 680–697 (2007)
4. Tian, H., et al.: A new method of virtual material hypothesis-based dynamic modeling on fixed joint interface in machine tools. Int. J. Mach. Tools Manuf. **51**, 239–249 (2011)
5. Fang, Z., et al.: Multi-body contact dynamic modeling of gear transmission. J. Mech. Trans. **33**(1), 15–18 (2009)
6. Zhang, X., et al.: Fractal model of normal contact stiffness between rough surfaces. Chin. J. Appl. Mech. **17**(2), 31–35 (2000)
7. Wen, S., et al.: Fractal model and simulation of normal contact stiffness of joint interfaces and its simulation. Trans. Chin. Soc. Agric. Mach. **40**(11), 197–202 (2009)
8. Majumdar, A., et al.: Fractal model of elastic-plastic contact between rough surfaces. J. Tribol. Trans. ASME **113**(1), 1–11 (1991)

9. Zhao, H., et al.: Fractal model of normal contact stiffness between two cylinders' joint interfaces. J. Mech. Eng. **47**(7), 53–58 (2011)
10. Li, X., et al.: Fractal prediction model for tangential contact damping of joint surface considering friction factors and its simulation. J. Mech. Eng. **48**(23), 46–50 (2012)
11. Wang, S., et al.: A fractal theory of the interfacial temperature distribution in the slow sliding regime: Part I-Elastic contact and heat transfer analysis. ASME J. Tribol. **116**, 812–823 (1994)
12. Wang, S., et al.: A fractal theory of the interfacial temperature distribution in the slow sliding regime: Part II-Multiple domains, elastoplastic contacts and applications. ASME J. Tribol. **116**, 824–832 (1994)
13. Ge, S. et al.: Tribological Fractal. China Machine Press, Beijing (2005)
14. Zhu, Y., et al.: The elastic elastoplastic and plastic fractal contact models for rough surface. J. Xi'an Inst. Technol. **21**(2), 150–157 (2001)
15. Johnson, K.-L. et al.: Contact Mechanics. Cambridge University Press, Cambridge (1985)
16. Huang, K., et al.: Research of fractal contact model on contact carrying capacity of two cylinders surface. Tribology **28**(6), 529–533 (2008)
17. Wen, S., et al.: Fractal model of tangential contact stiffness of joint interfaces and its simulation. Trans. Chin. Soc. Agric. Mach. **40**(12), 223–227 (2009)
18. Chen, Q.: Research on Gear Contact Strength Analysis of Automobile Gearbox by Fractal Theory. Hefei University of Technology, Hefei (2010)

Modular System with Varying Contact Elements for a Reconfigurable Parallel Robot

S. Kurtenbach, M. Hüsing and B. Corves

Abstract This paper introduces a modular system consisting of a 3 DOF robotic local structure with revolute joints. Several wrist joints are presented where the contact element, the end-effector, is systematically varied. The range of objects to be handled is analysed with the aim to identify a minimum of contact principles. Realizing these principles with one gripper each and defining general interfaces leads to a modular system, which warrants the versatility, flexibility and reconfigurability required for the introduced parallel handling system.

Keywords Modular system · Robotic local structure · Contact principle for end-effectors · Reconfigurable parallel robot

1 Introduction

Now-a-days parallel handing devices are used in various application areas. The Department of Mechanisms Theory and Dynamics of Machines (IGM) at RWTH Aachen University developed a new and innovative handling system based on a reconfigurable structure with modular layout [1–4]. The handling system moves the object with six degrees of freedom in space, using a parallel kinematic structure. This results only through the integration of the object using several robotic arms and makes it a part of the parallel kinematic structure. Figure 1 shows the entire handling system with all three robotic arms adapted to the object. This handling system is called PARAGRIP. A single robotic arm has six degrees of freedom; three

S. Kurtenbach (✉) · M. Hüsing · B. Corves
RWTH Aachen University, Aachen, Germany
e-mail: kurtenbach@igm.rwth-aachen.de

M. Hüsing
e-mail: mhuesing@igm.rwth-aachen.de

B. Corves
e-mail: corves@igm.rwth-aachen.de

S. Bai and M. Ceccarelli (eds.), *Recent Advances in Mechanism Design for Robotics*, Mechanisms and Machine Science 33,
DOI 10.1007/978-3-319-18126-4_29

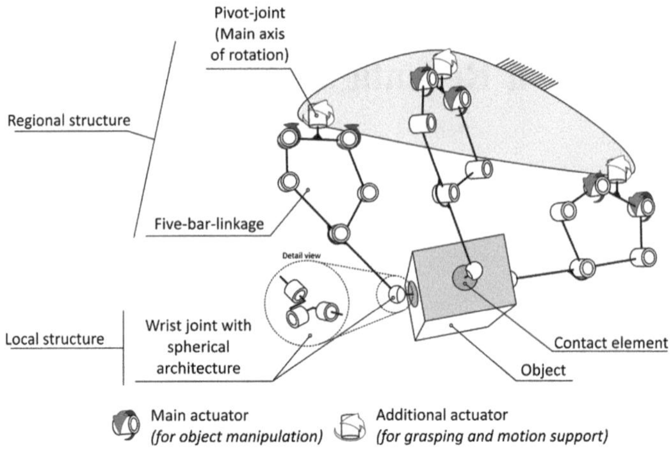

Fig. 1 Parallel kinematic handling system PARAGRIP

rotational degrees of freedom in the robotic regional structure and further three in the robotic local structure. The regional structure itself represents a hybrid structure. A planar five-bar linkage can be rotated around a previously located main axis of rotation. Drives (servo motors) are provided solely for the three degrees of freedom of the regional structure. The wrist joint is operated in a purely passive form. This is to reduce the number of drives for the entire handling system to nine servo motors to move an object with three translational and three rotational degrees of freedom. Similar applications for three serial industrial robots require 18 drives.

The current local structure of the robotic arms comprises a spherical arrangement of the wrist joint. Figure 2a schematically shows the spherical structure with the notation of various axes. The first joint (joint R) points with its axis to the object and therefore it is the x-direction. The second joint (joint S) is orthogonal to the joint R. Since y- and z-direction will be passed from joint S anyway during a rotation around the x-axis both arrangements are equal. The third joint (joint T) is again positioned directing to the object (x-direction). Thus, the spherical structure

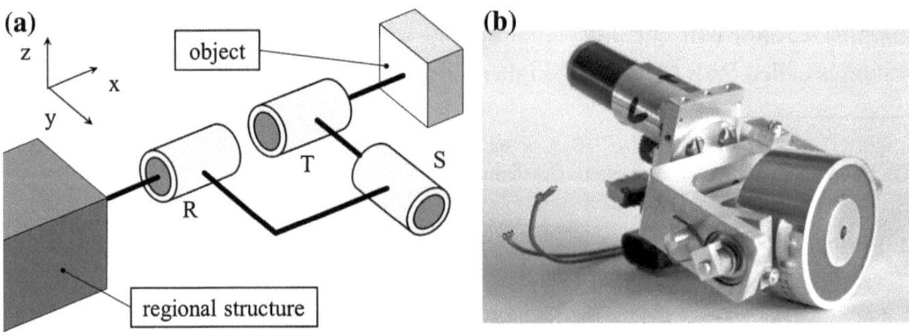

Fig. 2 Schematic representation (**a**) and image (**b**) of the current wrist joint

has 3 DOF and the type becomes XYX (or XZX, equal structure rotated around π/2 in singularity position). The singularity is achieved when both the first axis and the third one are positioned in x-direction. This arrangement must be avoided performing a motion due to high thrusts and jerks being avoided. Since the wrist joint is a passive one, the initial position before adapting to an object must be met precisely. The joint S uses a torque spring to keep the end-effector in its initial position. To warrant the initial position for the joint R, a small stepping motor is used which is switched of directly after reaching the initial position.

The integration of the object into the kinematic structure has been achieved in previous developments by using an electromagnet. This magnet represents the end-effector (Fig. 2b). The electrical energy is passed through the entire spherical structure by collector ring contacts, which are provided on the corresponding axes of rotation. In order to avoid a collision between the wrist joint and the object during docking, the R-axis is equipped and actuated by a stepping motor, which serves merely to pre-position the wrist joint. It is switched passive immediately after the object has been gripped.

The S-axis has a balancing and restoring spring element that holds the solenoid in the centre position while not in contact with the object. A disadvantage of the current robotic local structure is the focussing in exclusively ferromagnetic objects and so material properties of the object.

In this paper, the development of several spherical linkages with varying contact elements for a robotic local structure is presented with regard to the introduced handling system.

2 Range of Objects Handled

This chapter presents the range of different objects being grasped and manipulated. The aforementioned wrist joint using an electromagnet for integrating the object into the kinematic structure exclusively can handle ferromagnetic objects. In general, the object's properties and the kind of motion are decision making instruments for the selection of the respective gripper. This is the major challenge within the field of robotics where a certain object must be handled. These properties can be separated generally in the both categories outer object properties and inner object properties (Table 1). For both categories certain groups of contact elements can be preferred e.g. an object with grasping lashes can be grasped preferentially using mechanical grippers like a dual jaw gripper.

Depending on the inner properties like porosity it could be grasped with a vacuum gripper, too. Thus, is becomes a sufficient condition but not an essential one. This table basically introduces the wide range of different objects to be manipulated. For each object the best gripper can be tailor made and integrated in the PARAGRIP which is not practicable due to high costs and production effort. Thus, it is more advantageous to focus on a limited set of universal contact principles basically presented in [5]. There different end-effectors basing on mechanical

Table 1 Categories of material properties (according to [5])

Outer object properties	Inner object properties
Object's size (small/large)	Mass
Object's shape (Cube, Sphere, thin sheet etc.)	Material
Grasping lashes (existing/not existing)	Stiffness (stiff/flexible)
Number of grasping lashes (1, 2, 3, etc.)	Magnetizability (yes/no)
Form of the gripping surface (rectangular, spherical, etc.)	Porosity (impervious/porous)
Shape of the gripping surface (planar, curved, etc.)	Surface roughness (small/high)
Position of the gripping surface (inside/outside)	Temperature
Distance of gripping surfaces	

principles (finger, claw and jaw grippers), magnet principles, vacuum principles adhesive principles, hooking principles and clamping principles.

In the following chapter a modular system is presented consisting of the different end-effectors combined with the introduced robotic local structure.

3 Design of the Robotic Local Structure

For the PARAGRIP a modular system is to be realized within the Cluster of Excellence to manipulate a wide range of different objects. For the previously mentioned kinds of contact principles, the following end-effectors are selected (Table 2):

That ensures manipulating a wide range of different objects due to covering the most important contact principles. Furthermore, a variant for realizing the robotic local structure basing on compliant joints is presented where the already existing structure is entirely replaced through a compliant one. Thus a manipulation of objects in different fields of application is enabled.

Each gripper is mounted at the end of the robotic arm. In case of the PARAGRIP handling system, each gripper is positioned (robotic regional structure) and oriented (robotic local structure) with three degrees of freedom each. The energy supply for the end-effector must be guided through the entire robotic structure towards the end-effector. Since the complete robotic local structure consisting of serial spherical structure and end-effector is exchangeable, all supply lines (electricity, compressed air/vacuum and water) must be available at the interface between regional and local

Table 2 Determined end-effectors

Contact principles		End-effectors
Magnet principle	➔	Electromagnet
Vacuum principle	➔	Suction pads
Mechanical principle	➔	Planar compliant gripper
Adhesive principle	➔	Freezing gripper
Hooking principle	➔	Needle gripper
Clamping principle	➔	Dual jaw gripper

structure to supply the connected local structure with the required energy. Thus, each local structure must be able to transport the energy directly to the end-effector requiring this energy and parallel to that enable performing the motion. This task must be solved for every wrist joint each due to always changing boundary conditions and requirements and is presented in the following subsection beside the introduction of the current gripper.

3.1 Modular System

The following Figures (Figs. 3, 4, 5, 6, 7, 8) present six different robotic local structures which can be connected to the robotic regional structure. The mechanism

Fig. 3 Magnet gripper

Fig. 4 Vacuum gripper

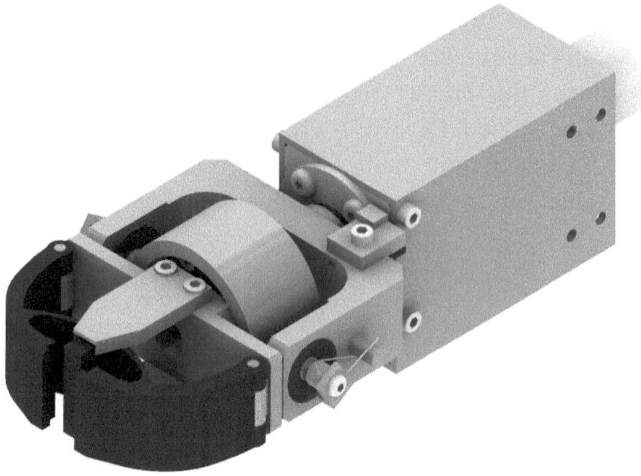

Fig. 5 Planar compliant mechanism

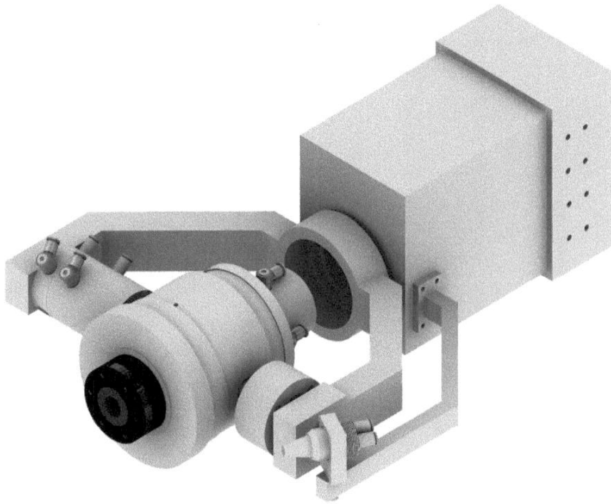

Fig. 6 Freezing gripper

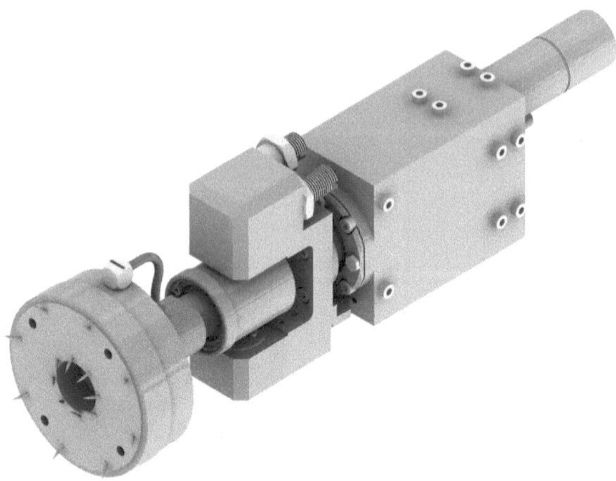

Fig. 7 Needle gripper

Fig. 8 Dual jaw gripper

bases on the serial spherical structure shown in Fig. 2a forming an XYX-arrangement (Joint R points in x-direction, joint S points in y-direction, joint T points in x-direction).

Figure 3 shows an electromagnet selected as end-effector. Using this kind of gripper, exclusively magnetisable objects with a plane face, where the connecting face of the magnet can connect to, can be grasped. The electric power must be guided through the entire mechanism towards the magnet using collector ring contacts in every revolute joint. Figure 4 introduces a vacuum gripper using a round, flat suction pad. In this case the power supply is pneumatic (vacuum). Thus, an air pipe must be integrated into the mechanism which is more difficult than electricity. Using straight rotary distributors which are integrated into every revolute joint the air line is guided towards the suction pad. The detailed development of this gripper can be taken from [6].

Figure 5 shows the mechanical gripping contact principle realized with a planar compliant gripper. This gripper can dock on to grasping lashes. The major difference is the passive contact element. Using the robotic arm, the gripper is pressed into lash whereby both flexible jaws are passive locking mechanisms and grasp the object that way. Figure 6 shows the most complex of all presented local structures. The contact to the object is realized by freezing together both contact partners. The adhesive medium enabling the freezing is water. The freezing element requires electricity for freezing and compressed air for cooling the element. Due to that both media must be provided at the end-effector. The rotary distributors used in this case are much bigger and more complex than in any other introduced structure. This makes the freezing local structure to the largest one.

Figure 7 shows the needle gripper which bases on a hooking principle. The robotic local structure completely is the same like the one of the vacuum gripper. Since the air line and the rotary distributor can be used for compressed air and for vacuum just the contact element and an adaption module must be exchanged. The needle gripper pushes its needles into the object which is e.g. a textile. Figure 8 completes the modular system presenting the dual jaw gripper as a pneumatic clamping principle. The dual jaws are actively driven and can dock on to grasping lashes. The systematic development of this wrist joint is explained in [7].

Since all wrist joints are passive ones, especially the S-axis must be restored by a spring. Otherwise it would bend down and thus not be in a defined position for adapting to the object. Each s-axis of every wrist joint is equipped with restoring springs.

Starting from this modular system the structures can be enhanced through a slight change regarding the joints used. Since the robotic local structure must orientate the end-effector, exclusively orientating joints can be used. Recently the flexible hinges become more important. Thus, it is possible to exchange the structure with revolute joints used as before through a completely compliant robotic local structure with large twist angles. This structure can be taken from Fig. 9. Its detailed development is explained in [8]. The single joints are dimensioned to realize large twist angles which make the structure very large itself compared to the conventional one. The cruciform hinges placed on the both first axes perform a twist angle of 40° and the third one basing on leaf springs performs 50°.

Using this kind of structure the field of activity can be increased where motion tasks from e.g. food industry can be realized. Usually robots in these fields of robotics must satisfy very high requirements according to cleaning and maintenance.

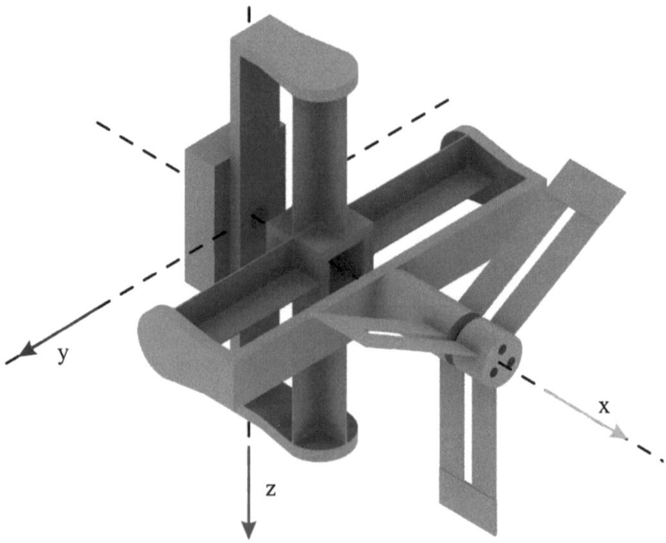

Fig. 9 3 DOF compliant mechanism

4 Conclusions

This paper introduces the development of a modular system for a parallel handling system. Several passively and actively driven end-effectors can be connected to the robotic regional structure and enable handling of a wide range of different objects. The selection of end-effectors bases on a systematic analysis of object's properties where subsequently the contact principles can be derived.

Since all robotic local structures should be integrated into the self-optimizing handling system which is to realize within the Cluster of Excellence at RWTH Aachen University the next step is to develop a changing station. This station can store all wrist joints and provide them in case of use.

Acknowledgments As part of this work was carried out at the Cluster of Excellence at RWTH Aachen "Integrative production technology for high-wage countries (EXC 128)"; the authors would like to thank The German Research Foundation for their support.

References

1. Müller, R., Riedel, M., Vette, M., Corves, B., Esser, M., Hüsing, M.: Reconfigurable self-optimising handling system. In: Ratchev, S. (ed.) IPAS 2010, IFIP AICT, vol. 315, pp. 255–262. Springer, Heidelberg (2010) ISBN 978-3-642-11597-4
2. Riedel, M., Nefzi, M., Huesing, M., Corves, B.: An adjustable gripper as a reconfigurable robot with a parallel structure. In: Proceedings of the Second International Workshop on Fundamental Issues and Future Research Directions for Parallel Mechanisms and Manipulators, pp. 253–260 (2008)
3. Riedel, M., Mannheim, T., Corves, B.: More flexibility and versatility in automated handling processes with an alterable parallel manipulator. In: ASME 2011 International Design Engineering Technical Conferences and Computers and Information in Engineering Conference (DETC 2011), Washington, DC, 28–31 Aug 2011
4. Riedel, M., Nefzi, M., Corves, B.: Performance analysis and dimensional synthesis of a six DOF reconfigurable parallel manipulator. In: IFToMM Symposium on Mechanism Design for Robotics 2012, Mexico City, Mexico, 28–30 Sept 2010
5. Müller, J.: Entwicklung eines Expertensystems zur Auswahl von Greifern. Dissertation, RWTH Aachen University, Aachen (1997)
6. Kurtenbach, S., Kochniss, M., Cousin, A., Corves, B.: Development of a pneumatic End-effector for a 3-DOF robotic local structure. In: Proceedings of the 2nd Conference on Mechanisms, Transmissions and Applications—MeTrApp, Bilbao, Spain, 2–4 Oct 2013
7. Kurtenbach, S., Kochniss, M., Cousin, A., Corves, B.: Structural and dimensional synthesis of pneumatical-mechanical end effectors for a 3-DOF robotic local structure. In: Proceedings of Mechatronik 2013, Aachen, Germany, 7–8 March 2013
8. Kurtenbach, S., Siebrecht, J., Schoenen, D., Corves, B.: Development of a 3-DOF compliant robotic local structure with large twist angle. In: Proceedings of MAMM 2014, Timisoara, Romania, 2–4 Oct 2014

Study on Nonlinear Dynamics of RV Transmission System Used in Robot Joints

Lijun Shan and Weidong He

Abstract The nonlinear dynamic characteristics of RV transmission device was investigated according to dynamics theory, mass centralized method was used to establish the nonlinear dynamical model of RV transmission device, which was composed of inertial component, elastic element and damping element. The influence of the nonlinear factors as meshing stiffness varying with time (though the variation in the number of meshing teeth), backlash of gear pairs and errors were considered. Movement differential equations of system were derived, and these equations were changed to non-dimension, which were solved by Runge-Kutta method. Influences of excitation frequency and damping on system dynamic characteristics were analyzed. As the excitation frequency increased, the system would vibrate from single periodic motion to harmonic motion, chaotic motion. Along with the reduced damping, chaos vibration occurs gradually after a doubling periodic of bifurcation, as the system gradually gets into the chaotic state. The impact is also changed from coexistence of non-impact and the unilateral impact to coexistence of the non-impact, unilateral impact and bilateral impact.

Keywords RV planetary transmission · Dynamics · Nonlinear differential equation

1 Introduction

RV transmission device is a new-type transmission device which is based on the planetary gear drive with few tooth difference, it has a wide range of transmission ratio, single stage transmission ratio range of 31–171, transfer torque per volume is

L. Shan (✉) · W. He
School of Mechanical Engineering, Dalian Jiaotong University, Dalian 116028, China
e-mail: slj@djtu.edu.cn

W. He
e-mail: hwd5870@163.com

S. Bai and M. Ceccarelli (eds.), *Recent Advances in Mechanism Design for Robotics*, Mechanisms and Machine Science 33,
DOI 10.1007/978-3-319-18126-4_30

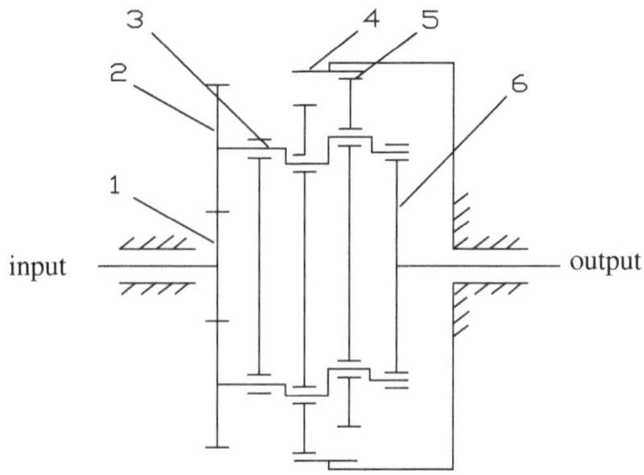

Fig. 1 Constriction of RV transmission drive. *1* Sun gear. *2* Planet gear. *3* Crank shaft. *4* Pin gear. *5* Cycloidal gear. *6* Planet carrier

large, cycloid is the multi-tooth meshing of hard tooth surface (up to 5–6 meshing teeth), which makes per volume can transfer large torque. In rated torque, elastic retrace tolerance is less than 6′, the transmission efficiency is high, usually 0.85–0.92. As RV transmission has above characteristics, it had been used in the transmission of industrial robot joints. RV reducer's transmission construction is shown as Fig. 1.

A lot of research has conducted on RV transmission up to now. The study of stress analysis and tooth profile optimization has reached a high theoretical level [1–3], however the study of its dynamic characteristics is less, and the nonlinear dynamic characteristics of the system often are the main factors that influence the performance of the whole machine. In the RV transmission system, contains many nonlinear factors such as backlash, bearing clearance, transmission error and time-varying meshing stiffness, these nonlinear factors are main excitations which cause the abnormal vibration of system. Therefore, it is necessary to establish a nonlinear dynamic model which considering time-varying meshing stiffness, backlash and transmission error and analysis the dynamic characteristics of the system [1–3].

2 The Establishment of RV Transmission System Dynamics Model

According to the RV transmission principle in Ref. [3], a nonlinear dynamic model of RV transmission system was established with lumped-mass method [4–6], as shown in Fig. 2.

3 Establishment and Solution of System Dynamics Differential Equation

Motion differential equation of the system was established according to system dynamics model, as shown in Fig. 2. Movement differential equations of system were derived which can be seen in reference [7]. These Movement differential equations were changed to second-order nonlinear differential equations with 15° of freedom by non-dimensional treatment as shown in formula (1).

$$
\begin{aligned}
&\ddot{\overline{X}}_{IS} - \overline{C}_{IS}\dot{\overline{X}}_{IS} + \overline{k}_{IS}\overline{X}_{IS} - \frac{M_{IS}}{M_s}\sum_{i=1}^{3}\overline{C}_{spi}\dot{\overline{X}}_{spi} - \frac{M_{IS}}{M_S}\overline{k}_{spi}\,f\left(\overline{X}_{spi},\overline{b_{spi}}\right) = -\frac{M_{IS}}{M_I}\overline{F}_I \\
&\ddot{\overline{X}}_{spi} - \frac{M_{spi}}{M_s}\overline{C}_{IS}\dot{\overline{X}}_{IS} - \frac{M_{spi}}{M_s}\overline{k}_{IS}\overline{X}_{IS} + \frac{M_{spi}}{M_s}\sum_{i=1}^{3}\overline{C}_{spi}\dot{\overline{X}}_{spi} + \frac{M_{spi}}{M_s}\sum_{i=1}^{3}\overline{k}_{spi}\,f\left(\overrightarrow{X}_{spi,}\overline{b}_{spi}\right) \\
&+\frac{M_{spi}}{M_{hpi}}\overline{C}_{spi}\dot{\overline{X}}_{spi} + \frac{M_{spi}}{M_{hpi}}\overline{k}_{spi}\,f\left(\overline{X}_{spi},\overline{b_{spi}}\right) - \frac{M_{spi}}{M_{hpi}}\sum_{j=1}^{2}\overline{C}_{hpiqj}\dot{\overline{X}}_{hpiqj} \\
&-\frac{M_{spi}}{M_{hpi}}\sum_{j=1}^{2}\overline{k}_{hpiqj}\overline{X}_{hpiqj} - \frac{M_{spi}}{M_{hpi}}\overline{C}_{hpio}\dot{\overline{X}}_{hpio} - \frac{M_{spi}}{M_{hpi}}\overline{k}_{hpio}\overline{X}_{hpio} = -M_{spi}\ddot{\overline{e}}_{spi} \\
&\ddot{\overline{X}}_{hpiqj} - \frac{M_{hpiqj}}{M_{hpi}}\overline{C}_{spi}\dot{\overline{X}}_{spi} - \frac{M_{hpiqj}}{M_{hpi}}\overline{k}_{spi}f\left(\overline{X}_{spi},\overline{b}_{spi}\right) + \frac{M_{spi}}{M_{hpi}}\sum_{j=1}^{2}\overline{C}_{hpiqj}\dot{\overline{X}}_{hpiqj} \\
&+\frac{M_{spi}}{M_{hpi}}\sum_{j=1}^{2}\overline{k}_{hpiqj}\overline{X}_{hpiqj} + \frac{M_{hpiqj}}{M_{hpi}}\overline{C}_{pio}\dot{\overline{X}}_{pio} + \frac{M_{hpiqj}}{M_{hpi}}\overline{k}_{pio}\overline{X}_{pio} \\
&+\frac{M_{hpiqj}}{M_{qj}}\sum_{i=1}^{3}\overline{C}_{hpiqj}\dot{\overline{X}}_{hpiqj} + \frac{M_{hpiqj}}{M_{qj}}\sum_{i=1}^{3}\overline{k}_{hpiqj}\overline{X}_{hpiqj} - \frac{M_{hpiqj}}{M_{qj}}\overline{C}_{qj1}\dot{\overline{X}}_{qj1} \\
&-\frac{M_{hpiqj}}{M_{qj}}\dot{k}_{qj1}f\left(\overline{X}_{qj1},\dot{b}_{qj1}\right) = 0 \\
&\ddot{\overline{X}}_{qj1} - \sum_{i=1}^{3}\overline{C}_{hpiqj}\dot{\overline{X}}_{hpiqj} - \sum_{i=1}^{3}\overline{k}_{hpiqj}\overline{X}_{hpiqj} + \overline{C}_{qj1}\dot{\overline{X}}_{qj1} + \overline{k}_{qj1}f\left(\overline{X}_{qj1},\overline{b_{qj1}}\right) = -M_{qj1}\ddot{\overline{e}}_{qj1} \\
&\ddot{\overline{X}}_{hpio} - \frac{M_{hpio}}{M_{hpi}}\overline{C}_{spi}\dot{\overline{X}}_{spi} - \frac{M_{hpio}}{M_{hpi}}\overline{k}_{spi}f\left(\overline{X}_{spi},\overline{b}_{spi}\right) + \frac{M_{hpio}}{M_{hpi}}\sum_{j=1}^{2}\overline{C}_{hpiqj}\dot{\overline{X}}_{hpiqj} \\
&+\frac{M_{hpio}}{M_{hpi}}\sum_{j=1}^{2}\overline{k}_{hpiqj}\overline{X}_{hpiqj} + \overline{C}_{hpio}\dot{\overline{X}}_{hpio} + \overline{k}_{hpio}\overline{X}_{hpio} = \frac{M_{hpio}}{M_o}\overline{F}_o
\end{aligned}
\tag{1}
$$

$i = 1, 2, 3;\ j = 1, 2$

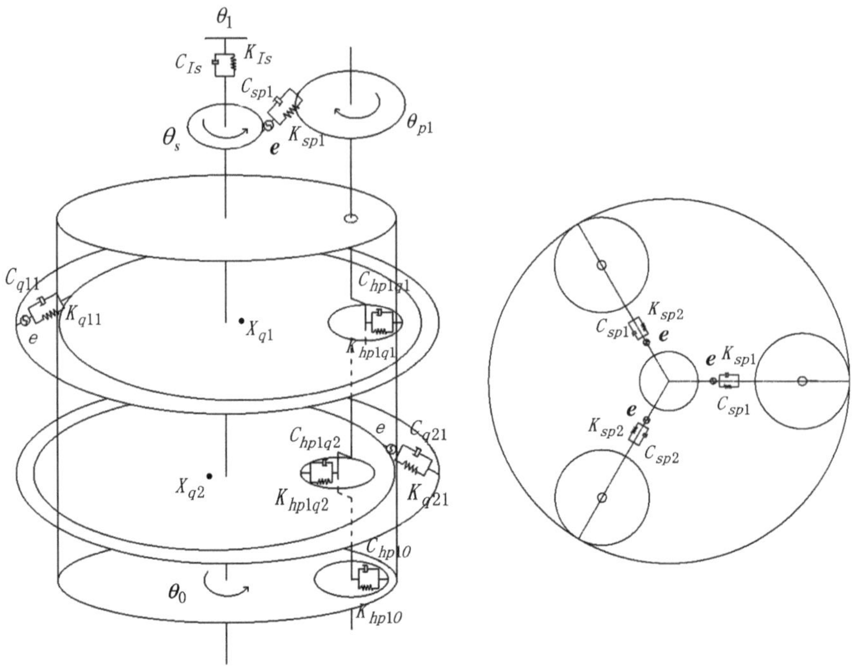

Fig. 2 The mechanical model of RV transmission

where: $M_{qji} = M_{qj}$

$$M_{Is} = \frac{M_I M_s}{M_I + M_s},\ M_{spi} = \frac{M_s M_{hpi}}{M_s + M_{hpi}},\ M_{hpiqj} = \frac{M_{hpi} M_{qj}}{M_{hpi} + M_{qj}},\ M_{hpiO} = \frac{M_{hpi} M_O}{M_{hpi} + M_O}$$

All equations above is represented as equation set (1), written in matrix form:

$$M\ddot{q} + c\dot{q} + kf(q) = F \tag{2}$$

4 The Steady-State Response of the System

4.1 Influences of Excitation Frequency on the Dynamic Characteristics of the Nonlinear System

Taking cycloid gear and pin gear mesh as an example,the influence of the system response by changing the frequency was analyzed. Select parameters: non-dimensional the average excitation $Fm = 0.1$, non-dimensional alternating excitation

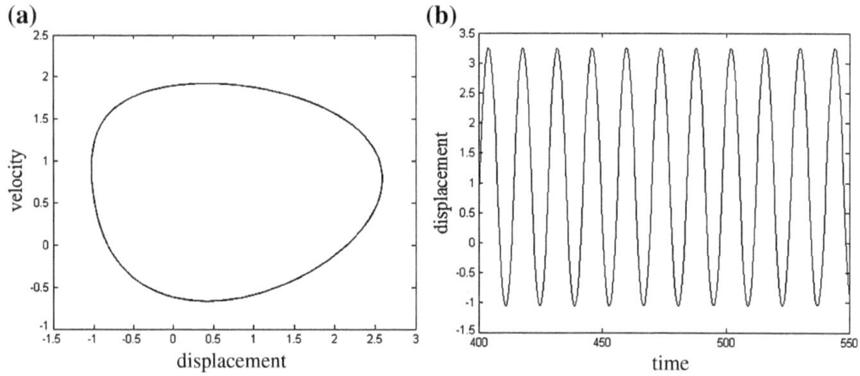

Fig. 3 Dynamic response of the system when $\Omega = 0.35$. **a** Phase diagram. **b** Time history

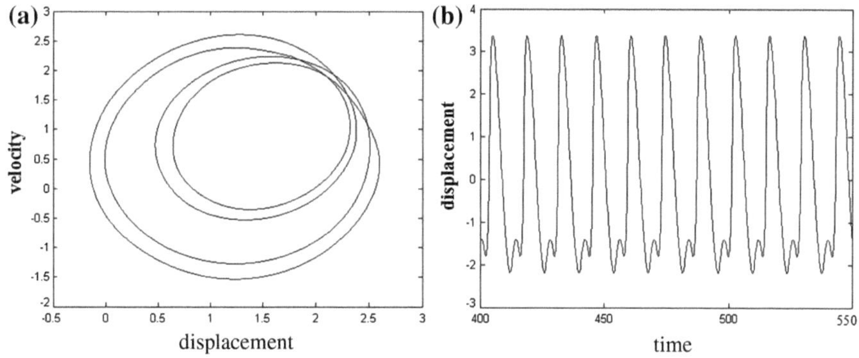

Fig. 4 Dynamic response of the system when $\Omega = 0.73$. **a** Phase diagram. **b** Time history

amplitude $Fa = 0.2$, non-dimensional gear backlash $b = 1$, non-dimensional damping $\xi = 0.03$, non-dimensional meshing stiffness $K = 0.1$. The non-dimensional frequency is selected respectively of 0.35, 0.73, 0.86. The adaptive Runge-Kutta method in Matlab software is used to calculate Eq. (2).

By analyzing Figs. 3, 4 and 5, it can be found that in the case of other parameters constant, when changing the excitation frequency, the system will occur from the non-harmonic single periodic motion which under the normal meshing condition to single periodic motion, harmonic motion, chaotic motion which under the condition out of normal meshing. The impact state of system is changed from periodic motion without impact into chaotic motion with bilateral impact.

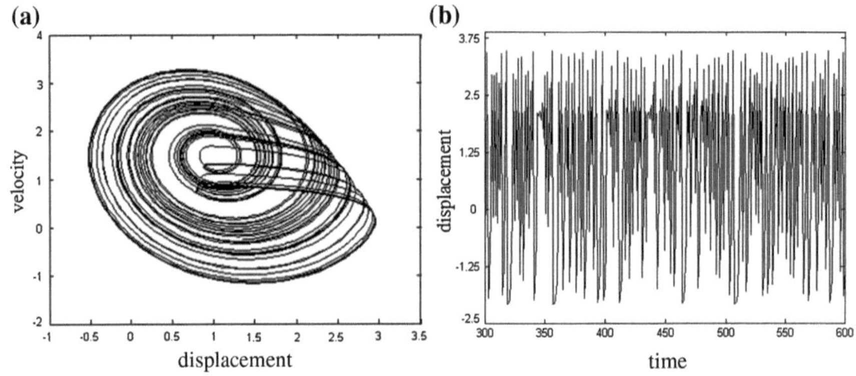

Fig. 5 Dynamic response of the system when $\Omega = 0.86$. **a** Phase diagram. **b** Time history

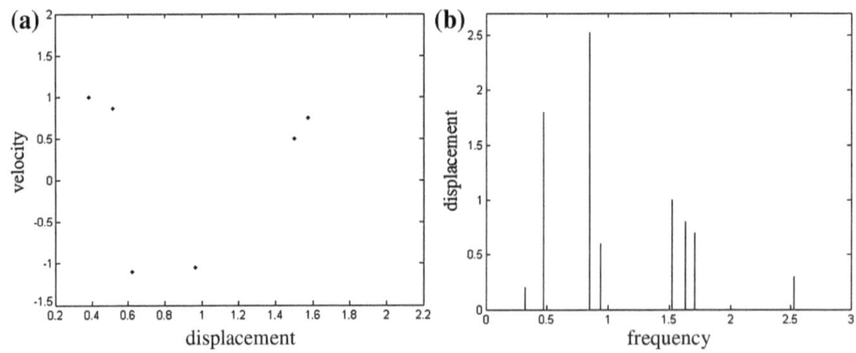

Fig. 6 Dynamic response of the system when $\xi = 0.12$. **a** Poincare section. **b** FFT spectrum

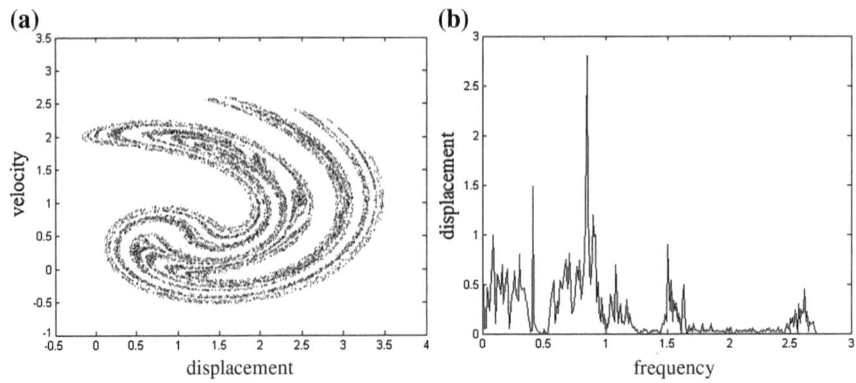

Fig. 7 Dynamic response of the system when $\xi = 0.03$. **a** Poincare section. **b** FFT spectrum

4.2 Infections of Damping on the Dynamic Characteristics of the Nonlinear System

Taking the meshing of sun gear and a planet gear for example to analysis, first determine the basic parameters: the non-dimensional average excitation $Fm = 0.1$, the non-dimensional alternating excitation amplitude $Fa = 0.2$, non-dimensional gear backlash $b = 1$, non-dimensional excitation frequency $\Omega = 0.85$. The non-dimensional damping ξ is selected respectively of 0.16, 0.12 and 0.03. Poincare section and FFT spectrum of different damping were obtained using the adaptive Runge-Kutta method.

By analyzing Figs. 6 and 7, it can be found that system vibrate from periodic of bifurcation into the chaotic state along with the damping reduced. Its impact has also changed gradually from the state of coexistence non-impact and unilateral impact into a state of coexistence non-impact, unilateral impact and bilateral impact.

5 Conclusions

(1) Nonlinear dynamics model of RV transmission system was established by considering infections of time-varying meshing stiffness, error and backlash etc. factors. Movement differential equations of system which derived were changed to second-order nonlinear differential equations by non-dimensional treatment.

(2) The equation was solved by numerical integral method. Influences of excitation frequency and damping on system dynamic characteristics were analyzed. With the variation of the excitation frequency, the motion state of the system has experienced from periodic motion without impact to chaotic motion with bilateral impact and finally to periodic motion with unilateral impact.

References

1. Seiki, T.: RV-AII Reduction Gear for High Precision Control[J], pp. 1–19. Teijin Seiki Co.Ltd, Japan (1996)
2. Li, D., Wang, Y., Fan, Y.: On the behavior of instantaneous center and mechanical characteristics of a RV reducer used in robot. Mech. Sci. Technol. **20**(6), 860–861 (2001)
3. He, W., Li, L., Li, X.:New optimized tooth-profile of cycloidal gear of high precision rv reducer used in robot. Chin. J. Mech. Eng. **36**(3), 51–55 (2000)
4. Zhang, D., Wang, G., Huang, T.: Dynamic formulation of RV reducer and analysis of structural parameters. Chin. J. Mech. Eng. **37**(1), 69–74 (2001)

5. Liu, J., Sun, T., Qi, H.: Research on natural frequency and dynamic model of RV reducer. Chin. J. Mech. Eng. **10**(4), 381–386 (1998)
6. Yan, X., Zhang, Ce., Z., Li, Y.: Natural frequency of torsional vibration of rotary vector reducer and its influencing factors. Mech. Sci. Technol. **23**(8) (2004)
7. Shan, L., Fang, X., He, W.: Analysis of nonlinear dynamic accuracy on RV transmission system. Adv. Mater. Res. (2012)

Compliance Modeling and Error Compensation of a 3-Parallelogram Lightweight Robotic Arm

Guanglei Wu, Sheng Guo and Shaoping Bai

Abstract This paper presents compliance modeling and error compensation for lightweight robotic arms built with parallelogram linkages, i.e., Π joints. The Cartesian stiffness matrix is derived using the virtual joint method. Based on the developed stiffness model, a method to compensate the compliance error is introduced, being illustrated with a 3-parallelogram robot in the application of pick-and-place operation. The results show that this compensation method can effectively improve the operation accuracy.

Keywords Parallelogram robot · Lightweight robotic arm · Error compensation · Stiffness

1 Introduction

Light-weight robotic arms with high payload capacity are desirable in food and pharmaceutical industries as well as assistive robots for the application of pick-and-place operations, such as [1–3]. Their motions are usually generated via the robotic controllers by virtue of the inverse kinematic model to compute the input signals for

G. Wu (✉) · S. Bai
Department of Mechanical and Manufacturing Engineering,
Aalborg University, Aalborg, Denmark
e-mail: gwu@m-tech.aau.dk

S. Bai
e-mail: shb@m-tech.aau.dk

S. Guo
Beijing Jiaotong University, Beijing, China
e-mail: shguo@bjtu.edu.cn

S. Bai and M. Ceccarelli (eds.), *Recent Advances in Mechanism Design for Robotics*, Mechanisms and Machine Science 33,
DOI 10.1007/978-3-319-18126-4_31

actuators corresponding to the desired end-effector position, where the compliance errors are ignored. However, under certain external load, the kinematic control becomes non-applicable due to the changes in the end-effector location caused by the limited strength of the robot components, namely, the actual trajectory will shift away from the desired path, resulting in the decreased product quality where high precision is needed in the applications.

The compliance error can be compensated through the calibration [4], but this technique is sometimes expensive. An economic way to solve the problem of error compensation is the modification of the robot control scheme [5] that defines the prescribed trajectory in Cartesian space: based on the error model, the loaded input trajectory is regenerated to achieve the coincidence of the output trajectory and the desired one, while input trajectory differs from the target one. This paper is focused on the modification of the input trajectory based on the compliance error model. The stiffness matrix for the serial robotics was first derived by Salisbury [6], where only the actuation compliance described by one-dimensional linear springs was considered. In this approach, the derivation of the stiffness is on the basis of the assumption that the manipulator is at an unloaded equilibrium configuration. In practice, the external loads directly influence on the manipulator equilibrium configuration and may modify the stiffness properties [7, 8]. Consequently, the structural compliance and the robot geometry change due to external loads should be considered [9, 10].

This paper deals with the compliance error analysis and compensation of a 3-parallelogram lightweight robotic arm. The Cartesian stiffness matrix is computed for loaded configurations and the effect of the parallelogram link (Π joints) is investigated. A method to modify the input trajectory is presented for error compensation. This method is numerically illustrated with the pick-and-place operation and the results show the effectiveness of the error compensation approach.

2 Kinematics of the Robot Under Study

The lightweight robot, shown in Fig. 1, has three degrees of freedom (dof), with one at the base and two at shoulder. Moreover, with an additional rotational degree of freedom on the end-effector, the robot performs the typical Schönfiles motion. The parallelogram links (Π joints) are used to reduce the inertia in motion. CANopen (Controller Area Network) bus is adopted for the communications between motors and EPOS controllers for real-time control. A USB-CAN interface is used to establish the communications between CANopen bus and PC.

Following the Denavit–Hartenberg (D–H) convention [11], Cartesian coordinate systems are established for each link of the robotic arm as shown in Fig. 2, with D–H parameters listed in Table 1. The transformation matrix in forward kinematics of the end-effector in reference frame is given as

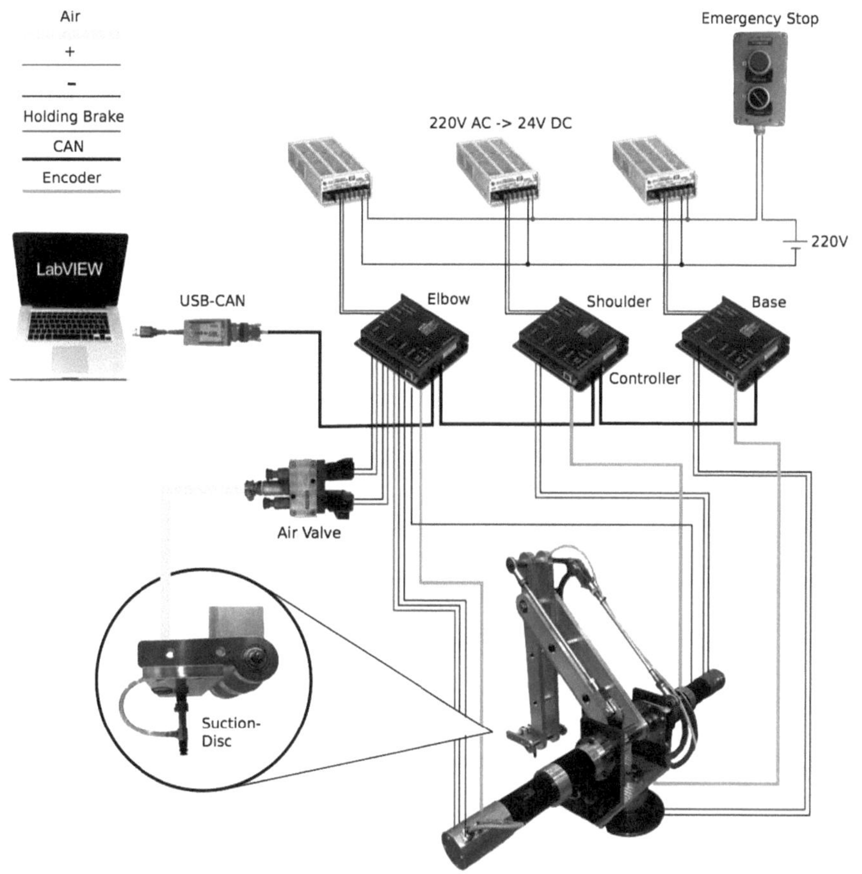

Fig. 1 The prototype of the SCARA robot and controller architecture

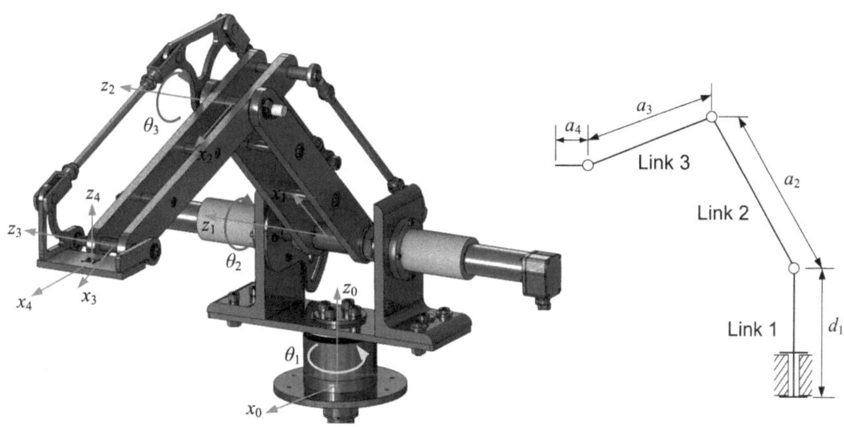

Fig. 2 The coordinates and parameterizations of the parallelogram robot

Table 1 D–H parameters of the robotic arm

Joint i	α_i	a_i (mm)	d_i (mm)	θ_i
1	$\pi/2$	0	250	θ_1
2	0	400	0	θ_2
3	0	400	0	θ_3
4	$-\pi/2$	100	0	$-(\theta_2+\theta_3)$

$$^0\mathbf{A}_4 = \begin{bmatrix} \mathbf{R} & \mathbf{q} \\ \mathbf{0} & 1 \end{bmatrix} = \prod_{i=1}^{4} {}^{i-1}\mathbf{A}_i; \quad {}^{i-1}\mathbf{A}_i = \begin{bmatrix} {}^{i-1}\mathbf{R}_i & \mathbf{q}_i \\ \mathbf{0} & 1 \end{bmatrix} \tag{1}$$

with

$$^{i-1}\mathbf{R}_i = \mathbf{R}(z_{i-1}, \theta_i)\mathbf{R}(x_i, \alpha_i); \quad \mathbf{q}_i = [a_i \cos\alpha_i \; a_i \sin\alpha_i \; d_i]^T \tag{2}$$

Let the end-effector position in the reference frame xyz be $\mathbf{q} = [x\,y\,z]^T$, with the consideration of the joint motions, the inverse geometric problem is solved below

$$\theta_1 = \tan^{-1}\frac{y}{x} \tag{3a}$$

$$\theta_3 = \cos^{-1}\frac{a_2^2 + a_3^2 - \|\mathbf{q} - \mathbf{a}_4 - \mathbf{d}_1\|^2}{2a_2a_3} - \pi; \quad \mathbf{a}_4 = a_4[\cos\theta_1 \; \sin\theta_1 \; 0]^T, \mathbf{d}_1 = d_1\mathbf{k} \tag{3b}$$

$$\theta_2 = -2\tan^{-1}\frac{a_3\sin\theta_3 + \sqrt{a_2^2 + a_3^2 + 2a_2a_3\cos\theta_3 - (z-d_1)^2}}{z - d_1 + a_2 + a_3\cos\theta_3} \tag{3c}$$

3 Stiffness Modeling

The compliance error computation pertains to the stiffness modeling [12]. Figure 3 shows the virtual joint method (VJM) model of the robotic arm, where **F** stands for the external loads and the gravity force is ignored in the stiffness modeling.

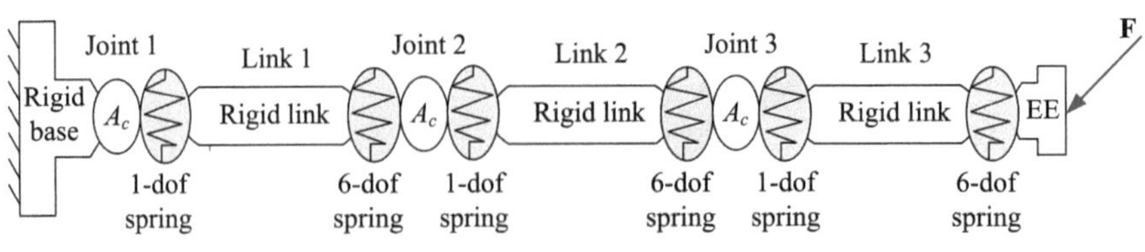

Fig. 3 Virtual-spring model, where A_c stands for the actuator, and EE for end-effector

Let $\boldsymbol{\theta}$, $\boldsymbol{\theta}'$ be the original and the deformed displacements of the virtual springs, respectively, For external force $\mathbf{F}$ and internal forces $\boldsymbol{\tau}_\theta$, their equilibrium equation is derived as below

$$\boldsymbol{\tau}_\theta = \mathbf{J}_\theta^T\mathbf{F}; \quad \boldsymbol{\tau}_\theta = \mathbf{K}_\theta(\boldsymbol{\theta}' - \boldsymbol{\theta}) \tag{4}$$

where $\mathbf{J}_\theta$ is the Jacobian matrix given in Appendix and $\mathbf{K}_\theta$ is the stiffness matrix in the joint space, namely,

$$\mathbf{K}_\theta = \mathrm{diag}\begin{bmatrix} K_{act,1} & \mathbf{K}_{L1} & K_{act,2} & \mathbf{K}_{L2} & K_{\theta 3} & \mathbf{K}_{L3} \end{bmatrix} \tag{5}$$

To compute the stiffness matrix of the loaded mode, let assume that a neighborhood of the loaded configuration, where the external loads, the joint and end-effector locations are supposed to be incremented by some small values $\delta\mathbf{F}$, $\delta\boldsymbol{\theta}$ and $\delta\mathbf{t}$, still satisfies the equilibrium conditions and linearized kinematic constraint:

$$(\mathbf{J}_\theta + \delta\mathbf{J}_\theta)^T(\mathbf{F} + \delta\mathbf{F}) = \mathbf{K}_\theta(\boldsymbol{\theta}' - \boldsymbol{\theta} + \delta\boldsymbol{\theta}) \tag{6a}$$

$$\delta\mathbf{t} = \mathbf{J}_\theta\delta\boldsymbol{\theta} \tag{6b}$$

Expanding Eq. (6a) and ignoring the high-order terms yields

$$\mathbf{J}_\theta^T\delta\mathbf{F} + \mathbf{H}_\theta^T \otimes \mathbf{F}\delta\boldsymbol{\theta} = \mathbf{K}_\theta\delta\boldsymbol{\theta}; \quad \mathbf{H}_\theta = \partial\mathbf{J}_\theta/\partial\boldsymbol{\theta} \tag{7}$$

where the symbol $\otimes$ represents the Kronecker product between matrices. Combining Eqs. (6b) and (7), the kinetostatic model of the robotic arm is reduced to

$$\begin{bmatrix} \mathbf{0} & \mathbf{J}_\theta \\ \mathbf{J}_\theta^T & \mathbf{H}_\theta^T \otimes \mathbf{F} - \mathbf{K}_\theta \end{bmatrix} \begin{bmatrix} \delta\mathbf{F} \\ \delta\boldsymbol{\theta} \end{bmatrix} = \begin{bmatrix} \delta\mathbf{t} \\ \mathbf{0} \end{bmatrix} \tag{8}$$

From $\delta\mathbf{F} = \mathbf{K}\delta\mathbf{t}$, the Cartesian stiffness matrix $\mathbf{K}$ of the robot is calculated as,

$$\mathbf{K} = \left(\mathbf{J}_\theta\left(\mathbf{K}_\theta - \mathbf{H}_\theta^T \otimes \mathbf{F}\right)^{-1}\mathbf{J}_\theta^T\right)^{-1} \tag{9}$$

Based on the above stiffness matrix, the compliance error can be computed from

$$\begin{bmatrix} \Delta\boldsymbol{\varphi}^T & \Delta\mathbf{p}^T \end{bmatrix}^T = \mathbf{K}^{-1}\begin{bmatrix} \mathbf{m}^T & \mathbf{f}^T \end{bmatrix}^T \tag{10}$$

3.1 Compound Stiffness of the Active Joint

The robot's third joint is driven by the actuator via the parallelogram link, as shown in Fig. 4a, consequently, its simplified stiffness $K_{\theta 3}$ is computed from

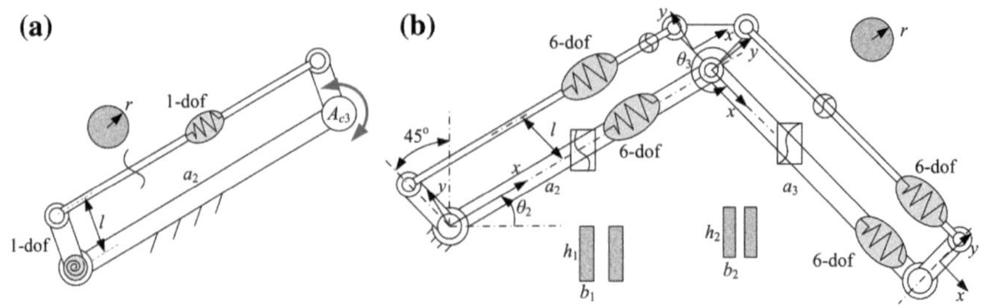

Fig. 4 Stiffness model of the parallelogram: **a** the third active joint; **b** two links

$$K_{\theta 3}^{-1} = K_{act,3}^{-1} + a_2/(EAl^2) \tag{11}$$

where $K_{\text{act},3}$ is the actuator stiffness, and E is the Young's modulus and A is the area of the cross section of the connecting link.

3.2 *Stiffness of the Π Joint*

With the virtual spring approach [10], the stiffness matrix $\mathbf{K}_{L2}$ of the upper arm illustrated in Fig. 4b is derived as

$$\begin{aligned} &\mathbf{K}_{L2} = \mathbf{K}_{a2} + \mathbf{K}_{s2}; \\ &\mathbf{K}'_{a2} = \begin{bmatrix} \mathbf{J}_{s21}\mathbf{K}_{l2}^{-1}\mathbf{J}_{s21}^T & \mathbf{J}_{q21} \\ \mathbf{J}_{q21}^T & 0 \end{bmatrix}^{-1}, \mathbf{K}'_{s2} = \begin{bmatrix} \mathbf{J}_{s22}\mathbf{K}_s^{-1}\mathbf{J}_{s22}^T & \mathbf{J}_{q22} \\ \mathbf{J}_{q22}^T & \mathbf{0}_2 \end{bmatrix}^{-1} \end{aligned} \tag{12}$$

with

$$\mathbf{J}_{s21} = \begin{bmatrix} \mathbf{i} & \mathbf{j} & \mathbf{k} & \mathbf{0}_3 & \mathbf{0}_3 & \mathbf{0}_3 \\ -\mathbf{r}_2 \times \mathbf{i} & -\mathbf{r}_2 \times \mathbf{j} & -\mathbf{r}_2 \times \mathbf{k} & \mathbf{i} & \mathbf{j} & \mathbf{k} \end{bmatrix}; \quad \mathbf{J}_{q21} = \begin{bmatrix} \mathbf{k} \\ -\mathbf{r}_2 \times \mathbf{k} \end{bmatrix} \tag{13}$$

$$\begin{aligned} &\mathbf{J}_{s22} = \begin{bmatrix} \mathbf{i} & \mathbf{j} & \mathbf{k} & \mathbf{0}_3 & \mathbf{0}_3 & \mathbf{0}_3 \\ \mathbf{r}_2 \times \mathbf{i} & \mathbf{r}_2 \times \mathbf{j} & \mathbf{r}_2 \times \mathbf{k} & \mathbf{i} & \mathbf{j} & \mathbf{k} \end{bmatrix}; \\ &\mathbf{J}_{q22} = \begin{bmatrix} \mathbf{k} & \mathbf{k} \\ (\mathbf{r}_2 - a_2\mathbf{i}) \times \mathbf{k} & \mathbf{r}_2 \times \mathbf{k} \end{bmatrix} \end{aligned} \tag{14}$$

and

$$\mathbf{r}_2 = l/2[-\cos(\theta_2 + \pi/4) \quad \sin(\theta_2 + \pi/4) \quad 0]^T \tag{15}$$

where $\mathbf{K}_{a2}$ and $\mathbf{K}_{s2}$ are the first six dimensional entries in $\mathbf{K}'_{a2}$ and $\mathbf{K}'_{s2}$, respectively. Moreover, $\mathbf{K}_{l2}$ is the stiffness matrix of the structural link and $\mathbf{K}_s$ is the stiffness matrix of the connecting link.

By the same token, the stiffness matrix $\mathbf{K}_{L3}$ of the forearm takes the similar formulation of Eq. (12), where $\mathbf{r}_3 = l/2[\cos(\theta_2 + \theta_3 - \pi/4) \quad \sin(\theta_2 + \theta_3 - \pi/4) 0]^T$ is to replace $\mathbf{r}_2$ in the Jacobian matrices.

The stiffness matrices of both the structural and connecting links are computed by a cantilever, i.e., $\mathbf{K}^{-1} = \mathbf{C}[c_{ij}]$, where the non-zero elements in the compliance matrix are:

$$\begin{aligned} &c_{11} = \frac{L}{GI_x};\ c_{22} = \frac{L}{EI_y};\ c_{33} = \frac{L}{EI_z};\ c_{44} = \frac{L}{EA};\ c_{55} = \frac{L^3}{3EI_z};\ c_{66} = \frac{L^3}{3EI_y} \\ &c_{26} = c_{62} = -\frac{L^2}{2EI_y};\ c_{35} = c_{53} = \frac{L^2}{2EI_z} \end{aligned} \tag{16}$$

where L is the link length, A is its cross-section area, I_y, I_z, and I_x are the quadratic and polar moments of inertia of the cross-section, and E and G are the Young's and shear modules, respectively.

4 Effect of the *Π* Joint

The effect of the *Π* joint to the manipulator stiffness can be evaluated by comparing of the stiffnesses for cases with/without the parallelogram link. The link properties are shown in Table 2 and the measured actuator stiffness are: $K_{act,1} = 8.24 \times 10^4$ Nm/rad, $K_{act,2} = K_{act,3} = 7.84 \times 10^4$ Nm/rad. Positioning errors at five workspace points over the workspace demonstrated in Fig. 5 are investigated [13]. With payload 3 kg, the positioning errors are shown in Fig. 6a.

When the robot is without Π joint, the stiffness matrix in the joint space of Eq. (17) is reduced to

$$\mathbf{K}_\theta = \mathrm{diag}[K_{act,1} \quad \mathbf{K}_{l1} \quad K_{act,2} \quad \mathbf{K}_{l2} \quad K_{act,3} \quad \mathbf{K}_{l3}] \tag{17}$$

The corresponding positioning errors are shown in Fig. 6a for comparison. The maximum difference between these two cases can reach up to 0.115 mm, which implies that the *Π* joint can improve the robot stiffness. When the parallelogram connecting link is made of steel instead, the compliance errors represented by the

Table 2 Link dimensions and properties of the parallelogram

l (mm)	r (mm)	b_1 (mm)	h_1 (mm)	b_2 (mm)	h_2 (mm)	E (GPa)	G (GPa)
120	6	15	60	15	50	69	26

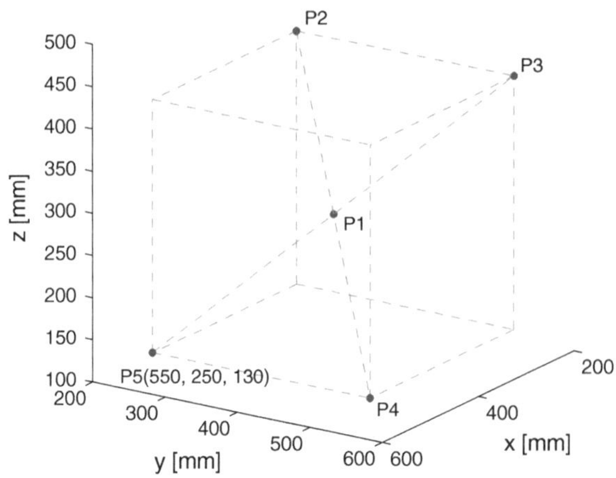

Fig. 5 The five positions within the workspace

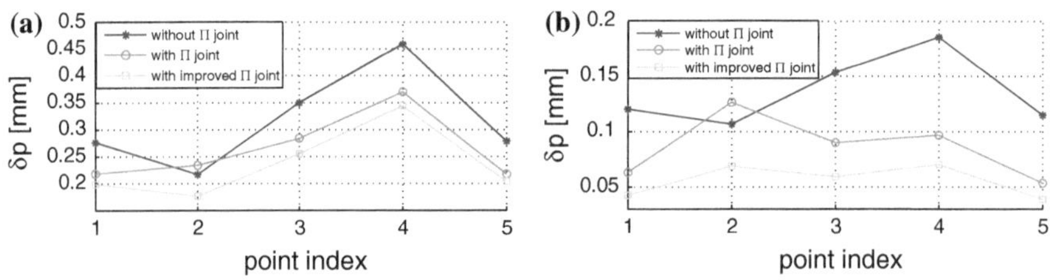

Fig. 6 The positioning error at five positions: **a** with compliant actuator; **b** with joint mechanically locked

green line can decrease 6–20 %, which means that the *Π* joint with higher stiffness leads to higher structural stiffness of the robot.

It is known that the manipulator stiffness mainly depends on the actuator and link stiffness. When the actuator is much stiffer than the link, supposed that $K_{act,i} = 10^6$ Nm/rad, $i = 1, 2, 3$, the position errors at the five positions are shown in Fig. 6b, from which it is seen that the *Π* joint with high rigidity can effectively improve the robot stiffness.

5 Error Compensation

5.1 Compensation Procedure

The motions of the robotic manipulators are usually generated according to the inverse kinematics, from which the input signals of the actuators $\boldsymbol{\theta}$ corresponding to the desired end-effector location $\mathbf{t}$ are computed. However, with the external loads

F applied to the end-effector, the kinematic control becomes non-applicable due to the compliance error $\delta\mathbf{t}$ of the end-effector, thus, the actual position is computed from the stiffness model

$$\mathbf{t}' = \mathbf{t} + \delta\mathbf{t}; \quad \delta\mathbf{t} = \mathbf{K}^{-1}\mathbf{F} \tag{18}$$

In order to make the end-effector be located in the desired position **t**, the compliant displacement from **t** to $\mathbf{t}'$, under the loads **F**, should be compensated. Assuming that the modified end-effector location be $\mathbf{t}_f = \mathbf{t} - \delta\mathbf{t}$, under the same loads **F**, the actual position after compensation $\mathbf{t}_c$ should be in the neighborhood of the desired position, namely,

$$\mathbf{t}_c = \mathbf{t}_f + \delta\mathbf{t}_f \approx \mathbf{t}; \quad \delta\mathbf{t}_f = \mathbf{K}_f^{-1}\mathbf{F} \tag{19}$$

where K_f is the stiffness evaluated at the deflected position. The modified end-effector location $\mathbf{t}_f$ can be calculated from the following iterative procedure,

$$\mathbf{t}_c' = \mathbf{t} + \lambda(\mathbf{t} - \mathbf{t}_c) \tag{20}$$

where the prime term corresponds to the next iteration, $\lambda \in (0, 1)$ is the scalar parameter to achieve the convergence. This iterative method, presented in Fig. 7, will stop until $\|\mathbf{t} - \mathbf{t}_c\| \leq \varepsilon$, where ε is an acceptable tolerance. Using this procedure to modify the reference trajectory used in the robotic control, it is possible to compensate compliance errors to follow accurately the desired trajectory.

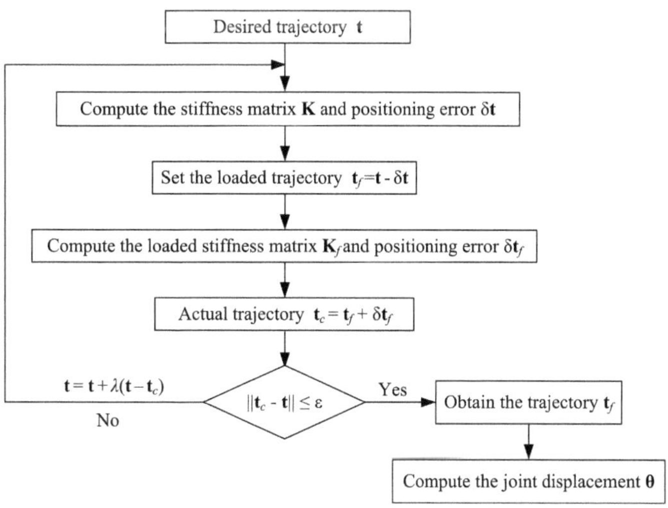

Fig. 7 Error compensation procedure for the robotic arm

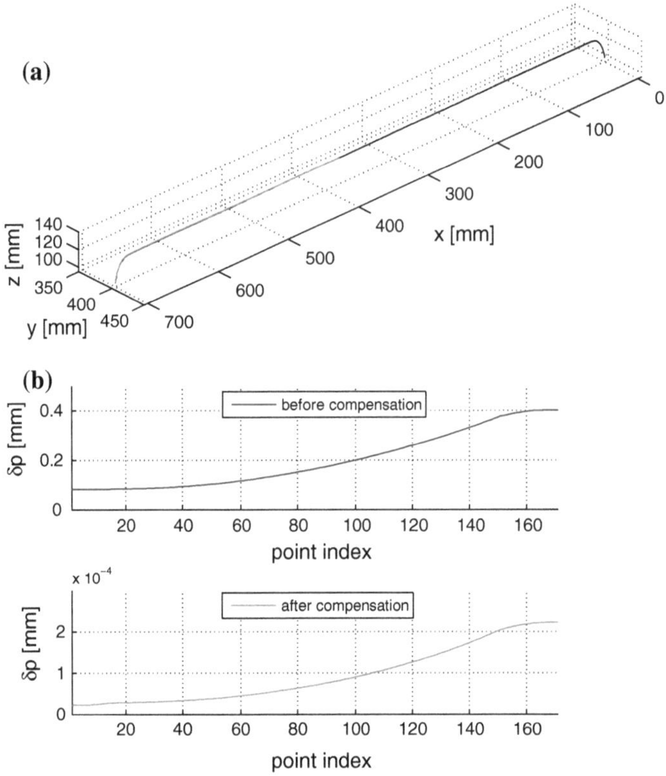

Fig. 8 Comparison of errors before/after compensation: **a** trajectory; **b** positioning error

5.2 Numerical Simulation

The foregoing mentioned error compensation procedure is illustrated with the parallelogram robot along a pick-and-place trajectory of 700 mm × 25 mm as displayed in Fig. 8a. Figure 8b shows the positioning error before and after compensation. It is seen that the maximum positioning error can reduce to $0.3\,\mu$m when the manipulator track the modified trajectory, which can be ignored as it is much smaller than the acceptable error. The comparison reveals that the approach of error compensation can effectively improve the pick-and-place operation accuracy.

6 Conclusions

This paper deals with the compliance error analysis and compensation for a 3-parallelogram robot with high payload ratio. Based on the derived stiffness of the Π joint, the Cartesian stiffness matrix is derived with the consideration of external loads. The Π joints can both reduce the inertia of the robotic arm in motion and improve the manipulator stiffness. An iterative error compensation method is introduced. The proposed approach implies that the manipulator accuracy can be effectively improved when the control strategy is based on the combination of the kinetostatic and kinematic models. The introduced approach is numerically illustrated with a trajectory of pick-and-place operation. The results show that the error compensation procedure can effectively improve the operation precision to make the manipulator track the desired trajectory within acceptable positioning errors.

Acknowledgments The authors would like to thank Palle Huus, Dennis Andersen, Nikolai Svalebæk, Nikolai Hansen, Mathias Kristensen and Mathias Jungersen for prototyping the robot.

Appendix

The force Jacobian matrix is expressed as

$$\mathbf{J}_\theta = \begin{bmatrix} \vec{\hat{\$}}_{\theta,1} & \vec{\hat{\$}}_{\theta,2} & \vec{\hat{\$}}_{21} & \cdots & \vec{\hat{\$}}_{26} & \vec{\hat{\$}}_{\theta,3} & \vec{\hat{\$}}_{31} & \cdots & \vec{\hat{\$}}_{36} \end{bmatrix} \in \vec{R}^{6\times 15} \tag{21}$$

with the unit screws

$$\hat{\$}_{\theta,1} = \begin{bmatrix} \mathbf{k} \\ -\mathbf{q}\times\mathbf{k} \end{bmatrix}; \hat{\$}_{\theta,2} = \begin{bmatrix} \mathbf{z}_1 \\ \mathbf{q}_1\times\mathbf{z}_1 \end{bmatrix}; \hat{\$}_{\theta,3} = \begin{bmatrix} \mathbf{z}_2 \\ \mathbf{q}_2\times\mathbf{z}_2 \end{bmatrix}; \hat{\$}_{n1} = \begin{bmatrix} \mathbf{x}_{n-1} \\ \mathbf{q}_n\times\mathbf{x}_{n-1} \end{bmatrix}$$
$$\hat{\$}_{n2} = \begin{bmatrix} \mathbf{y}_{n-1} \\ \mathbf{q}_n\times\mathbf{y}_{n-1} \end{bmatrix}; \hat{\$}_{n3} = \begin{bmatrix} \mathbf{z}_{n-1} \\ \mathbf{q}_n\times\mathbf{z}_{n-1} \end{bmatrix}; \hat{\$}_{n4} = \begin{bmatrix} \mathbf{0} \\ \mathbf{x}_{n-1} \end{bmatrix}; \hat{\$}_{n5} = \begin{bmatrix} \mathbf{0} \\ \mathbf{y}_{n-1} \end{bmatrix}; \hat{\$}_{n6} = \begin{bmatrix} \mathbf{0} \\ \mathbf{z}_{n-1} \end{bmatrix} \tag{22}$$

and

$$\mathbf{x}_n = \mathbf{R}_n\mathbf{i}; \quad \mathbf{y}_n = \mathbf{R}_n\mathbf{j}; \quad \mathbf{z}_n = \mathbf{R}_n\mathbf{k}; \quad \mathbf{q}_n = \mathbf{p}_n - \mathbf{q} \tag{23}$$

where $\mathbf{R}_n$ and $\mathbf{p}_n$, $n = 1, 2, 3$, respectively, are the rotation matrix and position vector extracted from $\prod_{i=1}^{n} {}^{i-1}\mathbf{A}_i$ of Eq. (1). Moreover, vectors $\mathbf{i}$, $\mathbf{j}$ and $\mathbf{k}$ represent the unit vectors of x-, y- and z-axis, respectively.

References

1. http://www.universal-robots.com/GB/Cases.aspx
2. Ivlev, O., et al.: Rehabilitation robots FRIEND-I and FRIEND-II with the dexterous lightweight manipulator. Technol. Disabil. **17**(2), 111–123 (2005)
3. Bien, Z., et al.: Integration of a rehabilitation robotic system (KARES II) with human-friendly man-machine interaction units. Auto. Robots **16**(2), 165–191 (2004)

4. Zhuang, H.: Self-calibration of parallel mechanisms with a case study on Stewart platforms. IEEE Trans. Robot. Autom. **13**(3), 387–397 (1997)
5. Klimchik, A., et al.: Compliance error compensation technique for parallel robots composed of non-perfect serial chains. Robot. Comput.-Int. Manuf. **29**(2), 385–393 (2013)
6. Salisbury, J.: Active stiffness control of a manipulator in cartesian coordinates. In: 19th IEEE Conference on Decision and Control Including the Symposium on Adaptive Processes, vol. 19 pp. 95–100. New Mexico (1980)
7. Chen, S.-F., Kao, I.: Conservative congruence transformation for joint and Cartesian stiffness matrices of robotic hands and fingers. Inter. J. Robot. Res. **19**, 835–847 (2000)
8. Alici, G., Shirinzadeh, B.: Enhanced stiffness modeling, identification and characterization for robot manipulators. IEEE T. Robot. **21**(4), 554–564 (2005)
9. Quennouelle, C., et al.: Stiffness matrix of compliant parallel mechanisms. In: Lenarčič, J., Wenger, P. (eds.) Advances in Robot Kinematics: Analysis and Design, pp. 331–341. Springer, Netherlands (2008)
10. Pashkevich, A., et al.: Enhanced stiffness modeling of manipulators with passive joints. Mech. Mach. Theory **46**(5), 662–679 (2011)
11. Denavit, J., Hartenberg, R.S.: A kinematic notation for lower-pair mechanisms based on matrices, Trans. ASME. J. Appl. Mech. **22**, 215–221 (1995)
12. Wu, G., et. al.: Mobile platform center shift in spherical parallel manipulators with flexible limbs. Mech. Mach. Theory. **75**, 12–26 (2014)
13. International Standard: Manipulating Industrial Robots-Performance Criteria and Related Test Methods ISO 9283: 1998

A Design Method of Thin-Walled and Four-Point Contact Ball Bearings of Industrial Robots

Xiujuan Zhang, Shuangchun Luo, Ronghua Li and Yi Tian

Abstract Thin-walled and four-point contact ball bearings are key elements for industrial robots, since they play a critical role in the action accuracy, running stability and flexibility, and service life for the main engine of robots. In this paper, a new design method which involves structure optimization, finite element analysis and accurate service life calculation has been developed. FPXU408-2RZ bearing is applied as an example to describe the design method.

Keywords Structure optimization · Improved genetic algorithm based on the family tree · Finite element analysis · Accurate calculation of service life

1 Introduction

Thin-walled and four-point contact ball bearings are key elements for industrial robots, since they play a critical role in the action accuracy, running stability and flexibility, and service life for the main engine of robots. The effective design method of bearings is very important for industrial robots. Scientists and engineers have given their contribution to the performance requirements, applications and structure characteristics of the thin-walled ball bearings [1]. Some design problems, such as the effect of support stiffness on the deformation of roller bearings, were mentioned without the introduction of design theory and method [2]. Chinese researchers have done some qualitative research on the effect of support stiffness on the deformation of roller bearings. But the influence of support stiffness on the inner load distribution has not been considered [3]. The structure optimization of the thin-walled ball bearings has been performed without considering the load-carrying capacity and stiffness performances [4]. Thus, the analysis results are not suitable

X. Zhang (✉) · S. Luo · R. Li · Y. Tian
School of Mechanical Engineering, Dalian Jiaotong University, Dalian, China
e-mail: zhangxiuj@djtu.edu.cn

S. Bai and M. Ceccarelli (eds.), *Recent Advances in Mechanism Design for Robotics*, Mechanisms and Machine Science 33,
DOI 10.1007/978-3-319-18126-4_32

for the design of thin-walled ball bearings. To solve this problem, this paper developed a new design method and FPXU408-2RZ bearing is applied as an example to further describe the new design method.

2 Design Method

The new design method goes through: (1) structure optimization. The optimization model is created after determining the objective function and constraints. The optimization is implemented using the improved Genetic Algorithm based on the family tree; (2) finite element analysis. The finite element analysis of whole shaft system is performed using the commercial application software ABAQUS. The equivalent stresses and displacements of inner and outer rings, separator, and rollers are obtained. The load distribution of bearing is acquired based on the finite element analysis results; and (3) accurate service life calculation. The accurate service life of bearings is then computed according to the theory of Lundberg and Palmgren. In this design method, FPXU408-2RZ bearing of industrial robots as shown in Fig. 1 is adopted as an example to further describe the calculation method. The ball bearing is 114.3 mm in inner diameter (d), 133.35 mm in outer diameter (D) and 12.7 mm in width (B).

2.1 Structure Optimization

2.1.1 Optimization Model

For the thin-walled and four-point contact ball bearings used in the industrial robots, the goal of structure optimization has two aspects. One is to pursue the high radial rated dynamic load capacity and obtain the long contact fatigue life. The other is to improve the high support stiffness and meet the requirements that the robot host should have the high repeat positioning accuracy. Thus, the radial basic

Fig. 1 Structure of bearing support system

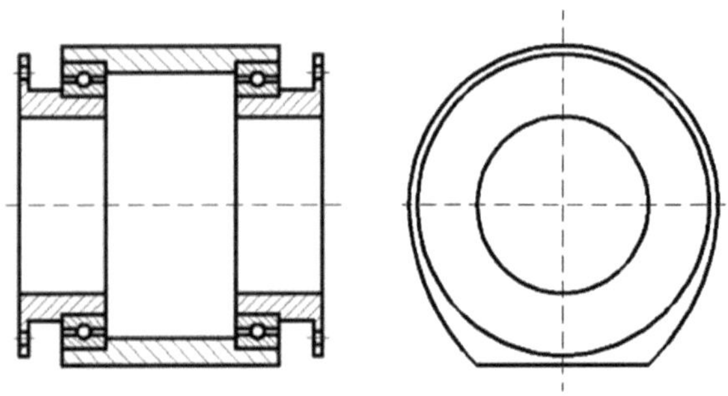

rated dynamic load and the radial stiffness are selected as the objective function for the structure optimization of bearings. According to Chinese standard on rated dynamic load and rated life of rolling bearings, the radial basic rated dynamic load for the balls whose diameter is not larger than 25.4 mm can be calculated as:

$$C_r = b_m f_c (i \cdot \cos\alpha)^{0.7} z^{2/3} D_W^{1.8} \tag{1}$$

where C_r is the radial basic rated dynamic load; b_m is the factor related with the fabrication method and is equals to 1.3; α is the nominal contact angle and is equals to 30°; z is the number of balls for a single column; D_w is the ball diameter; i is the column number of the bearing rollers and is equals to 2; and f_c is the factor relating to the geometric shape, manufacture accuracy, and materials of bearings. It can be computed as:

$$f_c = 39.93\lambda \frac{\gamma^{0.3}(1-\gamma)^{1.39}}{(1+\gamma)^{1/3}} \left(\frac{2f_i}{2f_i - 1}\right)^{0.41} \left\{ 1 + \left[1.04 \left(\frac{1-\gamma}{1+\gamma}\right)^{1.72} \left(\frac{f_i(2f_e - 1)}{f_e(2f_i - 1)}\right)^{0.41} \right]^{10/3} \right\}^{-0.3} \tag{2}$$

where λ is the coefficient of the machining error and is equals to 0.95; γ is the dimensionless geometrical parameter and is defined as $\gamma = D_w \cos\alpha / D_{wp}$; D_{wp} is the diameter of pitch circle of steel balls; f_i is the radius coefficient of raceway groove curvature for the inner ring of bearing and is equals to 0.55; f_e is the radius coefficient of raceway groove curvature for the outer ring of bearing and is equals to 0.55.

The stiffness of rolling bearings is the outer load for inner and outer rings of bearings to generate unit relative elastic displacement in the direction of the load. In the light of the relationship between the load and displacement, the radial stiffness of rolling bearings is:

$$R_r = 0.185 \times 10^4 \left(D_w F_r z^2 cos^5 \alpha\right)^{1/3} \tag{3}$$

where R_r is the radial stiffness of bearings and F_r is the radial outer load of bearings.

For the thin-walled and four-point contact ball bearings, the optimization variables should involve all diameters, number of balls, the lintel width of separator, and the center circle diameter of balls. For bearings, the nominal inner diameter d and the nominal outer diameter D have the relationship as:

$$D_w/(D - d) = K_w \tag{4}$$

where K_w is the factor of constraint condition of ball diameters and has the value of 0.25.

For the thin-walled and four-point contact ball bearings of industrial robots, their installation method is similar to that of deep groove ball bearings. The number of balls is limited by the ball filling angle. But the thickness of ring wall is thin and the range of the ball filling angle could be increased. Thus, the constraints of the number of balls can be written as:

$$\frac{\varphi_{\min}}{2\sin^{-1}D_{\mathrm{w}}/D_{\mathrm{wp}}}+1\leq z\leq\frac{\varphi_{\max}}{2\sin^{-1}D_{w}/D_{wp}}+1 \tag{5}$$

where $\varphi_{\min}$ is the minimum filling angle of ball and is equals to 186°; $\varphi_{\max}$ is the maximum filling angle of balls and is equals to 194°.

In order to increase the load-carrying capacity and stiffness, the number of balls should be selected as large as possible. But more balls would decrease the lintel width of separator. To ensure the sufficient strength of separator, the constraints of the lintel width of separator should be given and is expressed as:

$$\frac{\pi D_{wp}}{z}-1.05D_{w}\geq 1.5 \tag{6}$$

Because the wall thickness of the outer and inner rings for the thin-walled and four-point contact ball bearings is thin, generally, the center circle diameter of balls cannot be too large. The constraint of the center circle diameter of balls is:

$$D_{wp}=0.5(D+d) \tag{7}$$

2.1.2 Optimization Method

In this paper, the improved genetic algorithm based on family tree is selected as the optimization method [5]. Family Tree has a tree structure which can record the kinship between chromosomes and guide the calculation of inbreeding coefficients for them. Figure 2 shows an example of the family tree, in which dash lines represent mutation operations, solid lines represent crossover operations, and the chromosome with a cross means that the chromosome has been eliminated after a selection operation.

With the help of the family tree, an improved Genetic Algorithm [5] can be developed in order to solve the structure optimization of bearing. The developing procedure is introduced as follows: (1) determining encoding method; (2) determining initial population; (3) determining the degree of homology between chromosomes; (4) determining chromosome fitness; (5) determining selection method; (6) determining the method of genetic operations; (7) determining the method of reproduction; and (8) determining stop criterion. In this optimization method, the fitness of every chromosome is measured by the evaluation function:

$$f_i=\frac{1}{m_i}\sum_{j=1}^{m_i}K_j^{(i)}+k_1\sum_{j=1}^{n}\left[V_j\left(C_j+M_j\right)\right]^{-a}+P_i \tag{8}$$

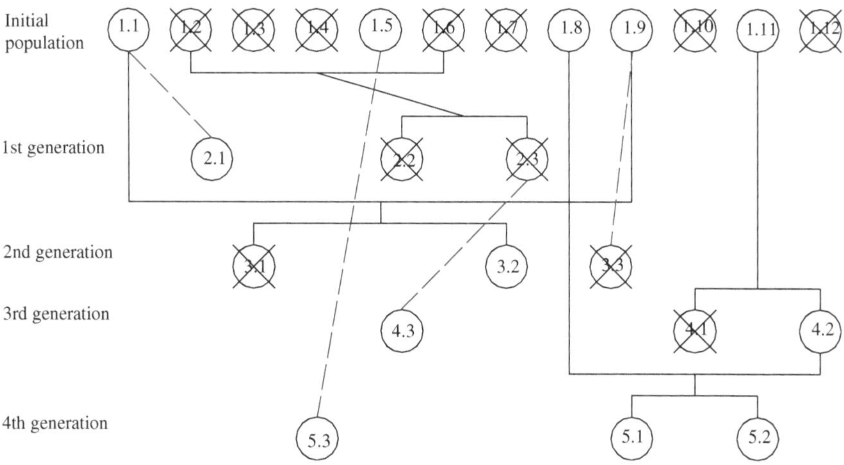

Fig. 2 Family tree of an evolution

In Eq. (8), the first item is for evaluating the material affinity of adjacent regions in the *i*th chromosome, where m_i is the number of boundaries of adjacent regions in the *i*th chromosome and $K_j^{(i)}$ is the material affinity value of the *j*th boundary in the *i*th chromosome. The second item is for evaluating the material cost and manufacturing cost, where C_j is the price per unit volume of material in the *j*th region and can be searched from the material database; V_j is the volume of the *j*th region; M_j is the manufacturing cost per unit volume for the *j*th region and can be searched from the database about manufacturability; and n is the number of material regions in the component. The coefficients k_1 and a are used to adjust the weight of items. The third item is for the penalty of not satisfying manufacturability constraints, where P_i is the penalty value of the *i*th chromosome.

Based on the above-mentioned optimization method, the optimization of FPXU408-2RZ bearing is performed. After the optimization, the optimum design variables are obtained. The ball diameter D_w is 4.762 mm, the number of balls z is 45, the center circle diameter of balls D_{wp} is 123.825 mm, and the radial basic rated dynamic load C_r is 11,633 N.

2.2 Finite Element Analysis

FPXU408-2RZ bearing of industrial robots is thin-walled bearing. Its installation shaft is hollow and the bearing housing is the thin-wall structure. The load distribution is flexible and the effect of the elastic deformation of shaft system on the load distribution should be considered. The typical theory of load distribution for bearings is performed based on the assumption that the outer and inner rings are rigid bodies. The influence of the elastic deformation of shaft system is not considered. Finite element method is developed based on the elastic theory in which the analysis object is considered as the elastic bodies. Thus, the load distribution problem of bearings with the flexible support can be solved well. In this paper, the finite element model of whole shaft system including the hollow shaft, bearings and bearing housing is created. The load distribution of whole shaft system is analyzed. As shown in Fig. 1, two sets of bearings are symmetrically arranged on the base of wrist joint. Two hollow shafts of bearings are connected with the robot arms using bolts, respectively. Due to the structure symmetry of the bearing support system, loads exerted on the two bearings are same when the external radial load is applied on the wrist joint base. Thus, only one bearing is analyzed in this paper. The wall thicknesses of bearing with the hollow shaft and wrist joint base are 4 mm.

Three-dimensional continuum element C3D8I is selected as the element type. Bearing raceway, rollers and hollow shaft are set as the positive surfaces. The inner and outer rings are set as the negative surface. The contact is set as ideal one and the friction coefficient between contact surfaces is 0. Material properties of whole shaft system are uniform. Young modulus of components is 200,000 MPa and Poisson's ratio is 0.3. Bearing mainly bears the radial load with the value of 2000 N. Since the hollow shaft is fixedly connected with the mechanical arm of robots, their motions along three axes are constrained, i.e., $U_x = U_y = U_z = 0$. The bearing housing has two sets of ball bearings which are respectively assembled on two hollow shafts. The motion of bearing housing along Y-axis is constrained, i.e., $U_z = 0$. Because the analysis model is a half of the entire shaft system, the symmetry constraint is imposed on the plane along the axial plane.

For the convenience of analysis, the angle of bottom ball is set to be 0°, the angle of balls from the bottom to the top are symmetrically set to be 0°, ±8°, ±16°, ±24°, ±32°, ±40°, ±48°, ±56°, ±64°, ±72°, ±72°, ±80° and ±88°, respectively.

As shown in Fig. 3, the displacement of lower part of outer ring is larger than that of upper part, which illustrates that the flexural elastic deformation of rings is generated. From the Fig. 4, the maximum equivalent stress for the whole shaft system is 553.0 MPa and generated on the roller with the position angle of 0°. The maximum equivalent stress for the outer ring is 553 MPa and generated on the roller with the position angle of 24°. The maximum equivalent stress for the inner ring is 335.3 MPa and generated on the roller with the position angle of 24°. Using the post processor of software, the contact loads between rollers and rings are obtained. The loads exerted on the rollers at different positions are calculated and given in Table 1. The load distribution between rollers and the raceways of inner and outer rings is changed due to the elastic deformation of shaft system. The imbalance of contact load is produced. Analysis results show that the load falls into the range from −64° to 64° and the number of loaded balls is 17. According to the hypothesis that the outer and inner rings are rigid bodies, the number of loaded balls is 23. It is obvious that the loaded region is narrowed due to the elastic deformation of shaft system.

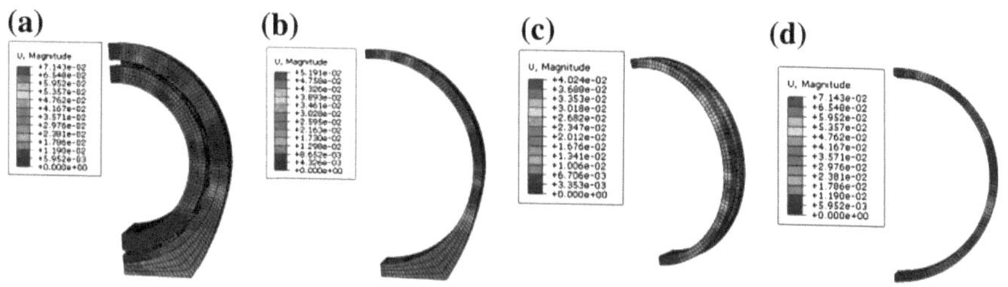

Fig. 3 Displacements of bearing, **a** entire model, **b** bearing separator, **c** inner ring of bearing, **d** outer ring of bearing

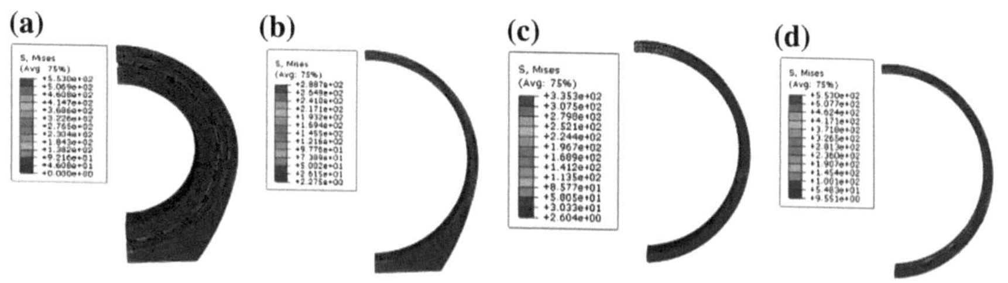

Fig. 4 Equivalent stresses of bearing, **a** entire model **b** bearing separator **c** inner ring of bearing **d** outer ring of bearing

2.3 Accurate Calculation of Service Life of Bearings

According to the accurate calculation method of bearings developed by Lundberg and Palmgren [6], the service life of bearings should be computed according to the actual working condition. The rated load and equivalent load of rollers of inner and outer rings should be computed at first. The rated life of single ring is calculated according to the relationship between the rated service life of rings and roller load. At last, the service life of whole bearing is the uniformly distributed roller load when the rated service life of rings is one million revolutions that 90 % of a group of bearings will compete or exceed before the failure criterion develops. For the point contact, the rated roller loads of outer and inner rings can be calculated as:

$$Q_{ci} = A\left(\frac{2R}{D_w}\frac{r_i}{r_i - R}\right)^{0.41}\frac{(1-\gamma)^{1.39}}{(1+\gamma)^{1/3}}\left(\frac{D_w}{d_m}\right)^{0.3} D_w^{1.8} Z^{-1/3} \tag{9}$$

$$Q_{ce} = A\left(\frac{2R}{D_w}\frac{r_c}{r_c - R}\right)^{0.41}\frac{(1+\gamma)^{1.39}}{(1-\gamma)^{1/3}}\left(\frac{D_w}{d_m}\right)^{0.3} D_w^{1.8} Z^{-1/3} \tag{10}$$

where Q_{ci} is the rated roller loads of inner ring; Q_{ce} is the rated roller loads of outer ring; A is the empirical parameter and is equals to 98.1; R is the radius curvature of the roller generatrix.

Table 1 Load distribution of inner right raceway

Force (N)	0°	±8°	±16°	±24°	±32°	±40°	±48°	±56°	±64°	±72°	±80°	±88°
Inner ring	129.6	147.8	172.1	181.2	165.9	119.3	60.3	35.3	14.6	5	0	0
Outer ring	281	244.8	211.8	198.2	167.3	129.6	75.3	40.2	16.4	9	0	0

After the calculation, Q_{ci} is 433 N and Q_{ce} is 486 N. The equivalent roller loads of rings are the uniformly distributed roller load on the hypothesis that the service life of rings is same as that of rings operating under the real load. It can be expressed as:

$$Q_{ei} = (\frac{1}{z}\sum_{j=1}^{z} Q_j^s)^{1/s} \tag{11}$$

$$Q_{ee} = (\frac{1}{z}\sum_{j=1}^{z} Q_j^{s\omega})^{1/\omega} \tag{12}$$

where Q_{ei} and Q_{ee} are the equivalent roller loads of inner and outer rings, respectively; s and ω are the load factors and can be expressed as s = 3 and ω = 10/3.

According to Eqs. (11) and (12), Q_{ei} is 80 N and Q_{ee} is 111 N. The rated service life of a single ring is the service life of the ring whose usage probability is 90 %. It turns out to be:

$$L_{10i} = (Q_{ci}/Q_{ei})^{\varepsilon} \tag{13}$$

$$L_{10e} = (Q_{ce}/Q_{ee})^{\varepsilon} \tag{14}$$

where L_{10i} and L_{10e} are the rated service life of inner and outer rings, respectively; ε is the service life index and is equals to 3.

After the calculation, L_{10i} is 159 and L_{10e} is 84. Thus, the service life of bearing can be calculated as:

$$L_{10} = (L_{10i}^{-e} + L_{10e}^{-e})^{-1/e} \tag{15}$$

where e is the Weibull distribution slope and is equals to 10/9.

The accurate service life of bearing is 5,800,000 rounds, which is smaller than the value of theoretical calculation. The elastic deformation of whole shaft system leads to the variation of the load distribution. The service life of bearing is decreased, which states that the effect of load distribution on the service life of bearing is great and the research on the flexible support system is very important for the load distribution.

3 Conclusions

A new design method for the thin-walled and four-point contact ball bearings of industrial robots is developed. FPXU408-2RZ bearing is selected as an example to introduce the design method. The analysis results state that the loaded region is from −64° to 64° and the number of loaded balls is 17. It is obvious that the loaded region is narrowed due to the elastic deformation of shaft system. The accurate service life of bearing is computed according to the load distribution of bearing obtained by the finite element analysis. The calculation result of service life shows that the rated service life of bearing is shorter than the theoretical calculation results due to the variation of load distribution.

References

1. Richard, F.: Evaluating bearing for robots. Mach. Des. **4**, 5–7 (1983)
2. Kenneth, W.: Design shape up with thin bearings. Mach. Des. **67**(1), 86–92 (1995)
3. Chen, L., Song, H.: Design and process analysis of thin-walled large size bearing with special use. Bearings **10**, 4–6 (2003) (in Chinese)
4. Zeng, Z.: Study on design technology of thin walled four point contact ball bearings used for industrial robots. Dissertation, Henan University of Science and Technology (2011) (in Chinese)
5. Zhang, X.J., Chen, K.Z., Feng, X.A.: Material selection using an improved genetic algorithm for material design of components made of a multiphase material. Mater. Des. **29**(5), 972–981 (2008)
6. Harris, T.A., Kotzalas, M.N.: Rolling Bearing Analysis-Essential Concepts of Bearing, Technology. Taylor & Francis, UK (2007)

On the Elastostatics of Spherical Parallel Machines with Curved Links

A. Cammarata and R. Sinatra

Abstract The paper presents the elastostatics analysis of a class of lower-mobility Parallel Kinematic Machines: the Spherical Parallel Machines. These robots usually recur to curved links in their structure to satisfy geometric constraints deriving from mobility reason. In fact, to make the mobile platform move with spherical motion all links or a part of these are constrained to have spherical motions too. This condition is generally obtained employing curved links with revolute pairs whose axes intersect at a common center of motion. Recurring to two-node Timoshenko's beam element with constant strain fields to simulate curved beams in space we adapt a methodology proposed by the same authors to study the elastostatics of Spherical Parallel Machines. The method is finally applied to study the error positioning analysis of the Agile Eye.

Keywords Spherical parallel robots · Elastostatics · Error positioning analysis · Curved Timoshenko's beam

1 Introduction

A Spherical Parallel Manipulator (SPM) is a reduced-mobility parallel robot that provides three degrees of freedom of pure rotations in which the mobile platform (MP) moves on the surface of a sphere centered at a point, named the rotation center of the spherical motions. While the kinematic analysis of SPMs has been largely investigated in the literature, the elastostatic analysis has received less attention. Probably the main reason to study stiffness of a SPM comes from the necessity to limit the positioning error of its end-effector. In fact, deformations jeopardize the

A. Cammarata (✉) · R. Sinatra
Department of Industrial Engineering, Catania University, Catania, Italy
e-mail: acamma@dii.unict.it

R. Sinatra
e-mail: rsinatra@dii.unict.it

S. Bai and M. Ceccarelli (eds.), *Recent Advances in Mechanism Design for Robotics*, Mechanisms and Machine Science 33,
DOI 10.1007/978-3-319-18126-4_33

mode of operation of these robots, specially when the latter are based on over-constrained architectures.

In [11] Liu et al. considers only the actuation compliance of a 3-RRR SPM to develop stiffness maps and atlases of a global stiffness index. A first structural optimization has been developed in [1] in which the shape of the curved links is optimized by means of the concept of *curve synthesis*. In [2, 3] Callegari et al. recur to a FEM approach to study the kineto-elasto-static synthesis and the design of a 3-CRU spherical wrist for miniaturized assembly tasks.

While the previous paper deals with different approaches based on the Jacobian matrix and FEM in [8] the authors use the strain energy and Castiglianos theorem to introduce the structural compliance of a 3-RRP SPM. In [15] Wu extend the same topic including the strain energy due to shear forces to study the stiffness analysis of a co-axial SPM.

In recent years Palmieri [13] and Wu [16] recur to FEM approach and virtual-spring method, respectively, to study the error positioning analysis of SPMs.

Here, we extend the method developed in [4] to study SPM with curved beams. The latter are modeled using two-node Timoshenko's spatial beams. The final elastostatic model is applied to investigate the error positioning analysis of the Agile Eye [9, 10].

2 Elastostatic Model

2.1 Small Displacements and Joint-Matrices

Here we extend the formulation developed by the same authors in [4, 5]. The proposed method is based on small displacements and Euler angles to describe rotations, while links are modeled by means of matrix structural elements. To study curved links here we recur to spatial curved beams derived applying Timoshenko's beam theory.

The following recall some concepts useful to understand the elastostatics analysis. A nodal displacement array is a six-dimensional array with three translational followed by three rotational displacements: $\mathbf{u} = \left[u_x \; u_y \; u_z \; u_\varphi \; u_\theta \; u_\psi\right]^T$. The number of nodes will depend on the degree of refinement of the structural mesh.

Joints introduce kinematic constraints between bodies of a mechanism that must be rewritten in terms of constraint equations between adjacent nodes. In particular, considering two nodes, belonging to two consecutive links bonded by a joint, it is possible to express the kinematic constraint provided by the joint through an equation among displacements:

$$\mathbf{u}_1^2 = \mathbf{u}_2^1 + \mathbf{H}\theta \tag{1}$$

in which subscripts of nodal displacements are referred to the link, while superscripts to one of the two end-nodes of the link. The joint-matrix **H** and the array θ depend on the joint type: the former containing unit vectors indicating geometric axes, the latter including joint displacements and rotations. Below we write the joint-matrices of some common kinematic pair:

Revolute joint:

$$\mathbf{h}_R = \begin{bmatrix} \mathbf{0} \\ \mathbf{e} \end{bmatrix}, \quad \theta_R = \theta \tag{2}$$

where **e** is the unit vector along the axis of the revolute joint R and θ is the angle of rotation about the said axis.

Universal joint:

$$\mathbf{H}_U = \begin{bmatrix} \mathbf{0} & \mathbf{0} \\ \mathbf{e}^1 & \mathbf{e}^2 \end{bmatrix}, \quad \theta_U = \begin{bmatrix} \theta^1 \\ \theta^2 \end{bmatrix} \tag{3}$$

where $\mathbf{e}^1$ and $\mathbf{e}^2$ are the unit vectors along the axes of the universal joint U and θ^1 and θ^2 are the angles of rotation about the axes of U.

Spherical joint:

$$\mathbf{H}_S = \begin{bmatrix} \mathbf{0} & \mathbf{0} & \mathbf{0} \\ \mathbf{e}^1 & \mathbf{e}^2 & \mathbf{e}^3 \end{bmatrix}, \quad \theta_S = \begin{bmatrix} \theta^1 \\ \theta^2 \\ \theta^3 \end{bmatrix} \tag{4}$$

where $\mathbf{e}^1$, $\mathbf{e}^2$ and $\mathbf{e}^3$ are the unit vectors along the axes of the spherical joint S and θ^1, θ^2 and θ^3 are the angles of rotation about the axes of S.

2.2 *Stiffness Matrix of a Curved Beam*

Before proceeding with the description of the method we haver to find the stiffness matrix of a curved beam. Here we employ a two-node Timoshenko's beam with constant strain fields, described in [7], in which strain functions are assumed independently in order to avoid locking phenomena. Similar formulation, based on modified isoparametric elements, can be found in the works of Prathap et al. [14] or Palaninathan et al. [12].

We briefly recall the mathematical background needed to obtain the stiffness matrix $\mathbf{K}_p$ of a curved beam as expressed into its local frame; the reader is referred to [7] for a complete description. Let us consider a curved beam of mean radius R, cross-section area A, Young's modulus E, shear modulus G, shear correction factor β^2 and moments of inertia I_x, I_y and I_z as referred to the local curvilinear coordinate system $x_i - z_i$ shown in Fig. 1. We can express the stress-strain relations by means of the matrix d_s, defined as

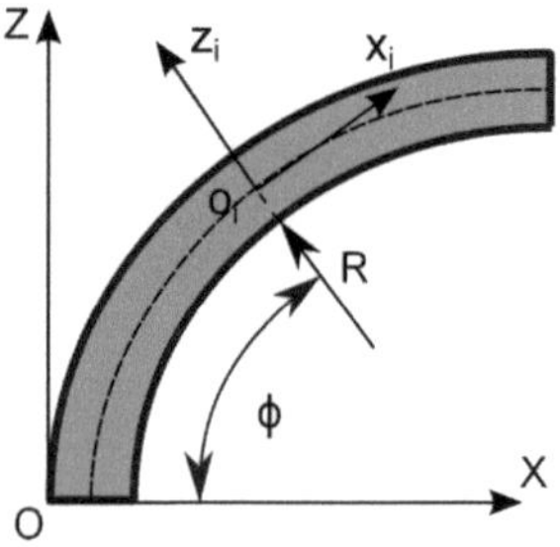

Fig. 1 Local Cartesian frame and curvilinear frame

$$\mathbf{D}_s = \mathbf{diag}\left(AE, \beta^2 GA, \beta^2 GA, GI_x, EI_y, EI_z\right). \tag{5}$$

Now, introducing the stain mode matrix Φ_1 and the rigid body mode matrix Φ_2:

$$\Phi_1 = \begin{pmatrix} 0 & -R^2\phi & R & 0 & 0 & 0 \\ 0 & 0 & 0 & 0 & R^2\phi & R\phi \\ R & R^2 & 0 & 0 & 0 & 0 \\ 0 & 0 & 0 & -R & 0 & 0 \\ 0 & R\phi & 0 & 0 & 0 & 0 \\ 0 & 0 & 0 & 0 & R & 0 \end{pmatrix} \tag{6}$$

$$\Phi_2 = \begin{pmatrix} 1 & -s(\phi) & c(\phi) & 0 & 0 & 0 \\ 0 & 0 & 0 & 1 & c(\phi) & s(\phi) \\ 0 & c(\phi) & s(\phi) & 0 & 0 & 0 \\ 0 & 0 & 0 & 0 & -\frac{c(\phi)}{R} & -\frac{s(\phi)}{R} \\ -\frac{1}{R} & 0 & 0 & 0 & 0 & 0 \\ 0 & 0 & 0 & 0 & -\frac{s(\phi)}{R} & \frac{c(\phi)}{R} \end{pmatrix} \tag{7}$$

the local stiffness matrix $\mathbf{K}_p$ can be written as

$$\mathbf{K}_p = R\left(\phi_f - \phi_i\right)\mathbf{T}_2\mathbf{A}_\phi^{-T}\mathbf{B}^T\mathbf{D}_s\mathbf{B}\mathbf{A}_\phi^{-1}\mathbf{T}_2^T \tag{8}$$

in which, denoting with ϕ_i and ϕ_f the initial and final angles of the curved beam, we have

$$\mathbf{A}_\phi = \begin{pmatrix} \Phi_1(\phi_i) & \Phi_2(\phi_i) \\ \Phi_1(\phi_f) & \Phi_2(\phi_f) \end{pmatrix},$$
$$\mathbf{B} = \begin{pmatrix} 1 & 0 & 0 & 0 & 0 & 0 & 0 & 0 & 0 & 0 & 0 & 0 \\ 0 & 0 & 0 & 0 & 0 & 1 & 0 & 0 & 0 & 0 & 0 & 0 \\ 0 & 0 & 1 & 0 & 0 & 0 & 0 & 0 & 0 & 0 & 0 & 0 \\ 0 & 0 & 0 & 0 & 1 & 0 & 0 & 0 & 0 & 0 & 0 & 0 \\ 0 & 1 & 0 & 0 & 0 & 0 & 0 & 0 & 0 & 0 & 0 & 0 \\ 0 & 0 & 0 & 1 & 0 & 0 & 0 & 0 & 0 & 0 & 0 & 0 \end{pmatrix} \tag{9}$$

while, by introducing $\beta_i = \pi/2 - \phi_i$ and $\beta_f = \pi/2 - \phi_f$, we set the canonical y-axis rotation matrix $\mathbf{T}_{2i}$ to express the local curvilinear frame rotated about the Y-axis of the generic angle β with respect to the local Cartesian frame $X - Z$ of Fig. 1, i.e.

$$\mathbf{T}_2 = \begin{pmatrix} \mathbf{T}_{2i}(\beta_i) & \mathbf{O} & \mathbf{O} & \mathbf{O} \\ \mathbf{O} & \mathbf{T}_{2i}(\beta_i) & \mathbf{O} & \mathbf{O} \\ \mathbf{O} & \mathbf{O} & \mathbf{T}_{2i}(\beta_f) & \mathbf{O} \\ \mathbf{O} & \mathbf{O} & \mathbf{O} & \mathbf{T}_{2i}(\beta_f) \end{pmatrix} \tag{10}$$

2.3 Stiffness Matrix of a Flexible Body Coupled to a Rigid Body

In deriving the stiffness matrix of a linkage we recall that the 12×12 stiffness matrix $\mathbf{K}_c$ of a flexible body coupled to a rigid body through a joint, with joint-matrix $\mathbf{H}$, has been obtained in [4]. Its final expression is as follows

$$\mathbf{K}_c = \begin{bmatrix} \mathbf{1} & \mathbf{O} \\ \mathbf{X}^1 & \mathbf{X}^2 \end{bmatrix}^T \begin{bmatrix} \mathbf{K}_1^{1,1} & \mathbf{K}_1^{1,2} \\ \mathbf{K}_1^{2,1} & \mathbf{K}_1^{2,2} \end{bmatrix} \begin{bmatrix} \mathbf{1} & \mathbf{O} \\ \mathbf{X}^1 & \mathbf{X}^2 \end{bmatrix} \tag{11}$$

where

$$\mathbf{X}^1 = -\mathbf{H}(\mathbf{H}^T\mathbf{K}_1^{2,2}\mathbf{H})^{-1}\mathbf{H}^T\mathbf{K}_1^{2,1} \tag{12a}$$

$$\mathbf{X}^2 = \left(\mathbf{1} - \mathbf{H}(\mathbf{H}^T\mathbf{K}_1^{2,2}\mathbf{H})^{-1}\mathbf{H}^T\mathbf{K}_1^{2,2}\right)\mathbf{G} \tag{12b}$$

The matrix $\mathbf{K}_1$ is the stiffness matrix of the flexible body while the matrix $\mathbf{G}$ extends the concept of rigid-body displacement to a virtual rigid body, i.e.

$$\mathbf{G} = \begin{bmatrix} \mathbf{1} & -\mathbf{D} \\ \mathbf{O} & \mathbf{1} \end{bmatrix} \tag{13}$$

with $\mathbf{1}$ and $\mathbf{O}$, respectively, denoting the 3×3 identity- and zero-matrices while $\mathbf{D}$ is the Cross-Product Matrix of the vector $\mathbf{d}$ pointing from the reference node of a virtual rigid body towards the joint node of the flexible body. The concept of virtual rigid body is used to say that the length of $\mathbf{d}$ remains the same when deforming.

For instance, let us consider two flexible links coupled by means of a revolute joint of unit vector $\mathbf{e}$, as shown in Fig. 2. As can be observed the two parts composing the revolute pair are not consecutive but far-between of a fixed distance

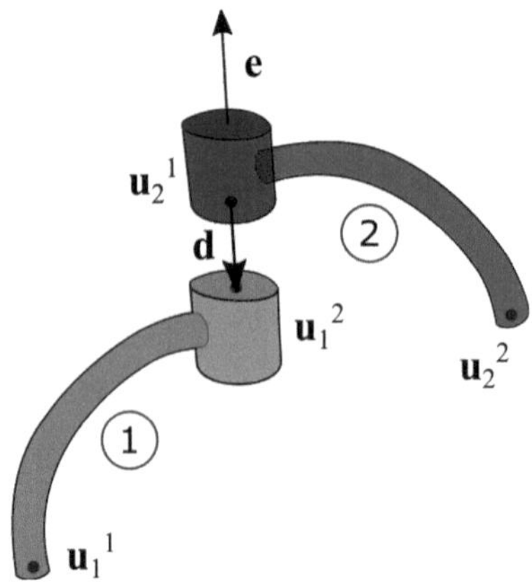

Fig. 2 Coupling of two curved beams through a revolute pair

individuated by the vector **d**. In this case the joint-matrix is a six-dimensional array $\mathbf{h}_R = [\mathbf{0}^T\, \mathbf{e}^T]^T$. Then upon substituting into Eq. (12), we obtain

$$\mathbf{X}^1 = -\frac{(\mathbf{h}_R\mathbf{h}_R^T)\mathbf{K}_1^{2,1}}{\mathbf{h}_R^T\mathbf{K}_1^{2,2}\mathbf{h}_R}, \quad \mathbf{X}^2 = \left(\mathbf{1} - \frac{(\mathbf{h}_R\mathbf{h}_R^T)\mathbf{K}_1^{2,2}}{\mathbf{h}_R^T\mathbf{K}_1^{2,2}\mathbf{h}_R}\right)\mathbf{G} \tag{14}$$

The expression of $\mathbf{K}_c$ can be derived by means of Eq. (11).

3 Case Study: Agile Eye

In this section we apply the elastostatics model developed in the previous sections to study the Agile Eye, a three DOF SPM developed in [9, 10]. The Agile Eye, shown in Fig. 3, has three limbs each made of two curved links, here named proximal and distal links, coupled by means of passive revolute joints. The proximal links are coupled to the BP by means of actuated revolute joints, while three further passive revolute joints connect the distal links to the MP. All the joints have intersecting axes at a common point, i.e. the center of the spherical motion.

3.1 Limb $\hat{R}\hat{R}\hat{R}$

Each limb of the Agile Eye is of type $\hat{R}\hat{R}\hat{R}$, meaning that it is made of three revolute pairs with intersecting axes. The *j*th proximal link of the *i*th limb is connected to the BP by means of an actuated revolute pair R_{i1}. To perform the elastostatic analysis we fix the actuated joints. This condition implies that the proximal links become cantilevers clamped to the fixed base. Let us consider the connection

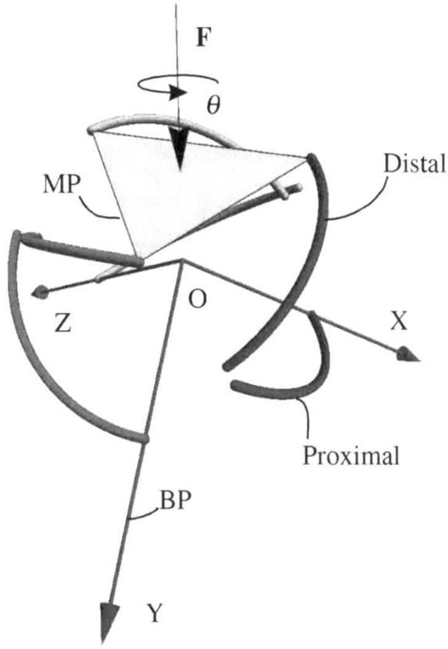

Fig. 3 Agile Eye in its home posture: $e = -\sqrt{3}/3[1, 1, 1]$, $\theta = 60°$

proximal-distal, here we use Eq. (11) to write $\mathbf{K}_{lp_i} = \mathbf{K}_c(\mathbf{h}_{R_{i2}}, \mathbf{G}_{O_{i2}}, \mathbf{K}_{p_i})$, where: R_{i2} refers to the middle revolute joint of the ith limb, $\mathbf{G}_{O_{i2}}$ is derived through Eq. (13) considering a vector $\mathbf{d}$ going from the second node of the proximal to the first adjacent node of the distal link, and finally $\mathbf{K}_{p_i}$ is the stiffness matrix of the ith proximal link.

Then the ith distal link is connected to the MP through the revolute joint R_{i3}. Again we write $\mathbf{K}_{ld_i} = \mathbf{K}_c(\mathbf{h}_{R_{i3}}, \mathbf{G}_{O_{i3}}, \mathbf{K}_{d_i})$, in which $\mathbf{G}_{O_{i3}}$ is built considering the array going from the second node of the distal to the ith attachment node of the MP, and $\mathbf{K}_{d_i}$ is the stiffness matrix of the ith distal link. Notice that both $\mathbf{K}_{p_i}$ and $\mathbf{K}_{d_i}$ must be expressed into the global frame of reference.

4 Error Positioning Analysis of the Agile Eye

The generalized stiffness matrix of the Agile Eye can be found exploiting the results from the previous subsections; for the sake of conciseness, readers are invited to consult [6] for further details on the final assembling. In order to improve the stiffness behaviour of the curved link, as already outlined in [7], we recur to six curved elements for each curved link. Geometrical and structural parameters of the robot are reported in Table 1. To investigate the error positioning analysis we applied a vertical force **F** to the center of the MP, as shown in Fig. 3, measuring the displacement s and the rotation ψ of the MP due to the deformation [13]. The rotation of the MP is calculated by means of its natural invariants, i.e. the angle of rotation θ and the unit vector e along the axis of rotation. Results are reported in Table 2 for five different poses.

Table 1 Geometrical and structural parameters of the Agile Eye

r_p	191.5	mm	Mean radius of proximal links
r_d	166.5	mm	Mean radius of distal links
l	20.0	mm	Side length of the square cross section of the curved links
α	90	°	Opening angle of the curved links
β	0.91		Shear correction factor
ν	0.33		Poisson's ratio: aluminum alloy
E	68.5	GPa	Elastic modulus of the links: aluminum alloy

Table 2 Error positioning analysis of the Agile Eye for a vertical force **F** = [10, 10, 10] [N]

Natural invariants	Displacement and deformation angle (mm, °)
(e, θ)	(s, ψ)
$(-\sqrt{3}/3[1, 1, 1], 0°)$	(0.095400 mm, 0.012061°)
$(-\sqrt{3}/3[1, 1, 1], 30°)$	(0.156513 mm, 0.028559°)
$(-\sqrt{3}/3[1, 1, 1], 60°)$	(0.100929 mm, 0.011687°)
$(-1/\sqrt{434}[20, 3, 5], 30°)$	(0.064678 mm, 0.002977°)
$(-1/\sqrt{434}[20, 3, 5], 60°)$	(0.108887 mm, 0.016022°)

We compared results to FEM to check the accuracy of the proposed method. First we compared the displacements obtained through FEM to those coming from the curved element developed in Sect. 2.2. The element is quite accurate for in-plane forces (relative error of 94.18 % w.r.t. FEM for the z-displacement when a force acts along the z-axis direction) while loses precision for out-of-plane wrenches (relative error of 86.50 % for the y-displacement when a force along the y-axis is applied). This gap is essentially due to different reasons: not slender element, square cross section, holes in the FEM model of the link to allow revolute connections, simplified assumptions of the theory such as constant strain fields inside elements.

To exclude errors in the calculation of the stiffness model we derived the 12 × 12 stiffness matrices of the proximal and distal link via FEM. Then these matrices have been included in the mathematical model as superelements.

We tested the Agile Eye at its home posture, i.e. when $e = -\sqrt{3}/3[1, 1, 1]$ and $\theta = 60°$, as shown in Fig. 4. In this particular configuration the first and third revolute pairs of a limb lie on a same plane passing for the vertical direction. As reported in Table 3 the accuracy of the model with superelements is high when compared to FEM with relative errors below 0.04 %. The same model with curved beams seems to be more rigid than FEM model. The relative errors, w.r.t. FEM, on displacement and rotation respectively are about 26 and 46 %. These high values comes from the inaccuracy on single curved elements.

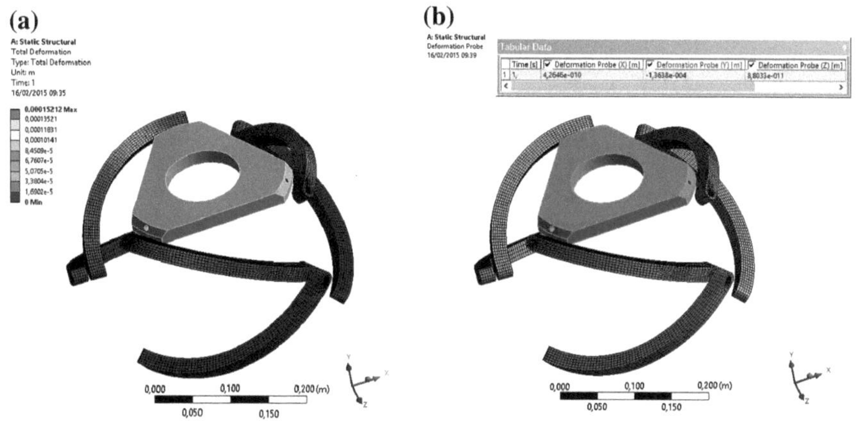

Fig. 4 FEM model of the Agile Eye in its home posture: $\mathbf{e} = -\sqrt{3}/3[1, 1, 1]$, θ = 60°. The maximum deformations occur at the revolute joints of the MP. **a** Total deformation, **b** MP displacement

Table 3 Comparison with FEA results of the Agile Eye at its home posture for a vertical force $F = [10, 10, 10]$ [N]

Curved beams	Superelements	FEM
Displacement s (mm)		
0.100929	0.136428	0.136380
Rotation ψ (°)		
0.011687	0.021576	0.021569

5 Conclusions

The elastostatic model of Spherical Parallel Machines with curved links has been derived by means of a method based on strain energy minimization and nodal constraints due to the joints. Curved beams have been modeled recurring to two-node curved elements based on Timoshenko beam theory. The procedure to create the stiffness matrix of curved beams has been recalled, as also the way to obtain the stiffness matrix of bodies coupled through joints. The proposed method has been applied to the Agile Eye robot to study the positioning error resulting from the application of a vertical static force on the end-effector. Results have been compared to FEM for validation. To exclude mistakes in the stiffness modeling we also implemented superelements, obtained by FEM, to model the flexible links. The discrepancy of results w.r.t. FEM is due to several aspects such as: not slender element, square cross section, different geometric features in FEM model, simplified assumptions of the theory. Even if the method seems to be more suitable for planar cases, it is a starting point to improve the formulation, based on stiffness matrices, of three dimensional curved beams.

References

1. Bidault, F., Teng, C.P., Angeles, J.: Structural optimization of a spherical parallel manipulator using a two-level approach. In: Proceedings of ASME 2001 Design Engineering Technical Conferences (2001)
2. Callegari, M., Cammarata, A., Gabrielli, A., Ruggiu, M., Sinatra, R.: Analysis and design of a spherical micromechanism with flexure hinges. J. Mech. Des. **131**(5), 051003 (2009)
3. Callegari, M., Gabrielli, A., Ruggiu, M.: Kineto-elasto-static synthesis of a 3-cru spherical wrist for miniaturized assembly tasks. Meccanica **43**(4), 377–389 (2008)
4. Cammarata, A.: On the stiffness analysis and elastodynamics of parallel kinematic machines. Kinematic Dynamics and Control, Serial and Parallel Robot Manipulators (2012)
5. Cammarata, A., Condorelli, D., Sinatra, R.: An algorithm to study the elastodynamics of parallel kinematic machines with lower kinematic pairs. J. Mech. Rob. **5**(1), 011004 (2013)
6. Cammarata, A., Rosario, S.: Elastodynamic optimization of a 3t1r parallel manipulator. Mech. Mach. Theory **73**, 184–196 (2014)
7. Choi, J.k., Lim, J.k.: General curved beam elements based on the assumed strain fields. Comput. Struct. **55**(3), 379–386 (1995)
8. Enferadi, J., Tootoonchi, A.A.: Accuracy and stiffness analysis of a 3-rrp spherical parallel manipulator. Robotica **29**(2), 193–209 (2011)
9. Gosselin, C.M., Hamel, J.F.: The agile eye: a high-performance three-degree-of-freedom camera-orienting device. In: Proceedings of the IEEE International Conference on Robotics and Automation, pp. 781–786 (1994)
10. Gosselin, C.M., Lavoie, E.: On the kinematic design of spherical three-degree-of-freedom parallel manipulators. Int. J. Rob. Res. **12**(4), 394–402 (1993)
11. Liu, X.J., Jin, Z.L., Gao, F.: Optimum design of 3-dof spherical parallel manipulators with respect to the conditioning and stiffness indices. Mech. Mach. Theor. **35**(9), 1257–1267 (2000)
12. Palaninathan, R., Chandrasekharan, P.: Curved beam element stiffness matrix formulation. Comput. Struct. **21**(4), 663–669 (1985)
13. Palmieri, G.: On the positioning error of a 2 dof spherical parallel wrist with flexible links and joints—a fem approach. Submitted to Mechanical Sciences
14. Prathap, G., Babu, C.R.: An isoparametric quadratic thick curved beam element. Int. J. Numer. Meth. Eng. **23**(9), 1583–1600 (1986)
15. Wu, G.: Stiffness analysis and optimization of a co-axial spherical parallel manipulator. Model. Ident. Control **35**(1), 21–30 (2014). doi:10.4173/mic.2014.1.2
16. Wu, G., Bai, S., Kepler, J.: Mobile platform center shift in spherical parallel manipulators with flexible limbs. Mech. Mach. Theor. **75**, 12–26 (2014)

Accuracy Analysis of a Tripod Parallel Grinder

P. Zou, F. Liu, X.J. Gao, X.L. Yang and M.Z. Ai

Abstract This paper presents the accuracy analysis of a tripod parallel grinder. The manipulator structure was introduced and its inverse geometric problem was studied. The inverse Jacobian matrix of the tripod parallel grinder was obtained upon the differentiation of the loop-closure displacement equations. The influence of the geometric parameters to the accuracy of the tripod parallel grinder was investigated. Through simulation, the influences of link length error to accuracy were analyzed, which is the basis of error compensation and control for the machine tool.

Keywords Parallel machine tool · Inverse position solution · Accuracy analysis · Simulation

1 Introduction

The parallel machine tool, compared with traditional serial mechanism, has many advantages, such as, high stiffness, high carrying capacity, and so on. However, it is reported from the literature that the parallel machine tool is not necessarily more accurate than a serial machine with the same manufacturing and assembling

P. Zou (✉) · F. Liu · X.J. Gao · X.L. Yang
School of Mechanical Engineering & Automation, Northeastern University, Shenyang, China
e-mail: pzou@me.neu.edu.cn

F. Liu
e-mail: ysulf@163.com

X.J. Gao
e-mail: gxj7976@126.com

X.L. Yang
e-mail: xlyang@me.neu.edu.cn

M.Z. Ai
Liaoning Provincial Institute of Measurement, Shenyang, China
e-mail: aammzz2006@163.com

S. Bai and M. Ceccarelli (eds.), *Recent Advances in Mechanism Design for Robotics*, Mechanisms and Machine Science 33,
DOI 10.1007/978-3-319-18126-4_34

precision [1]. Low accuracy remains a bottleneck for further industrial applications of parallel machine tool.

To enhance the accuracy of parallel machine tool, it is important to evaluate the end-effector's accuracy in the design phase. There are many ways to establish error model [2–4]. This paper concentrates on a tripod parallel grinder whose error model is derived by differentiating the loop-closure displacement equation. Analysis result is visualized by means of MATLAB, from which the influence of geometric errors to the accuracy of the parallel machine tool can be observed.

2 Structure Descriptions

As shown in Fig. 1, this machine tool has five degrees of freedom (dofs). Translations along X and Y can be achieved by two motors independently through ball screws. Tilt orientation of tripod parallel bench can be achieved by two leg motors. Rotation of spindle can be achieved by spindle motor. The tool head is fixed on the mobile platform of the tripod parallel universal wrist, which can generate an unlimited rotation under a certain tilt angle. Thus, the machine can be used for drilling, grinding and milling. Particularly, it can be used for complex helical drill point grinding.

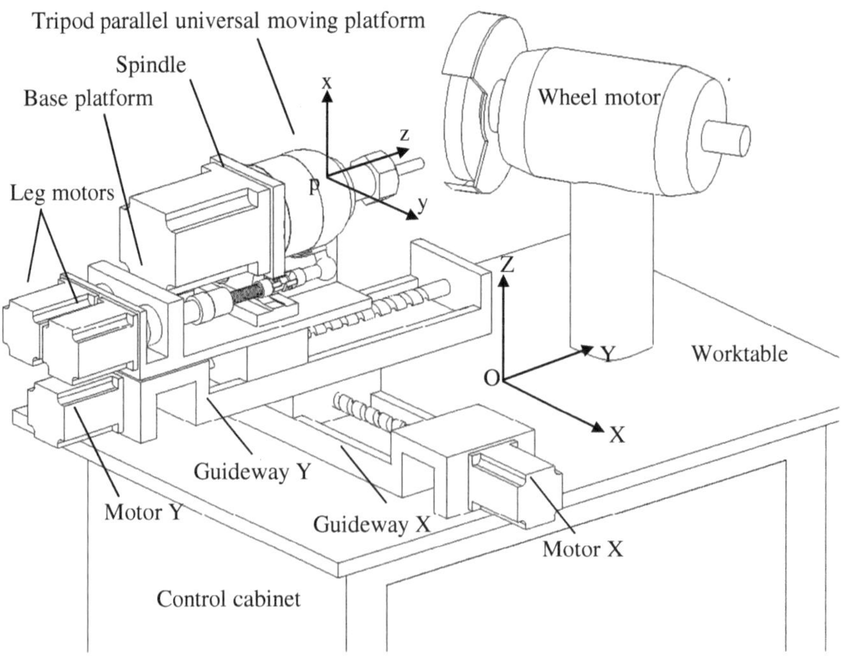

Fig. 1 Structure of the tripod parallel grinder

3 Position Reverse Solution

As shown in Fig. 2, the reference coordinate system $o_1\text{-}x_1y_1z_1$ is fixed in the reference platform, with the origin o_1 located at the geometric center of the reference platform. The moving coordinate system $p\text{-}xyz$ is attached to the moving platform with the origin p located on the geometric center of the moving platform, and N is the reference point of end-effectors. Positive stop is used to restrict moving platform rotation around Z.

The moving platform is connected with the reference platform by three links, among them two links are telescopic, and one constant-length is spindle. A_i and B_i ($i = 1, 2$) are the centers of spherical and revolute joint, respectively. $\Delta o_1A_1A_2$ and ΔpB_1B_2 are assumed to be equilateral triangles. R and r denote radiuses of reference platform and moving platform respectively. h denotes the length of the spindle.

The transformation matrix can be formulated as

$$[T] = \begin{bmatrix} C\phi_y & 0 & S\phi_y \\ S\phi_y S\phi_x & C\phi_x & C\phi_y S\phi_x \\ -S\phi_y C\phi_x & S\phi_x & C\phi_y C\phi_x \end{bmatrix} \tag{1}$$

where "S" and "C" denote sine and cosine functions, respectively. ϕ_x, ϕ_y are angles of moving platform rotating around its X axis and Y axis.

An arbitrary vector $\vec{R}'$ in $p\text{-}xyz$ can be transformed to $\vec{R}$ in $o_1\text{-}x_1y_1z_1$ as follows.

$$\vec{R} = [T]\vec{R}' + \vec{P} \tag{2}$$

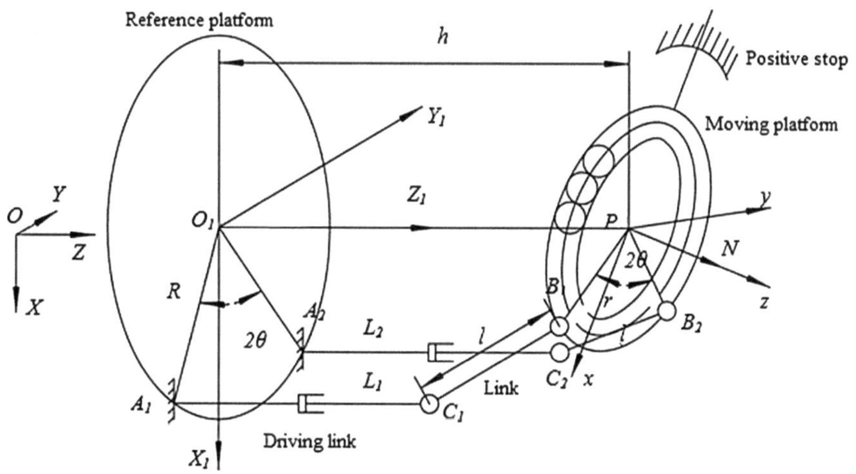

Fig. 2 Tripod parallel universal wrist coordinate systems

where $\vec{P} = \begin{bmatrix} 0 & 0 & h \end{bmatrix}^T$ is the translation vector of $p\text{-}xyz$ respect to $o_1\text{-}x_1y_1z_1$.

$$\vec{A_1} = \begin{bmatrix} A_{1x} \\ A_{1y} \\ A_{1z} \end{bmatrix} = \begin{bmatrix} RC\theta \\ -RS\theta \\ 0 \end{bmatrix}, \vec{A_2} = \begin{bmatrix} A_{2x} \\ A_{2y} \\ A_{2z} \end{bmatrix} = \begin{bmatrix} RC\theta \\ RS\theta \\ 0 \end{bmatrix} \tag{3}$$

$\vec{A_1}$, $\vec{A_2}$ are position vectors of A_i (i = 1, 2) in $o_1\text{-}x_1y_1z_1$.

$$\vec{B'_1} = \begin{bmatrix} B'_{1x} \\ B'_{1y} \\ B'_{1z} \end{bmatrix} = \begin{bmatrix} rC\theta \\ -rS\theta \\ 0 \end{bmatrix}, \vec{B'_2} = \begin{bmatrix} B'_{2x} \\ B'_{2y} \\ B'_{2z} \end{bmatrix} = \begin{bmatrix} rC\theta \\ rS\theta \\ 0 \end{bmatrix} \tag{4}$$

$\vec{B'_1}$, $\vec{B'_2}$ are position vectors of B_i (i = 1, 2) in $p\text{-}xyz$.

Transformation of position vector of point Bi from frame xyz to frame $x_1y_1z_1$ can be formulated as

$$\vec{B_i} = [T]\vec{B'_i} + \vec{P} \tag{5}$$

Let L_i be the length of telescopic link A_iC_i, the coordinate of C_i in reference coordinate frame can be formulated as

$$\vec{C_i} = \vec{A_i} + \begin{bmatrix} 0 \\ 0 \\ L_i \end{bmatrix} \tag{6}$$

Let l be the length of fixed-length link, the following kinematic constraint exists:

$$l^2 = |B_iC_i|^2 \tag{7}$$

As a result, with know ϕ_x and ϕ_y, the length of L_i can be calculated as follows.

$$\begin{cases} L_1 = -rC\theta S\phi_y C\phi_x - rS\theta S\phi_x + h - \\ \qquad \sqrt{l^2 - \left(rC\theta C\phi_y - RC\theta\right)^2 - \left(rC\theta S\phi_y S\phi_x - rS\theta C\phi_x + Rs\theta\right)^2} \\ L_2 = -rC\theta S\phi_y C\phi_x + rS\theta S\phi_x + h - \\ \qquad \sqrt{l^2 - \left(rC\theta C\phi_y - RC\theta\right)^2 - \left(rC\theta S\phi_y S\phi_x + rS\theta C\phi_x - Rs\theta\right)^2} \end{cases} \tag{8}$$

4 Error Modeling Method

Equation (8) can be expressed as objective function of variables as below:

$$\begin{aligned} f_1(\phi_x, \phi_y, L_1, r, R, h, l, \theta) &= L_1 + rC\theta S\phi_y C\phi_x + rS\theta S\phi_x - h + a = 0 \\ f_2(\phi_x, \phi_y, L_2, r, R, h, l, \theta) &= L_2 + rC\theta S\phi_y C\phi_x - rS\theta S\phi_x - h + b = 0 \end{aligned} \tag{9}$$

which can be simply described as

$$\begin{cases} f_1(\phi_x, \phi_y, L_1, r, R, h, l, \theta) = 0 \\ f_2(\phi_x, \phi_y, L_2, r, R, h, l, \theta) = 0 \end{cases} \tag{10}$$

where

$$\begin{aligned} a &= \sqrt{l^2 - (rC\theta C\phi_y - RC\theta)^2 - (rC\theta S\phi_y S\phi_x - rS\theta C\phi_x + RS\theta)^2} \\ b &= \sqrt{l^2 - (rC\theta C\phi_y - RC\theta)^2 - (rC\theta S\phi_y S\phi_x + rS\theta C\phi_x - RS\theta)^2} \end{aligned}$$

Differentiating Eq. (10) leads to

$$\begin{bmatrix} \frac{\partial f_1}{\partial \phi_x} & \frac{\partial f_1}{\partial \phi_y} \\ \frac{\partial f_2}{\partial \phi_x} & \frac{\partial f_2}{\partial \phi_y} \end{bmatrix} \begin{bmatrix} d\phi_x \\ d\phi_y \end{bmatrix} = \begin{bmatrix} -\frac{\partial f_1}{\partial L_1} dL_1 - \frac{\partial f_1}{\partial l} dl - \frac{\partial f_1}{\partial h} dh - \frac{\partial f_1}{\partial R} dR - \frac{\partial f_1}{\partial r} dr - \frac{\partial f_1}{\partial \theta} d\theta \\ -\frac{\partial f_2}{\partial L_2} dL_2 - \frac{\partial f_2}{\partial l} dl - \frac{\partial f_2}{\partial h} dh - \frac{\partial f_2}{\partial R} dR - \frac{\partial f_2}{\partial r} dr - \frac{\partial f_2}{\partial \theta} d\theta \end{bmatrix} \tag{11}$$

The orientation error of the mobile platform can be cast in the matrix form, namely,

$$\begin{bmatrix} d\phi_x \\ d\phi_y \end{bmatrix} = \begin{bmatrix} \frac{\partial f_1}{\partial \phi_x} & \frac{\partial f_1}{\partial \phi_y} \\ \frac{\partial f_2}{\partial \phi_x} & \frac{\partial f_2}{\partial \phi_y} \end{bmatrix}^{-1} \begin{bmatrix} -\frac{\partial f_1}{\partial L_1} dL_1 - \frac{\partial f_1}{\partial l} dl - \frac{\partial f_1}{\partial h} dh - \frac{\partial f_1}{\partial R} dR - \frac{\partial f_1}{\partial r} dr - \frac{\partial f_1}{\partial \theta} d\theta \\ -\frac{\partial f_2}{\partial L_2} dL_2 - \frac{\partial f_2}{\partial l} dl - \frac{\partial f_2}{\partial h} dh - \frac{\partial f_2}{\partial R} dR - \frac{\partial f_2}{\partial r} dr - \frac{\partial f_2}{\partial \theta} d\theta \end{bmatrix} \tag{12}$$

where, the inverse Jacobian matrix of the machine tool is described as

$J^{-1} = \begin{bmatrix} \frac{\partial f_1}{\partial \phi_x} & \frac{\partial f_1}{\partial \phi_y} \\ \frac{\partial f_2}{\partial \phi_x} & \frac{\partial f_2}{\partial \phi_y} \end{bmatrix}$, mapping the orientation errors and geometric parameters.

5 Accuracy Analysis

Structure error of parallel machine is one of the most important sources of error which influences machining accuracy. Structure error includes manufacturing error, assembly error, link length error and error caused by temperature change and elastic deformation. In the paper, link length error is addressed [5, 6].

For the L_1 and L_2 are symmetrically arranged, the length error of L_1 and L are discussed in the paper.

When the length of L_1 has error, the difference matrix of Eq. (10) can be described as

$$\begin{bmatrix} \frac{\partial f_1}{\partial \phi_x} & \frac{\partial f_1}{\partial \phi_y} \\ \frac{\partial f_2}{\partial \phi_x} & \frac{\partial f_2}{\partial \phi_y} \end{bmatrix} \begin{bmatrix} d\phi_x \\ d\phi_y \end{bmatrix} = \begin{bmatrix} -\frac{\partial f_1}{\partial L_1} dL_1 \\ 0 \end{bmatrix} \tag{13}$$

That is,

$$\begin{bmatrix} d\phi_x \\ d\phi_y \end{bmatrix} = \begin{bmatrix} \frac{\partial f_1}{\partial \phi_x} & \frac{\partial f_1}{\partial \phi_y} \\ \frac{\partial f_2}{\partial \phi_x} & \frac{\partial f_2}{\partial \phi_y} \end{bmatrix}^{-1} \begin{bmatrix} -\frac{\partial f_1}{\partial L_1} dL_1 \\ 0 \end{bmatrix} \tag{14}$$

When the parallel manipulator is not at the singular configuration, namely the $|J^{-1}|$ is far away from the zero, if the size parameters and error values are known, the pose error of end-effectors can be calculated.

The geometric parameters and error values are given in Table 1.

When the error tolerance is $d_{L1} = 0.1$ mm, $-80 < \phi_x < 80$, $-80 < \phi_y < 80$, the result can be simulated in Fig. 3.

Similarly, when the length of L has error, the first-order Eq. (10) can be described as

$$\begin{bmatrix} \frac{\partial f_1}{\partial \phi_x} & \frac{\partial f_1}{\partial \phi_y} \\ \frac{\partial f_2}{\partial \phi_x} & \frac{\partial f_2}{\partial \phi_y} \end{bmatrix} \begin{bmatrix} d\phi_x \\ d\phi_y \end{bmatrix} = \begin{bmatrix} -\frac{\partial f_1}{\partial l} dl \\ -\frac{\partial f_2}{\partial l} dl \end{bmatrix} \tag{15}$$

Table 1 Size parameters and error of the parallel machine tool

L (mm)	h (mm)	r (mm)	R (mm)	θ (°)
80	190	65	65	30
0.01	0.5	0.01	0.01	0.025

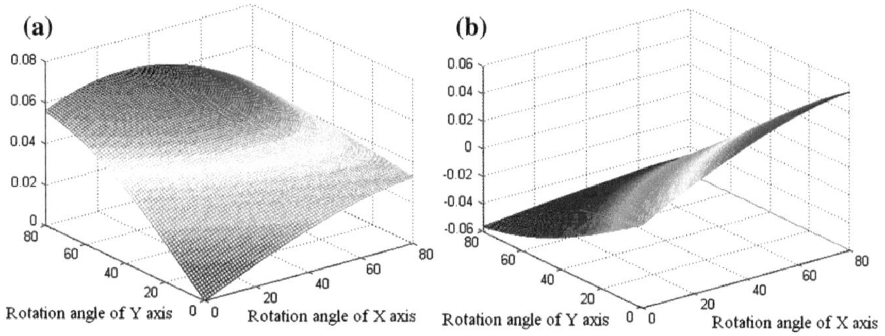

Fig. 3 Error of link L_1 on the influence of the moving platform position, **a** angle error $\Delta\phi_x$, **b** angle error $\Delta\phi_y$

The above equation has the alternative expression:

$$\begin{bmatrix} d\phi_x \\ d\phi_y \end{bmatrix} = \begin{bmatrix} \dfrac{\partial f_1}{\partial \phi_x} & \dfrac{\partial f_1}{\partial \phi_y} \\ \dfrac{\partial f_2}{\partial \phi_x} & \dfrac{\partial f_2}{\partial \phi_y} \end{bmatrix}^{-1} \begin{bmatrix} -\dfrac{\partial f_1}{\partial l} dl \\ -\dfrac{\partial f_2}{\partial l} dl \end{bmatrix} \tag{16}$$

where, $\frac{\partial f_1}{\partial l} = \frac{l}{a}$, $\frac{\partial f_2}{\partial l} = \frac{l}{b}$.

When the parallel manipulator is not at the singular configuration, namely $|J^{-1}|$ is far away from the zero, if the size parameters and error values are known, $[d\phi_x, d\phi_x]^T$ can be simulated as shown in Fig. 4.

According to the simulation results, when the length of L_1 has error, angle error $\Delta\phi_x$ is increased with the increasing rotation angle around X axis and Y axis. On the other hand, angle error $\Delta\phi_y$ is increased with the increasing rotation angle around X

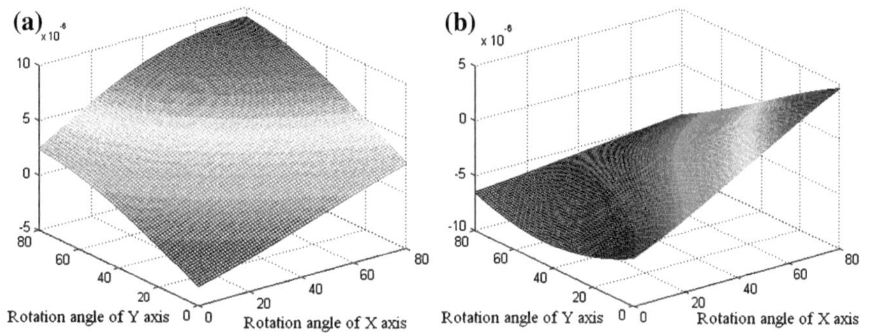

Fig. 4 Error of link L on the influence of the moving platform position, **a** angle error $\Delta\phi_x$, **b** angle error $\Delta\phi_y$

axis, but decreases with the increasing rotation angle around Y axis. When the length of L has error, angle error in the X axis of rotation is increased with the increasing rotation angles around X axis and Y axis, and angle error $\Delta\phi_y$ is increased with the increasing rotation angle around X axis significantly.

6 Conclusions

In this paper, a new type of tripod parallel grinder and its working principle were introduced. Inverse position solution was obtained. Error modeling was established through differential displacement equation. Accuracy analysis was studied under the condition that the lengths of telescopic link and fixed link have errors. In addition, the simulation of accuracy analysis was conducted. According to the simulation results, length error of fixed link has greater effect on the pose accuracy of moving platform, which should be strictly controlled. Conclusions mentioned above can provide theoretical basis for error compensation and control.

References

1. Wang, J.: On the accuracy of a Stewart platform—Part I. The effect of manufacturing tolerances. IEEE international conference on robotics and automation, Atlanta, GA, USA, pp. 114–120, 1993
2. Ropponen, T.: Accuracy analysis of a modified steward platform manipulator. IEEE international conference on robotics and automation, pp. 521–525, 1995
3. Pate1, A.J.: Volumetric error analysis of a stewart platform—based machine tool. Ann. CIRP **46** (1), 287–290 (1997)
4. Soons, J.A.: Error analysis of a hexapod machine tool. In: Laser Metrology and Machine Performance, pp. 346–358. Computational Mechanics Publishing House, Boston (1997)
5. Sebastien.: Accuracy analysis of 3-DOF planar parallel robots. Mech. Mach. Theory **43**(2008), 445–458 (2007)
6. Sebastien.: Accuracy analysis of 3TIR fully-parallel robots. Mech. Mach. Theory **45**(2010), 695–706 (2009)

General Purpose Software to Solve the Inverse Dynamics and Compute the Internal Efforts of Non-redundant Planar Mechanisms

Erik Macho, Victor Petuya, Monica Urizar and Alfonso Hernandez

Abstract GIM is a general purpose kinematics simulation program which was initially developed to compute and visualize the motion of mechanisms of any number of DOFs. In this motion simulation, at each mechanism position, the velocities and accelerations results are obtained and stored. This paper presents a new computation module recently added to the GIM program which is able to read those kinematic results in order to build the inertial efforts and simulate the inverse dynamics of the mechanism. All joint efforts are computed across the motion, so the free solid diagram of any link can be depicted at a specific position, as well as animated over time. In addition, for straight axis elements of the mechanism, the inertial forces are considered as distributed loads, and the internal efforts diagrams (axial and shear forces and bending moment) are traced. These results allow a further cross section dimensional design for such elements.

Keywords Inverse dynamics · Kinematics · Planar linkages · Simulation software

1 Introduction

This paper presents a general methodology for solving the inverse dynamics of planar mechanisms. This methodology is integrated in a general purpose mechanisms analysis and design software named GIM [1, 2]. There are many well-known

E. Macho · V. Petuya · M. Urizar · A. Hernandez (✉)
Faculty of Engineering, University of the Basque Country UPV/EHU, Bilbao, Spain
e-mail: a.hernandez@ehu.es

E. Macho
e-mail: erik.macho@ehu.es

V. Petuya
e-mail: victor.petuya@ehu.es

M. Urizar
e-mail: monica.urizar@ehu.es

S. Bai and M. Ceccarelli (eds.), *Recent Advances in Mechanism Design for Robotics*, Mechanisms and Machine Science 33,
DOI 10.1007/978-3-319-18126-4_35

and reliable programs that are able to deal with the simulation of mechanisms [3–5], but usually in all of these commercial packages, due to their wide range of applications, the generation of the geometrical model to be analyzed requires a considerable effort. The main reason is that when mechanical loads have to be considered in the motion analysis, as in a dynamic simulation, the real solid shape of each element has to be used. This is required to take into account their real inertia, as well as to compute stresses and deformations in the material.

In the approach presented in this paper, mechanism elements are considered as perfectly rigid bodies, so the material properties do not need to be defined, neither the solid shape. The main advantage of doing this is that the definition of the geometrical model can be done in an quicker way. The software GIM provides a geometry module optimized for generating the kinematic schemes of mechanisms in a quite easy process.

Instead of computing stresses, for the mechanism prismatic elements, internal efforts diagrams are obtained. This information can be used in an early design stage to define the cross section needed for such elements.

Target mechanisms can be composed of any number of elements. The most common kinematic joints are included, i.e. revolute joints as well as articulated and rigid sliders. Also line-disk and disk-disk rolling and cam contacts can be included. The mechanism also can has any number of degrees of freedom. Ideal joints are modeled so no gap or friction is considered. Elements are modeled as perfectly rigid bodies, so elastic deformations are also not taken into account. This means that no deformation compatibility equation can be posed, that leads to the main limitation of this methodology, which is that redundant mechanical systems cannot be analyzed.

2 Motion Simulation

In order to compute the inertial forces, at as first step, the motion of the mechanism has to be simulated. The objective is to obtain, for each moving element, the gravity center acceleration and the angular acceleration. This must be done at each mechanism position during the motion. In the motion simulation module of GIM the position problem is posed in a general way to provide an algorithm capable of solving all types of mechanisms.

The mechanical system is translated into a set of geometrical constraints [6]. There are several types of constraints, which relate the coordinates of the mechanism nodes. Each constraint is one equation of the position problem, whose variables are the coordinates of all nodes. Being n the number of unknowns, and m the number of independent constraints, if the mechanical system has no redundancies, the number of degrees of freedom of the mechanism is $f = n - m$.

The user then needs to set f actuators to achieve a defined motion. Each actuator, like in the mechanism itself, is translated into a geometrical constraint, which is added to the set of constraints of the mechanism and a system of n nonlinear equations is completed. Once the position, velocity and acceleration functions (of time) are set, the motion of the system is simulated.

These are some of the main constraints:

- Distance constraint: The distance between two system nodes is a specific value. This is used to model solid elements, like bars, and to model piston-type and sliding actuators.
- Sliding constraint: A node must lie in the line defined by other two nodes. This one is used to define any type of slider.
- Absolute angle constraint: A segment defined by a couple nodes must form a specific angle with a fixed reference. This is used for example to define fixed rotatory actuators.
- Relative angle constraint: The angle between two couples of nodes is a specific valued. This one is used to model rigid elements and joints and to define floating rotatory actuators.

There are many other constraints used to model additional joints, like rollings, and the constraint collection could be enlarged in the future if necessary (for example, to deal with spatial mechanisms). Each constraint always needs to be formulated in an analytical way because it has to be derived two times with respect to time. The first derivative is needed not only to solve the velocity problem, but also to obtain the Jacobian matrix of the system of position equations. Such a Jacobian is used to solve iteratively the nonlinear position problem. The Newton-Raphson algorithm is applied. The second derivative is needed to solve the acceleration problem, whose results are the accelerations of all the mechanism nodes.

For illustrative purposes the simplest constraints, the distance one and the absolute angle one, will be developed. For the distance constraint, being q the distance between two points whose position vectors are $\mathbf{r_A}$ and $\mathbf{r_B}$, and being $\mathbf{a} = \mathbf{r_B} - \mathbf{r_A}$, the constraint equation can be set as:

$$C = \frac{a_x^2}{2} + \frac{a_y^2}{2} - \frac{q^2}{2} = 0 \tag{1}$$

The first derivative with respect time is:

$$\dot{C} = a_x \cdot \dot{a}_x + a_y \cdot \dot{a}_y - q \cdot \dot{q} = 0 \tag{2}$$

where the coefficients of the Jacobian matrix of the nonlinear position problem (which is also the coefficient matrix of the velocity problem, which is linear) and the independent term of the velocity problem can be clearly identified.

The second derivative with respect time is:

$$\ddot{C} = a_x \cdot \ddot{a}_x + a_y \cdot \ddot{a}_y - q \cdot \ddot{q} + \dot{a}_x^2 + \dot{a}_y^2 - \dot{q}^2 = 0 \tag{3}$$

where the coefficients and the independent term of the linear acceleration problem can be clearly identified. Inspecting this equation it is obvious the fact that the coefficients matrix is the same in the velocity and acceleration problems and that to solve the acceleration problem, the velocity problem needs to have been solved previously. If the distance q is constant, $\dot{q} = \ddot{q} = 0$.

For the absolute angle constraint, being q the angle of vector $\mathbf{a} = \mathbf{r_B} - \mathbf{r_A}$ with respect to the positive horizontal axis, where $\mathbf{r_A}$ and $\mathbf{r_B}$ are the position vectors of two mechanism nodes, the constraint equation can be set as:

$$C = \sin q \cdot a_x - \cos q \cdot a_y = 0 \tag{4}$$

The first derivative with respect time is:

$$\dot{C} = \sin q \cdot \dot{a}_x - \cos q \cdot \dot{a}_y + (\cos q \cdot a_x + \sin q \cdot a_y)\dot{q} = 0 \tag{5}$$

And the second one is:

$$\begin{aligned} \ddot{C} = \sin q \cdot \ddot{a}_x - \cos q \cdot \ddot{a}_y + (\cos q \cdot a_x + \sin q \cdot a_y)\, \ddot{q} \\ + 2(\sin q \cdot \dot{a}_x - \cos q \cdot \dot{a}_y) + 2(\cos q \cdot \dot{a}_x + \sin q \cdot \dot{a}_y)\dot{q} = 0 \end{aligned} \tag{6}$$

All the other types of constraints included in the general methodology are not described in detail here due to the space limitation, but follow a similar posing process.

Finally, the user needs to specify the total time of the desired motion, as well as the number of steps to perform during such time, since the position and acceleration problems (Fig. 1) are solved at a sequence of discrete instants. These results can be immediately transformed to those needed to obtain the inertial loads in each element (Fig. 2). To do this, the mass properties of solids must be set.

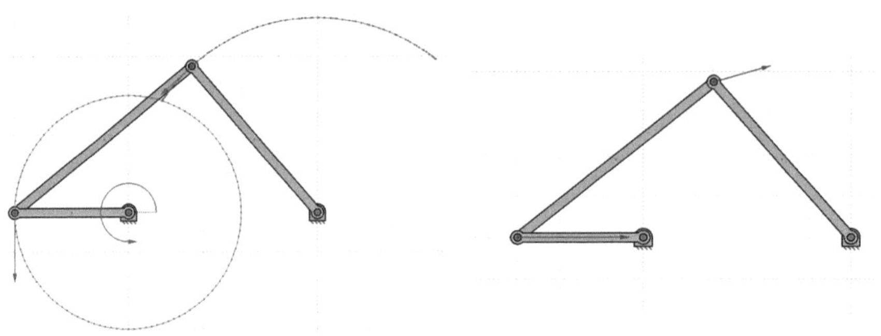

Fig. 1 Motion simulation. Nodes velocities and accelerations

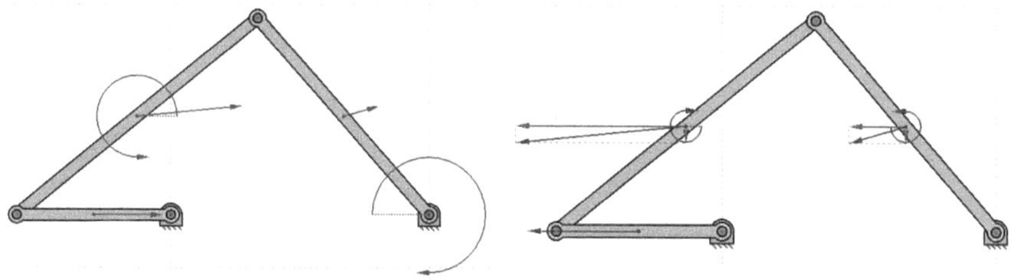

Fig. 2 Elements angular accelerations and gravity centers accelerations. Inertial loads

3 Force Analysis

The main goal at this step is to obtain the free solid diagram of each element. The full set of external forces and moments existing in an element are the data required to trace internal efforts diagrams. Once the inertial forces are known, in order to compute the efforts transmitted at kinematic joints, as well as the forces needed in the actuators to achieve the simulated motion, if the mechanical system has not redundant geometrical constraints, the mechanism can be analyzed as an isostatic structure.

Applying the D'Alembert principle, if inertial forces are considered as external loads, each element can be analyzed like being in a static equilibrium. Apart from the known inertial loads, the user in this step can add as many known loads as needed to the system. These additional applied loads can be punctual forces or moments, or linearly distributed forces in any direction.

The first unknowns of the inverse dynamic problem are the forces and/or moments transmitted among the mechanism elements due to the existence of joints. Each type of kinematic joint, depending on the relative motions that allows, introduces some specific load unknowns (Fig. 3):

- Revolute joint: Allows the rotation, so no moment is transmitted. The unknown is the force transmitted, which is unknown in module, but also in direction. Two unknown are then considered, which are precisely the two cartesian components of such a force.
- Articulated slider: Allows the rotation as well as the displacement along the sliding direction. The only unknown is the module of a force in the direction normal to the sliding one.
- Rigid slider: Allows only the displacement along the sliding direction. Two unknowns appear, the value of the force normal to the sliding direction and the moment responsible of constraining the rotation.
- Rigid union: When several elements are rigidly joined no relative motion is possible among them. In this case, for each element, at the rigid joint two perpendicular forces and a moment are posed as unknowns.

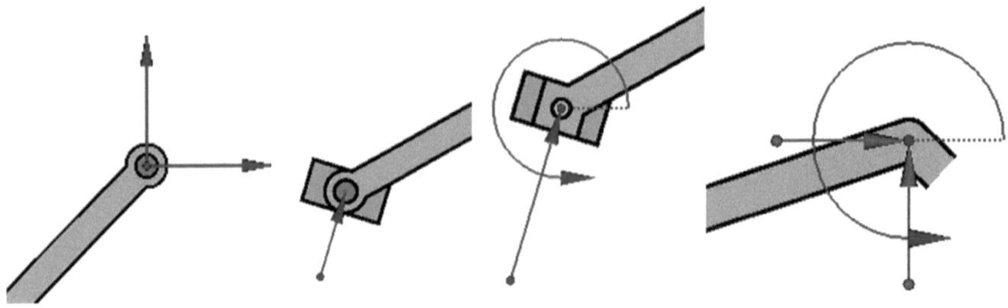

Fig. 3 Unknown transmitted loads in some types of joints

- Rolling contact: Two forces are posed, in the tangent and the normal directions to the contact. It is supposed that the contact is always capable of guaranteeing the rolling, so the user should check the value of the tangent force obtained.
- Cam contact: Frictionless contact is considered, so just one force normal to the contact direction is added to the set of unknowns.

In order to minimize the number of unknowns the action-reaction principle is applied in all joints where only two elements are connected and no external load exists. The dynamic equilibrium equations are always linear. The system of equations is formed as follows:

- For each element three equations are set, the equilibrium of forces in two different directions and the equilibrium of moments whit respect a certain point.
- For each joint where the action-reaction principle is not applicable, the two equations of equilibrium of forces are posed and if the joint is not articulated a third equation considering the equilibrium of moments is added.

Having done this, for an f degrees of freedom non-redundant mechanism, s equations with r unknowns (the reaction loads at joints) are achieved, where $s = r + f$. In order to reach a determined compatible system of equations, f unknowns more are needed. These are the loads, forces and/or moments, responsible of generating the specific motion previously simulated. By default, these are set in the joints actuated during the motion simulation, but the user at this point, has the option of changing them and locate the actuators in another joints. This ability is provided because sometimes the motion is simulated specifying the desired motion of the output element(s) instead of the actual actuated one(s).

Once the actuation loads are added to the linear system of s equations all the unknown are solved, the r reaction loads and the f actuation loads. At this point, the free solid diagrams of each mechanism element can be depicted. Obviously, this process must be performed at each discrete position across the motion.

Please note in Fig. 4 that when the mechanism is depicted assembled, just the equivalent resultants of the inertial loads distributions are drawn at each element (in fact, such resultants are used to pose and solve the aforementioned equilibrium

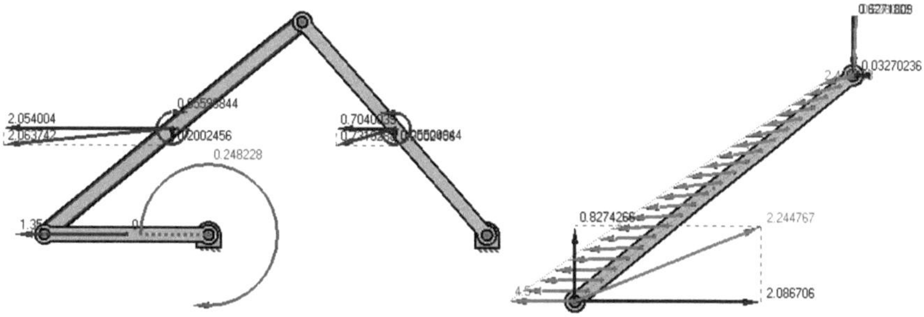

Fig. 4 Actuating loads. Joint reactions results, free solid diagram

equations), but when the complete free solid diagram of a straight axis element is depicted, the real inertial load distribution is obtained and drawn. This is required to compute the internal efforts, which is explained in the following section.

4 Internal Efforts

Internal efforts diagrams, i.e. axial (N_x) and shear $\left(V_y\right)$ forces and bending moment (M_z) only have sense in prismatic shape elements. According to the classical approaches [7, 8], when necessary, the whole beam is divided in spans and each span is analyzed separately. Span endings are given by the existence of loads in the element free solid diagram. Where a punctual load (force or moment), is applied, o where a distributed force starts or ends, there is a span ending. For the simplest case, a binary beam without external applied loads, the element does not need to be divided, due to the fact that all punctual loads are applied just in the two beam endings and the inertial load, as will be proved next, is a distributed force all across the beam length (starts and ends at the beam endings).

All punctual and distributed forces have to be projected along the beam axis direction (x) and the perpendicular direction (y). The forces projections along x direction will affect the axial force diagram, while the forces projections along y direction will affect the shear force and bending moment diagrams. Finally, punctual moments will affect only the bending moment diagram.

To compute the internal efforts, the inertial loads must be considered as distributed all across the beam element. The inertial force at each cross slice of the beam, is proportional to the opposite of the cross section center acceleration. The constant of proportionality is simply the mass per length of the beam. It can be proved in a very simple way that both projections of the inertial force distributions along x and y directions are linearly distributed forces (for a solid moving element, the extremes of the acceleration vectors of all points over a straight line are also aligned).

Being P a generic point on the moving beam, located a distance x of one extreme, point A, of the beam. $\mathbf{a_P}$ and $\mathbf{a_A}$ are the accelerations of both points, ω and α are the angular velocity and acceleration of the element. Then:

$$\mathbf{a_P} = \mathbf{a_A} + \boldsymbol{\alpha} \times \mathbf{AP} - \omega^2 \cdot \mathbf{AP} \tag{7}$$

Considering the dot product of all terms with the unit vector across the x direction, $\mathbf{u_x}$, i.e., the projection of all vectors in the equation along the beam axis:

$$\mathbf{a_P} \cdot \mathbf{u_x} = \mathbf{a_A} \cdot \mathbf{u_x} + (\boldsymbol{\alpha} \times \mathbf{AP}) \cdot \mathbf{u_x} - \omega^2 \cdot \mathbf{AP} \cdot \mathbf{u_x} \tag{8}$$

Since $\mathbf{AP}$ is parallel to $\mathbf{u_x}$ and $\boldsymbol{\alpha} \times \mathbf{AP}$ is perpendicular to $\mathbf{AP}$ and consequently to $\mathbf{u_x}$, and taking into account that $|\mathbf{AP}| = x$ and $|\mathbf{u_x}| = 1$:

$$(\mathbf{a} \times \mathbf{AP}) \cdot \mathbf{u_x} = 0 \quad \text{and} \quad \mathbf{AP} \cdot \mathbf{u_x} = x$$

So the linear force distribution along the x direction is proportional to the opposite of:

$$a_{Px} = a_{Ax} - \omega^2 \cdot x \tag{9}$$

In the same way, considering the dot product of all terms with the unit vector across the y direction, $\mathbf{u_y}$, i.e., the projection of all vectors in the equation along the perpendicular to the beam axis:

$$\mathbf{a_P} \cdot \mathbf{u_y} = \mathbf{a_A} \cdot \mathbf{u_y} + (\boldsymbol{\alpha} \times \mathbf{AP}) \cdot \mathbf{u_y} - \omega^2 \cdot \mathbf{AP} \cdot \mathbf{u_y} \tag{10}$$

Since $\mathbf{AP}$ is perpendicular to $\mathbf{u_y}$ and $\boldsymbol{\alpha} \times \mathbf{AP}$, being perpendicular to $\mathbf{AP}$, is parallel to $\mathbf{u_y}$, and taking into account that $|\mathbf{u_y}| = 1$ and $|\boldsymbol{\alpha} \times \mathbf{AP}| = \alpha \cdot x$ (because $\boldsymbol{\alpha}$ is perpendicular to $\mathbf{AP}$):

$$(\boldsymbol{\alpha} \times \mathbf{AP}) \cdot \mathbf{u_y} = \alpha \cdot x \quad \text{and} \quad \mathbf{AP} \cdot \mathbf{u_y} = 0$$

So the linear force distribution along the y direction is proportional to the opposite of:

$$a_{Py} = a_{Ay} + \alpha \cdot x \tag{11}$$

Once is known the fact that the inertial forces distributions in both directions are linear, instead of using the proportionality constants $-\omega^2$ and α, the projections of the acceleration of the second extreme of the beam, point B can be directly used. This is the most natural way of considering the inertial effect, because at the motion simulation the acceleration vectors of all mechanism nodes were computed.

Being m and L the mass and the length of the element respectively, the inertial forces distribution functions finally used are:

$$q_{in\,x,y}(x) = \left(-\mathbf{a_A} \cdot \mathbf{u_{x,y}} - \frac{\mathbf{a_B} \cdot \mathbf{u_{x,y}} - \mathbf{a_A} \cdot \mathbf{u_{x,y}}}{L} \cdot x\right)\frac{m}{L} \tag{12}$$

Considering now the well-known relations among the distributed forces functions ant the internal efforts diagrams functions:

$$q_x(x) = \frac{dN_x(x)}{dx} \quad q_y(x) = \frac{dV_y(x)}{dx} \quad V_y(x) = \frac{dM_z(x)}{dx} \tag{13}$$

It is concluded that $N_x(x)$ and $V_y(x)$ will be in general second order polynomials, while $M_z(x)$ will be a third order polynomial. This fact has the advantage that an analytical discussion (exact computation) of the internal efforts functions can be performed in an easy way to obtain all relative extremes (maximum values points) and roots (sign changing points).

Not only the internal efforts diagrams for each element (Fig. 5), but also the color maps of values for all the mechanical system (Fig. 6) can be depicted.

Obviously, when the user needs to take into account the own weights, although for making the computation of the joint reactions just a punctual force is added at the gravity center of the element, when obtaining the internal efforts in beam-type elements, the weigh is uniformly distributed.

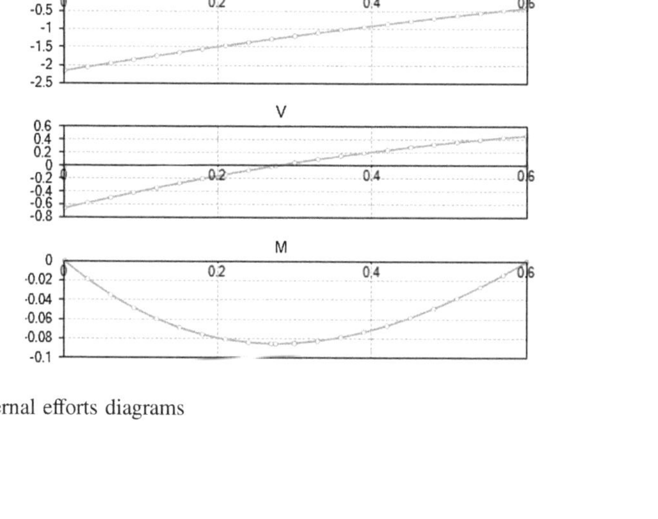

Fig. 5 Internal efforts diagrams

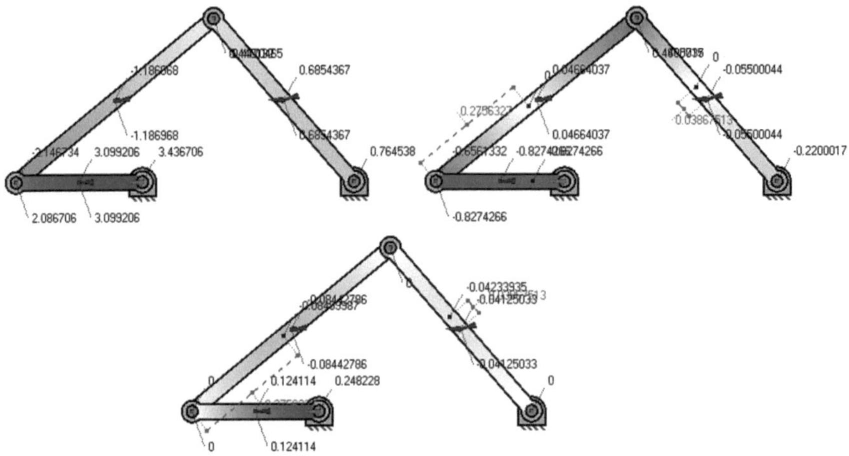

Fig. 6 Internal efforts maps. Axial forces, shear forces and bending moments

5 Conclusions

In this paper authors have presented a new general purpose software tool for solving the inverse dynamics of non-redundant planar mechanisms. A rigid solid approach is used, so no stresses or elastic deformations are computed. However, internal efforts diagrams are obtained. The main advantage is that the model pre-processing process can be completed in an extremely simple way. Furthermore, the efforts diagrams can be used to set the optimal shape of the elements cross section in an early designing step. Standard methodologies are applied in the different process steps, the motion simulation, the reactions computation, or the diagrams tracing. The main contribution is the smart combination of all those processes in a unified easy usable methodology which is now implemented in the software GIM.

Acknowledgments The authors wish to acknowledge the financial support received from the Spanish Government through the Ministerio de Economa y Competitividad (Project DPI2011-22955) and the Regional Government of the Basque Country through the Departamento de Educacin, Universidades e Investigacin (Project IT445-10) and UPV/EHU under program UFI 11/29.

References

1. GIM: www.ehu.eus/compmech/software/. CompMech Research Group, University of the Basque Country
2. Petuya, V., Macho, E., Altuzarra, O., Pinto, C., Hernandez, A.: Educational software tools for the kinematic analysis of mechanisms. Comput. Appl. Eng. Educ. **22**, 72–86 (2014)
3. ADAMS: www.mscsoftware.com. MSC Software Corporation

4. DADS: www.lmsintl.com. LMS International NV
5. Working Model: www.design-simulation.com. Design Simulation Technologies, Inc
6. Garcia de Jalon, J., Serna, M.A., Aviles, R.: Computer method for kinematic analysis of lower-pair mechanisms, velocities and accelerations, position problems. Mech. Mach. Theory **16**, 543566 (1981)
7. Waldron, K.J., Kinzel, G.L.: Kinematics, Dynamics and Design of Machinery. Wiley, New York (1999)
8. Beer, F.P., Johnston, E.R.: Vector Mechanics for Engineers. McGraw-Hill Book Company Inc., New York (1977)

Dynamic Modeling of Flexible Robot Manipulators: Acceleration-Based Discrete Time Transfer Matrix Method

Rasmus Srensen, Mathias Rahbek Iversen and Xuping Zhang

Abstract This paper presents a new and computationally efficient method for the modelling of flexible robot manipulators. The proposed method avoids the global dynamics by decomposing it to the component dynamics. The component dynamics is established, and is linearized based on the acceleration-based state vector. The transfer matrices for different type of components are created, and the systematic dynamics of a flexible robot manipulator is then established by transferring the state vector from the base to the end-effector without increasing the order of the system matrices. The numerical simulations of a flexible manipulator are conducted for verifying the proposed methodologies.

Keywords Flexible robot · Transfer matrix method · State vector · Dynamic modeling

1 Introduction

In contrast to the rigid manipulators, light weight manipulators offer advantages such as higher speed, better energy efficiency, improved mobility, and higher payload-to-arm weight ratio. However, at high operational speeds, inertial forces of moving components become quite large, leading to considerable deformation in the light links, and generating unwanted vibration phenomena. Hence, elastic vibrations of light weight links must be taken into account in the modelling, design, and control of the robot manipulators. In the past decades, significant progresses have been made into the dynamic modeling of manipulators with flexible components [1]. Different discretization techniques, such as the finite element method (FEM) [2, 3], the assumed mode method (AMM) [4, 5], and the lumped parameter method (LPM) [6, 7], have been reported extensively for modeling the dynamics of flexible

R. Srensen · M.R. Iversen · X. Zhang (✉)
Department of Engineering, Aarhus University, Inge Lehmanns Gade 10, 8000 Aarhus C, Denmark
e-mail: xuzh@ase.au.dk

S. Bai and M. Ceccarelli (eds.), *Recent Advances in Mechanism Design for Robotics*, Mechanisms and Machine Science 33,
DOI 10.1007/978-3-319-18126-4_36

robot manipulators. However, the matrix size of global dynamic model of a robot manipulator increases with the number of the Degree of Freedom, and therefore heavy computation of dynamic modelling is still a big concern in terms of real-time control.

Alternatively, the transfer matrix method can be used to model linear and continuous parameter systems without discretization [8, 9]. Using the integration procedure, the discrete-time transfer matrix method (DT-TMM) was presented to perform the dynamic analysis of large systems that consists of large subsystems, each of which is a simple dynamic element. The DT-TMM was further developed to model multi-body system dynamics using linearization and integral schemes. With the TD-MM, the local dynamics is transmitted and concatenated through the transfer matrix of each subsystem. As a result, the matrix size of the dynamic equations does not increase with the DOF of multi-body systems as the traditional procedure of establishing the global dynamic equations are avoided. Therefore, the computation efficiency can be significantly improved. In this work, the DT-transfer matrix method is extended to the dynamic modeling of lightweight robot manipulators with the consideration of link flexibility. The procedure of DT-TMM presented in this work for modeling the dynamics of flexible manipulators is: (1) a robot manipulator is decomposed into subsystems or elements: links, motors, and connections between a link and motors; (2) the dynamics model of each element or subsystem is established; (3)the state vector of the input board and the output board of the element is defined, the dynamics of each element is linearized in terms of the defined state vector of the inboard and the outboard, and the transfer matrix is formulated to transform the state vector of the element from the inboard to the outboard; (4) the boundary conditions between the links is applied, and the transformation of the state vector is generated from one link to the neighboring links so that the dynamics of links is connected with motors and fixed mounting; (5) the transformation of the dynamics of the robot manipulator are conducted from the base to the last link, and hence the global transfer matrix is obtained base on the transformation of the input state vector at the base and the output state vector at the tip of the manipulator. The global dynamics is avoided using the proposed method. The matrix size of the robot manipulator dynamics does not increase with the DOF of the robot manipulator, and hence significantly reduce the computation cost.

2 State Vectors and Transformation

The state vector is a column vector that represents the internal forces (force q_x, q_y, q_z, and moment M_x, M_y, M_z) and displacement (rigid motion $x, y, z, \theta_x, \theta_y, \theta_z$

and deformation $w^1, w^2, \ldots, w^n$) at a particular location within a system. In this work, the acceleration-based integral scheme is selected as

$$\mathbf{z} = [\overbrace{\ddot{x}, \ddot{y}, \ddot{z}}^{\text{acceleration}}, \underbrace{\ddot{\theta}_x, \ddot{\theta}_y, \ddot{\theta}_z}_{\text{angular acceleration}}, M_x, M_y, M_z, q_x, q_y, q_z, \overbrace{\ddot{w}^1, \ddot{w}^2, \ldots, \ddot{w}^n}^{\text{modal acceleration}}, 1]^T \quad (1)$$

The sign convention of the elements in the state vector related to the reference coordinate system is illustrated in Fig. 1. The position coordinates and orientation angles are defined as positive when they are in the positive directions of the coordinate axes. The inboard forces and outboard moments applied on the elements are positive if they are in the positive directions of the coordinate axes, and the outboard forces and inboard moments acted on the elements are negative if they are in the positive directions of the coordinate axes.

In the DT-TMM, a transfer matrix is formulated to transform the state vector form one end of the component to the other end of the component as

$$\underset{[n\times 1]}{\mathbf{z}^O} = \underset{[n\times m]}{\mathbf{U}} \underset{[m\times 1]}{\mathbf{z}^I} \quad (2)$$

For a robot manipulator system, the order of the system transfer matrix $\mathbf{U}_{sys}$ depends on the size of the boundary state vectors of the system.

$$\underset{[n\times 1]}{\mathbf{z}^O} = \mathbf{U}_j\mathbf{U}_{j-1}\cdots\mathbf{U}_2\mathbf{U}_1 \underset{[m\times 1]}{\mathbf{z}^I} = \underset{[n\times m]}{\mathbf{U}_{sys}} \underset{[m\times 1]}{\mathbf{z}^I} \quad (3)$$

In (3), j is the number of elements divided from the robot manipulator system. It is clear that the order of the system matrix does not increase when the degrees of freedom of the manipulator increase, which results in significant reduction of the computational time and storage requirements. Each component of a robot

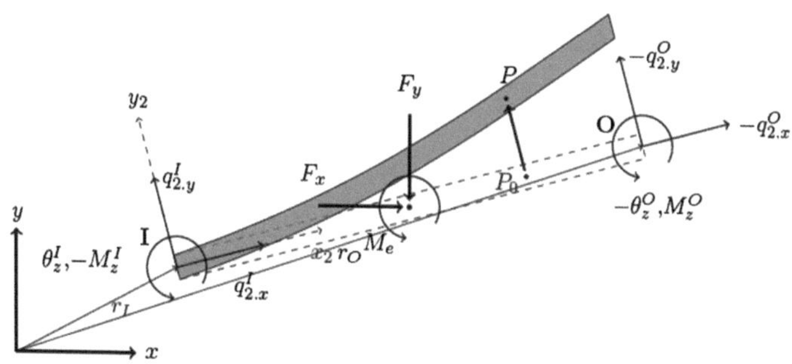

Fig. 1 State vectors and coordinate system

manipulator has a unique transfer matrix based on the physical behavior it models. The transfer matrices of different elements are established to propagate the state vectors of flexible robot manipulator components in Sect. 4.

3 Linearization and Integration Schemes

The nonlinearity terms may result from 1st and 2nd second order time derivatives, trigonometric terms, multiplications and couplings of variables related to the state vectors. This brings the challenges, if not possible, to formulate the linear transformation between the outboard state vector and the inboard state vector for a subsystem, as shown in Eq. (3). Additionally if no linearization is performed, the transfer matrix may contain some terms related to the current values of the stave vector variables. This would be incompatible with the TMM formulation, as transfer matrices are only allowed to have state vector variables related to the previous time steps. Mathematically, linearization is carried using the method of a Taylor series expansion. The linearization is not detailed in this work due to the limited space. The dynamic equations of the components usually contain terms of the displacement (linear and angular) variables, velocity (linear and angular), and accelerations (linear and angular). To transfer the state vectors with transfer matrices as in Eq. (3), the displacement and velocity variables need to be described as the functions of the variables at the previous time step when the acceleration-based integration scheme. This process can be conducted through an implicit integration method. The principle of an implicit integration is to find a solution by solving an equation consisting of both the current step variables and previous step variables. Assuming i−1 to be the previous time step, and i the current time step, the general form of an implicit scheme can be written as $\xi = \xi^{(i-1)} + f(\xi, t)$, where ξ and $\xi^{(i-1)}$ are the current and previous time stable variables respectively, and f is a function of the current time step variables. Using Newmark-β integration method [10], this implicit equation can be solved and rewritten in terms of TMM parameters as

$$\xi = \beta \Delta t^2 \ddot{\xi} + \xi^{(i-1)} + \dot{\xi}^{(i-1)} \Delta t + \left(\frac{1}{2} - \beta\right) \ddot{\xi}^{(i-1)} \Delta t^2 \tag{4}$$

$$\dot{\xi} = \gamma \Delta t \ddot{\xi} + \dot{\xi}^{(i-1)} + (1 - \gamma) \ddot{\xi}^{(i-1)} \Delta t \tag{5}$$

where the integration parameters γ and β are weighting coefficient in the Newmark method, and the two weighting parameters γ and β play a key role in the stability and convergence of analyses. It was proved that $\gamma = 0.5$, or else spurious damping will be added to the system. In general, the Newmark scheme has shown to be *unconditionally stable* with the following parameters $2\beta \geq \gamma \geq \frac{1}{2}$.

4 Transfer Matrix of Components

To establish the system transfer matrix of a robot manipulator, the robot manipulator system is broken down to motors and links. A local transfer matrix is then formulated for each link, motor, and connection between a motor and a link.

4.1 Dynamics and Transfer Matrix of a Flexible Link Element

The lightweight links are assumed as a slender beam, and the state vectors and coordinate system are defined as shown in Fig. 1. The linear transformation is formulated based on the kinematics, the force and moment balance of an Euler-Bernoulli beam. The elastic deformation u can then be described by the spatial mode shape function $\mathbf{N}(x)$ and the time-dependent modal coordinates $\mathbf{w}(t)$ as $u(x,t) = \mathbf{N}(x)\mathbf{w}(t)$. In this work, only the first three vibration modes are included to calculated the elastic deformation. The kinematic relation from inboard to outboard end is established, and then is differentiated twice with respect to time as

$$\begin{bmatrix} \ddot{x} \\ \ddot{y} \\ \ddot{\theta} \end{bmatrix}^{O} = \begin{bmatrix} \ddot{x}^{I} - \ddot{\theta}\sin\theta L - \dot{\theta}^{2}\cos\theta L - \ddot{u}\sin\theta - 2\dot{u}\dot{\theta}\cos\theta - u\ddot{\theta}\cos\theta_{z} + u\dot{\theta}_{z}^{2}\sin\theta_{z} \\ \ddot{y}^{I} + \ddot{\theta}\cos\theta L - \dot{\theta}^{2}\sin\theta L + \ddot{u}\cos\theta - 2\dot{u}\dot{\theta}\sin\theta - u\ddot{\theta}_{z}\sin\theta_{z} - u\dot{\theta}^{2}\cos\theta \\ \ddot{\theta}^{I} \end{bmatrix} \quad (6)$$

The force and moment balance equations of the flexible beam are conducted based on the Newton-Euler equations by including the impact of inertial forces and deformation (the force and moment equations are not detailed due to the limited space). The dynamic equations are then linearized, reorganized, and combined with Eq. (6) to generate the transfer formation equation of a flexible link as

$$\mathbf{z}^{O} = \mathbf{U}\mathbf{z}^{I}$$
$$= \begin{bmatrix} 1 & 0 & u_{1,3} & 0 & 0 & 0 & u_{1,7} & u_{1,8} & u_{1,9} & u_{1,10} \\ 0 & 1 & u_{2,3} & 0 & 0 & 0 & u_{2,7} & u_{2,8} & u_{2,9} & u_{2,10} \\ 0 & 0 & 1 & 0 & 0 & 0 & 0 & 0 & 0 & 0 \\ u_{4,1} & u_{4,2} & u_{4,3} & 1 & 0 & 0 & u_{4,7} & u_{4,8} & u_{4,9} & u_{4,10} \\ u_{5,1} & 0 & u_{5,3} & 0 & 1 & 0 & u_{5,7} & u_{5,8} & u_{5,9} & u_{5,10} \\ 0 & u_{6,2} & u_{6,3} & 0 & 0 & 1 & u_{6,7} & u_{6,8} & u_{6,9} & u_{6,10} \\ 0 & 0 & 0 & 0 & 0 & 0 & 1 & 0 & 0 & 0 \\ 0 & 0 & 0 & 0 & 0 & 0 & 0 & 1 & 0 & 0 \\ 0 & 0 & 0 & 0 & 0 & 0 & 0 & 0 & 1 & 0 \\ 0 & 0 & 0 & 0 & 0 & 0 & 0 & 0 & 0 & 1 \end{bmatrix} \begin{bmatrix} \ddot{x} \\ \ddot{y} \\ \ddot{\theta} \\ M_{z} \\ q_{x} \\ q_{y} \\ \ddot{w}^{1} \\ \ddot{w}^{2} \\ \ddot{w}^{3} \\ 1 \end{bmatrix}^{I} \quad (7)$$

The components of the transfer matrix are listed at the Ref. [12].

4.2 Dynamics and Transfer Matrix of a Rotational Motor Joint

Physically, the motor (driven joint) is used to generate the rotation between the mounted inboard and outboard links. The angular acceleration related to the two attached bodies is expressed as $\ddot{\theta}^I_{(\kappa+1)} = \ddot{\theta}^I_{(\kappa-1)} + \ddot{\theta}^D$ where $\ddot{\theta}^D$ is the angle driver. Combining the torque and force relationship at both ends of the motor, the transfer matrix of a motor is given as

$$\mathbf{z}^O = \mathbf{U}_\kappa \mathbf{z}^I = \begin{bmatrix} 1 & 0 & 0 & 0 & 0 & 0 & 0 \\ 0 & 1 & 0 & 0 & 0 & 0 & 0 \\ 0 & 0 & 1 & 0 & 0 & 0 & \ddot{\theta}^D_z \\ 0 & 0 & 0 & 1 & 0 & 0 & 0 \\ 0 & 0 & 0 & 0 & 1 & 0 & 0 \\ 0 & 0 & 0 & 0 & 0 & 1 & 0 \\ 0 & 0 & 0 & 0 & 0 & 0 & 1 \end{bmatrix} \begin{bmatrix} \ddot{x} \\ \ddot{y} \\ \ddot{\theta} \\ M_z \\ q_x \\ q_y \\ 1 \end{bmatrix}^I \tag{8}$$

4.3 Dynamics and Transfer Matrix of Fixed Mountings

In this work, the connection between a motor and its attached links is modeled as a fixed mounting. The mounting is denoted element κ, and inboard and outboard body are denoted $\kappa - 1$ and $\kappa + 1$, respectively. When both inboard and outboard links are rigid bodies, the transfer matrix of the fixed mounting is an identity matrix. However when a flexible beam is inboard and/or outboard body, the mounting becomes slightly more complex.

4.3.1 Fixed Mounting with Rigid Inboard Body and Flexible Outboard Beam

$$EI_{2.z}\frac{\partial^4 u}{\partial x_2^4} = \bar{f}_{2.y} - \frac{\partial}{\partial x_2}\bar{M}_e - \rho A\left(\ddot{y}_2{}^{O_2} + \ddot{\theta} x_2 - \dot{\theta}^2 u + \ddot{u}\right) \tag{9}$$

Integrating Eq. (9) along the beam length and including the effect of centrifugal stiffening, and then the equation can be rewritten written in a matrix format by the linearization, rearrangement, and matrix partition as

$$\mathbf{z}^O = \mathbf{U}\mathbf{z}^I = \begin{bmatrix} 1 & 0 & 0 & 0 & 0 & 0 & 0 \\ 0 & 1 & 0 & 0 & 0 & 0 & 0 \\ 0 & 0 & 1 & 0 & 0 & 0 & 0 \\ 0 & 0 & 0 & 1 & 0 & 0 & 0 \\ 0 & 0 & 0 & 0 & 1 & 0 & 0 \\ 0 & 0 & 0 & 0 & 0 & 1 & 0 \\ u_{7,1} & u_{7,2} & u_{7,3} & 0 & 0 & 0 & u_{7,7} \\ u_{8,1} & u_{8,2} & u_{8,3} & 0 & 0 & 0 & u_{8,7} \\ u_{9,1} & u_{9,2} & u_{9,3} & 0 & 0 & 0 & u_{9,7} \\ 0 & 0 & 0 & 0 & 0 & 0 & 1 \end{bmatrix} \begin{bmatrix} \ddot{x} \\ \ddot{y} \\ \ddot{\theta} \\ M \\ q_x \\ q_y \\ 1 \end{bmatrix}^I \tag{10}$$

All the items $u(i,j)$ in Eq. (10) are detailed in Ref. [12].

4.3.2 Fixed Mounting with a Flexible Inboard Body and a Flexible Outboard Body

The relationship between inboard and outboard angle of the inboard body $\kappa - 1$ needs to be included and can be described as

$$\begin{aligned} \boldsymbol{\theta}^O_{z(\kappa-1)} &= \boldsymbol{\theta}^I_{z(\kappa-1)} + \left.\frac{\partial u_{y(\kappa-1)}}{\partial x_2}\right|_{x_2=L} \\ &= \boldsymbol{\theta}^I_{z(\kappa-1)} + \mathbf{N}'_y\Big|^{(\kappa-1)}_{x_2=L} \mathbf{w}_{y(\kappa-1)} \\ &= \boldsymbol{\theta}^I_{z(\kappa+1)} \end{aligned} \tag{11}$$

The second order differentiation of Eq. (11) with respect to time is written as

$$\begin{aligned} \ddot{\theta}^O_{z(\kappa-1)} &= \ddot{\theta}^I_{z(\kappa-1)} + \mathbf{N}'_y\Big|_{x_2=L} \ddot{\mathbf{w}}_{y(\kappa-1)} \\ &= \ddot{\theta}^I_{z(\kappa+1)} \end{aligned} \tag{12}$$

The calculation of modal coordinates at the outboard link must be based on the actual outboard angle of the inboard link. Combining the equations of the elastic deformation, force, and moment, the transfer matrix is given as

$$\mathbf{z}^O = \mathbf{U}\mathbf{z}^I = \begin{bmatrix} 1 & 0 & 0 & 0 & 0 & 0 & 0 & 0 & 0 & 0 \\ 0 & 1 & 0 & 0 & 0 & 0 & 0 & 0 & 0 & 0 \\ 0 & 0 & 1 & 0 & 0 & 0 & u_{3,7} & u_{3,8} & u_{3,9} & 0 \\ 0 & 0 & 0 & 1 & 0 & 0 & 0 & 0 & 0 & 0 \\ 0 & 0 & 0 & 0 & 1 & 0 & 0 & 0 & 0 & 0 \\ 0 & 0 & 0 & 0 & 0 & 1 & 0 & 0 & 0 & 0 \\ u_{7,1} & u_{7,2} & u_{7,3} & 0 & 0 & 0 & u_{7,7} & u_{7,8} & u_{7,9} & u_{7,10} \\ u_{8,1} & u_{8,2} & u_{8,3} & 0 & 0 & 0 & u_{8,7} & u_{8,8} & u_{8,9} & u_{8,10} \\ u_{9,1} & u_{9,2} & u_{9,3} & 0 & 0 & 0 & u_{9,7} & u_{9,8} & u_{9,9} & u_{9,10} \\ 0 & 0 & 0 & 0 & 0 & 0 & 0 & 0 & 0 & 1 \end{bmatrix} \begin{bmatrix} \ddot{x} \\ \ddot{y} \\ \ddot{\theta}_z \\ M_z \\ q_x \\ q_y \\ \ddot{w}^1 \\ \ddot{w}^2 \\ \ddot{w}^3 \\ 1 \end{bmatrix}^I \tag{13}$$

All the items $u(i,j)$ in Eq. (13) are detailed in Ref. [12].

5 Case Studies and Simulations

The simulations of a single link manipulator are carried out using parameters identical to those used by Shabana and Berzeri in Ref. [11]. The comparison, as shown in Fig. 2, is made to verify the proposed model in this work. Figure 2 shows that the tip displacement of the rotating flexible beam calculated from the model in this work is very close the the result in Ref. [11]. Figure 3 compares the results based on the DT-TMM method and the traditional Lagrange-Euler method. The comparison demonstrates that the results of the two methods agree well with each other.

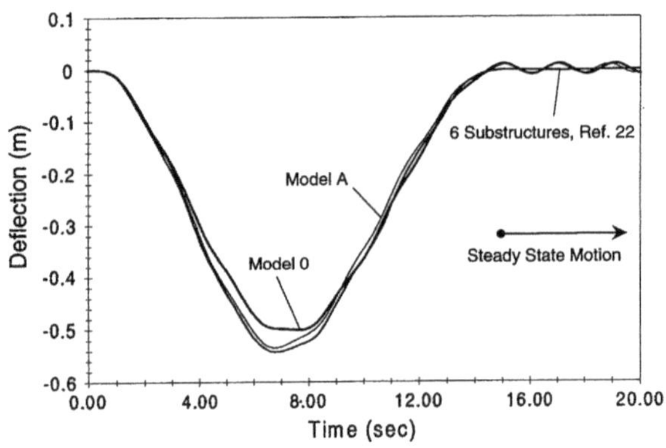

Fig. 2 Tip-displacement from a simulation by Shabana using ANCF elements with centrifugal stiffening [11]

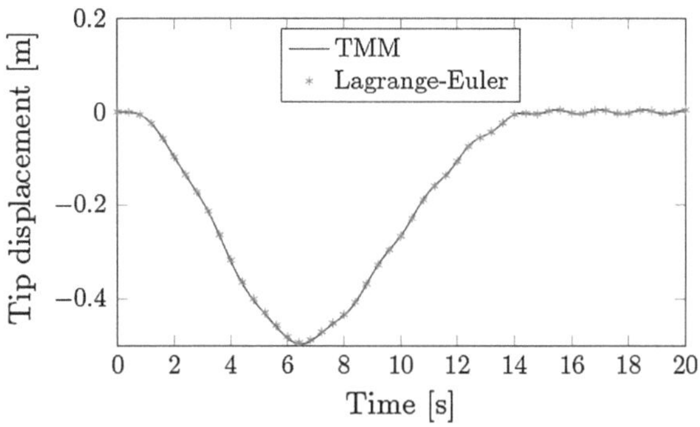

Fig. 3 Comparison of tip-displacement between TMM and Euler-Lagrange for simulations with $\Delta t = 0.001$ giving a Mean Absolute Percentage Error of MAPE $= 0.57\%$

6 Conclusion and Discussion

A computationally efficient dynamic modelling approach is presented in this research. With the proposed method, the computation of the large size dynamic equations is avoided by dividing the global dynamics to component dynamics. The methodology of defining state vectors, linearizing the component dynamics, and deriving transfer matrix for different types of components has been presented with details. Numerical simulations of flexible manipulators are conducted to verify the proposed dynamic modelling method.

References

1. Dwivedy, S.K., Eberhard, P.: Dynamic analysis of flexible manipulators-a literature review. Mech. Mach. Theory **41**, 749–777 (2006)
2. Zhang, X., Yu, Y.: A new spatial rotor beam element for modeling spatial manipulators with moint and link flexibility. Mech. Mach. Theory **35**(3), 403–421 (2000)
3. Usoro, P.B., Nadira, R., Mahil, S.S.: A finite element/Lagrangian approach to modeling lightweight flexible manipulators. ASME J. Dyn. Syst. Meas. Control **108**, 198205 (1986)
4. Book, W.J.: Recursive Lagrangian dynamics of flexible manipulator arms. Int. J. Rob. Res. **3** (3), 87–101 (1984)
5. Zhang, X., Mills, J.K., Cleghorn, W.L.: Dynamic modeling and experimental validation of a 3-PRR parallel manipulator with flexible intermediate link. J. Intell. Rob. Syst. **50**(4), 323–340 (2007)
6. Ge, S.S., Lee, T.H., Zhu, G.: A new lumping method of a flexible manipulator. In: Proceedings of the American Control Conference, Albuquerque, New Mexico, pp. 1412–1416, June 1997
7. Carbone, G.: Carbone: stiffness analysis and experimental validation of robotic systems. Front. Mech. Eng. **6**, 182–196 (2011)

8. Kitis, L.: Natural frequencies and mode shapes of flexible mechanisms by a transfer matrix method. Finite Elem. Anal. Des. **6**, 267–285 (1990)
9. Rui, X., He, B., Lu, Y., Lu, W., Wang, G.: Discrete time transfer matrix method for multibody system dynamics. Multibody Syst. Dyn. **14**, 317–344 (2005)
10. Newmark, N.M.: A method of computation for structural dynamics. J. Eng. Mech. Div. Proc. Am. Soc. Civil Eng. **85**, 67–94 (1959)
11. Berzeri, M., Shabana, A.A.: Study of the centrifugal stiffening effect using the finite element absolute nodal coordinate formulation. Technical Report MBS99-5-UIC (2001)
12. Srensen, R., Iversen, M.R.: Dynamic modeling for wind turbine instability analysis using discrete time transfer matrix method. Master Thesis, Department of Engineering, Aarhus University (2014)

Comparative Study on the Kinematic and Static Performance of Two 1T2R Parallel Manipulators

Binbin Lian, Hélène Chanal and Yan Jin

Abstract Focusing on the potential of one translational and two rotational (1T2R) parallel kinematic machines (PKMs) for high precision manufacturing, this paper carried out the first time a comparative study on the kinematic and static performance of two promising Exechon variants, named PAW and PAW-II. PAW has the topology with 2R$\underline{P}$U-S$\underline{P}$R while the topology of PAW-II is 2R$\underline{P}$U-R$\underline{P}$S. Herein, R, U, $\underline{P}$, S denote revolute joint, universal joint, actuated prismatic joint and spherical joint, respectively. After introducing their architectures and inverse kinematics, two key performance characteristics, i.e. workspace and stiffness were conducted for the two PKMs. Comparative results show that PAW has a smaller workspace but better stiffness performance comparing to PAW-II.

Keywords Parallel manipulators · Parallel kinematic machine · Workspace analysis · Stiffness modelling

B. Lian · Y. Jin (✉)
School of Mechanical and Aerospace Engineering,
Queen's University Belfast, Belfast, UK
e-mail: y.jin@qub.ac.uk

B. Lian
e-mail: b.lian@qub.ac.uk

B. Lian
School of Mechanical Engineering, Tianjin University, Tianjin, China

H. Chanal
Institut Pascal, Clermont Université, Clermont-Ferrand, France
e-mail: helene.chanal@ifma.fr

S. Bai and M. Ceccarelli (eds.), *Recent Advances in Mechanism Design for Robotics*, Mechanisms and Machine Science 33,
DOI 10.1007/978-3-319-18126-4_37

1 Introduction

In the past three decades, parallel kinematic machines (PKMs) have become a research hotspot among academia and industry, because of their higher stiffness, higher accuracy and higher dynamic capability comparing to serial manipulators. In the fields like aerospace manufacture and assembly, PKMs show huge potential as machine tools to meet the requirements of high precision, high working load and high productivity [1]. Equipped with three degree-of-freedom (3-DoF) PKMs which provide one translational and two rotational (1T2R) motion DoFs, 5-axis hybrid manipulators such as Tricept mechanism [2, 3] and Sprint Z3 head [4, 5] are successful applications of these PKMs.

Although a large number of 3-DOF PKMs have been produced [6, 7], the PKMs which can offer both translational and rotational DOFs have not been extensively studied. In particular, the PKMs with 1T2R DOFs have special importance in machine tools as they can possibly replace the existing problematic two serial rotational motion axes which bring negative effects on the response speed and stiffness of 5-axis machine tools. Among these existing 1T2R PKMs, axial symmetrical 3R$\underline{P}$S architecture and its variant 3$\underline{P}$RS architecture are the main focus. Research has been conducted on their kinematic analysis and optimization [8–11], dynamic performance [12, 13], accuracy [14] and stiffness [15]. It is obvious that 3R$\underline{P}$S PKM and 3$\underline{P}$RS PKM are constructed by the same set of kinematic joints but of different topology in each leg. Despite lots of work has been conducted on these two PKMs independently, there is no special focus on the effects of different joint arrangements to the performance of the whole mechanism, which is one of the important issues in designing PKMs.

Similar to the concept of 3$\underline{P}$RS PKM, Neumann [16] invented a promising machine tool called Exechon which consists of a novel 1T2R PKM and a 2-DOF serial wrist. The unique over-constrained architecture enables Exechon to achieve rather high stiffness and accuracy [17]. Inspired by the Exechon PKM, its variants PAW and PAW-II are recently proposed [18]. PAW has a 2R$\underline{P}$U-S$\underline{P}$R architecture and PAW-II has a 2R$\underline{P}$U-R$\underline{P}$S architecture. According to the mobility analysis, both PAW and PAW-II PKMs have 1T2R motion DOF and over-constrained architectures which led to simpler mechanisms with less number of links and joints [18]. It is believed that these two variants of Exechon are especially suitable for aerospace manufacturing tasks along curved surface. In order to further understand their performances, this paper carries out the comparative study of the two novel PKMs. Specially, as the only disparity of PAW and PAW-II is from the joint sequence of the leg, the comparative study will also shed light to the effect of varied joint arrangement, which will provide insightful knowledge in PKM design.

Having outlined the challenges mentioned above, the reminder of the paper is organized as follows. Section 2 describes the architecture of PAW and PAW-II and establishes their inverse kinematics. In Sect. 3, the orientation workspaces of the two PKMs are investigated with the same set of geometrical parameters, while their end-effector deflections with applied efforts are formulated by strain energy and Castigliano's theorem in Sect. 4. The conclusions are drawn in Sect. 5.

2 Architecture Description and Inverse Kinematics

As shown in Fig. 1a, PAW consists of fixed base, moving platform and three legs in between. Leg_1 and leg_3 are identical. Each of them has a R$\underline{\text{P}}$U chain, with an R joint connected to the fixed base, linked to a linear actuator and a U joint to the moving platform. The two axes connected to the moving platform of the two U joints from leg_1 and leg_3 are collinear, while the other two axes of the two U joints are in parallel with the R joints on the base of the two legs. The topology of leg_2 is S$\underline{\text{P}}$R and the R joint axis is parallel to the collinear axis of U joints. The three legs are all actuated by linear actuators. Figure 1b shows the architecture of PAW-II whose leg_2 is slightly different from PAW. Leg_2 of PAW-II is composed of one R joint on the base, followed by one linear actuator and one S joint connecting to the moving platform.

For the convenience to describe the motions, the same coordinate systems are defined for PAW and PAW-II. A_i and B_i $(i = 1, 2, 3)$ denote the joint centers to the base and moving platform respectively. The distance between point A_1 (B_1) to point A_3 (B_3) is l_1 (l_3). Point O is the middle point of A_1A_3 and the base coordinate system $\{O - X_bY_bZ_b\}$ is assigned to it, in which X_b is directed from point A_1 to point A_3, Y_b is from point O towards point A_2. Similarly, point O_e is the central point of B_1B_3. The platform coordinate system $\{O_e - X_eY_eZ_e\}$ is defined at point O_e, and X_e is

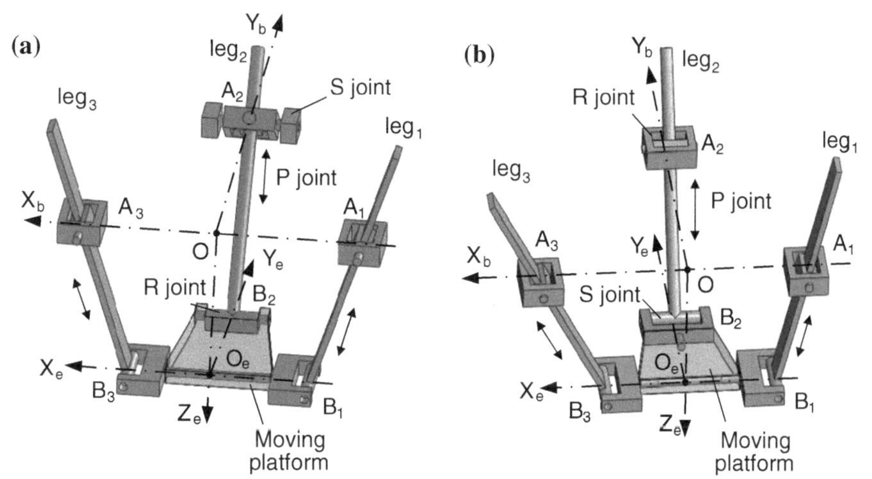

Fig. 1 Schematic diagram of **a** PAW (2RPU-1SPR) and **b** PAW-II (2RPU-1SPR)

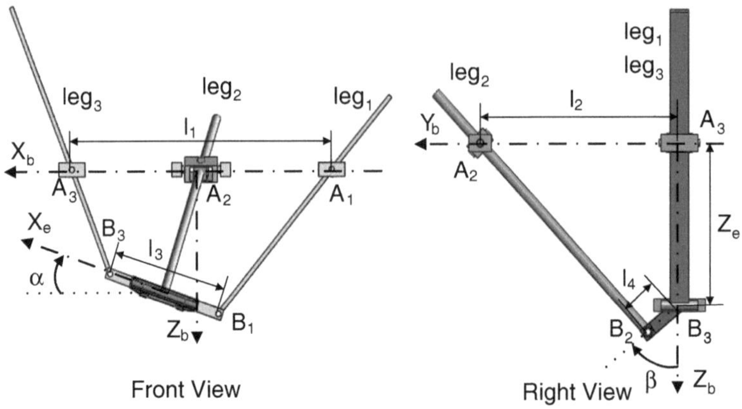

Fig. 2 Parameter description of PAW and PAW-II

directed from point B_1 to point B_3, Y_e from point O_e to point B_2. The distance between point O (O_e) to point A_2 (B_2) is l_2 (l_4).

Mobility analysis shows that both PAW and PAW-II have 1T2R DoF [18]. The posture of moving platforms can be expressed by three parameters α, β and z_e (See Fig. 2). α represents the angle of frame $\{O_e\}$ rotating about Y_b measured from X_b to X_e, β denotes the angle between the platform and the plane formed by points A_1, B_1, B_3 and A_3. Let ${}^{O}_{O_e}R$ be the rotation matrix of frame $\{O_e\}$ to frame $\{O\}$, and it is formulated as

$$ {}^{O}_{O_e}R = \begin{bmatrix} c\alpha & -c\beta \cdot s\alpha & -s\beta \cdot s\alpha \\ 0 & s\beta & -c\beta \\ s\alpha & c\beta \cdot c\alpha & s\beta c\alpha \end{bmatrix} \tag{1} $$

where s represents the sine function, and c the cosine function.

For PAW, since point O_e is always moves within the plane formed by points A_1, B_1, B_3, A_3, vector $\boldsymbol{OO}_e$ can be expressed as $\boldsymbol{OO}_e = [x_e \quad 0 \quad z_e]^T$ in the base frame. Given the constraint offered by leg_2, the determination of x_e for the two PKMs are different. For PAW, as X_b is perpendicular to the plane formed by points A_2, B_2, O_e, x_e is obtained as follow.

$$ x_e = -\tan\alpha \cdot z_e \tag{2} $$

For PAW-II, as the R joint axis of leg_2 is perpendicular to A_2B_2, x_e is calculated as

$$ x_e = \cos\beta \cdot \sin\alpha \cdot l_2 \tag{3} $$

Therefore, when α, β and z_e are known, the actuators' displacements are obtained by

$$q_i = |\boldsymbol{A}_i\boldsymbol{B}_i| = \left|\boldsymbol{OO}_e + {}^{O}_{O_e}R \cdot \boldsymbol{O}_e\boldsymbol{B}_i - \boldsymbol{OA}_i\right|, \quad (i = 1, 2, 3) \tag{4}$$

As a result, both PAW and PAW-II have a unique inverse kinematic solution.

3 Kinematic Performance

Given the same dimensions to the two PKMs, as shown in Table 1, the workspace of PAW and PAW-II can be defined as the set of α, β and z_e values, which can be obtained by examining the boundary formed by constraints via inverse kinematics.

For both PAW and PAW-II, the searching ranges of α, β and z_e are set as $[-90°, 90°]$, $[0, 180°]$ and $[0.9\,\text{m}, 1.4\,\text{m}]$. The constraints of three legs are $0.8\,\text{m} \le q_1 \le 1.5\,\text{m}$, $0.8\,\text{m} \le q_3 \le 1.5\,\text{m}$ $0.9\,\text{m} \le q_2 \le 1.6\,\text{m}$, and the rotating angles between R joints (on the base) and Z_b axis are within $[20°, 160°]$.

The orientation workspaces for the two PKMs are shown in Figs. 3 and 4. It is obvious that the motion range of β could reach from 0° to 180°, while the motion range of α will become smaller with increasing z_e value for both PAW and PAW-II.

Table 1 Nominal values of dimensional parameters

l_1	l_2	l_3	l_4
1000 mm	635 mm	353 mm	106 mm

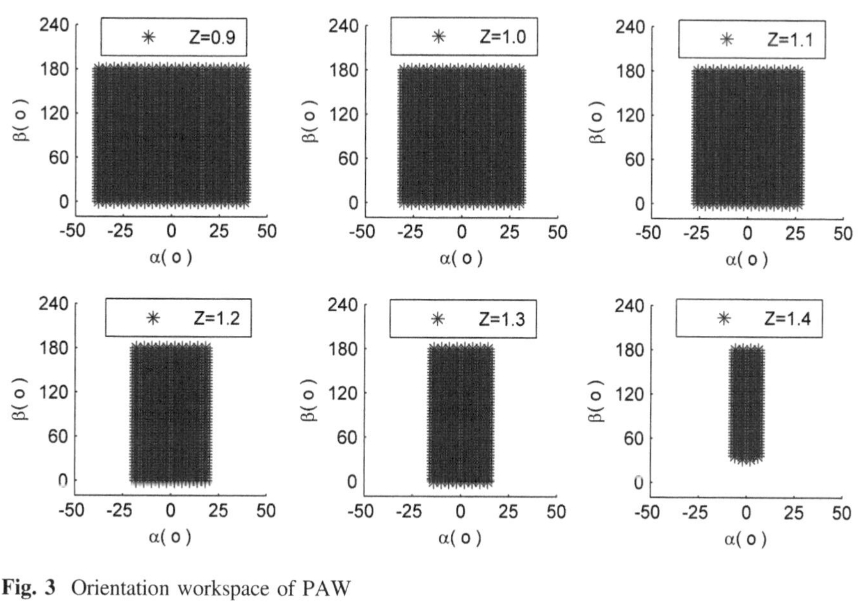

Fig. 3 Orientation workspace of PAW

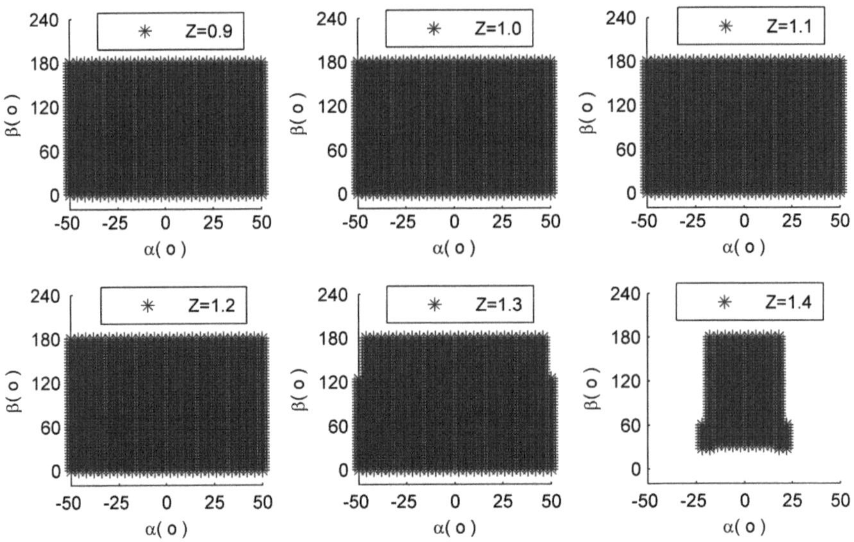

Fig. 4 Orientation workspace of PAW-II

Examining their architectures, the motion range of α is constrained by the lengths of leg_1 and leg_3, as well as the location of the spherical joint in leg_2. Since the spherical joint is on the moving platform rather than on the base, PAW-II has a much larger orientation capability in α than PAW. The orientation workspaces are symmetrical about $\alpha = 0$. This is because of their symmetrical architecture about the Y_bOZ_b plane at home position.

4 Static Performance

Bonnemains et al. [19, 20] proposed energy based method for stiffness modeling, taking into account joint and leg deflections for over-constrained PKMs. In this paper, the same method is applied to PAW and PAW-II. Related stiffness characteristics of the two PKMs are listed in Table 2. By applying 1kN to point O_e in x, y, z directions respectively, the displacements in every direction is shown in Figs. 4, 5, 6, 7, 8 and 9.

Table 2 Stiffness characteristics of PAW and PAW II PKMs

Element	Stiffness characteristic	Value
Bearing assembly	Radial stiffness	$1.04e^{9}$ N/m
Screw	Section	0.001 m^2
Rail	Moments of inertia about x_i y_i	$8.2448e^{-8}$ m^2

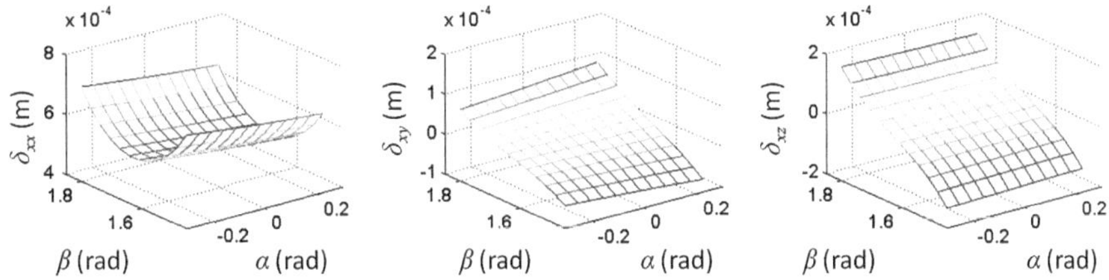

Fig. 5 Displacements of point O_e for a 1 kN effort in $\boldsymbol{x}$ direction and $z_e = 1$ m for PAW

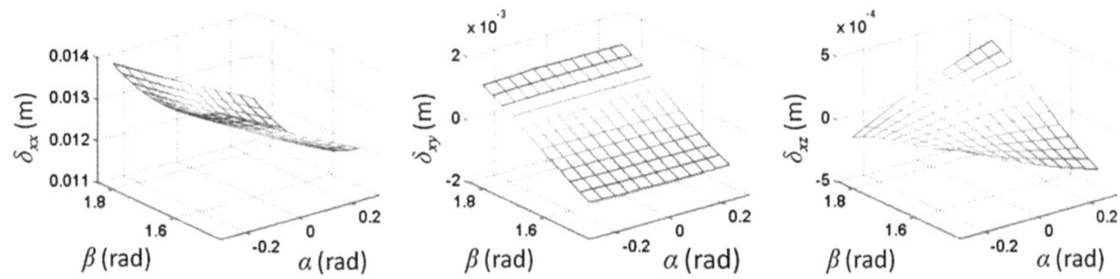

Fig. 6 Displacements of point O_e for a 1 kN effort in $\boldsymbol{x}$ direction and $z_e = 1$ m for PAW-II

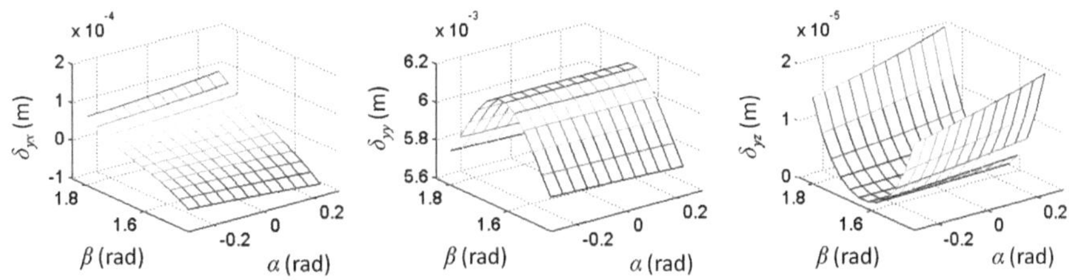

Fig. 7 Displacements of point O_e for a 1 kN effort in $\boldsymbol{y}$ direction and $z_e = 1$ m for PAW

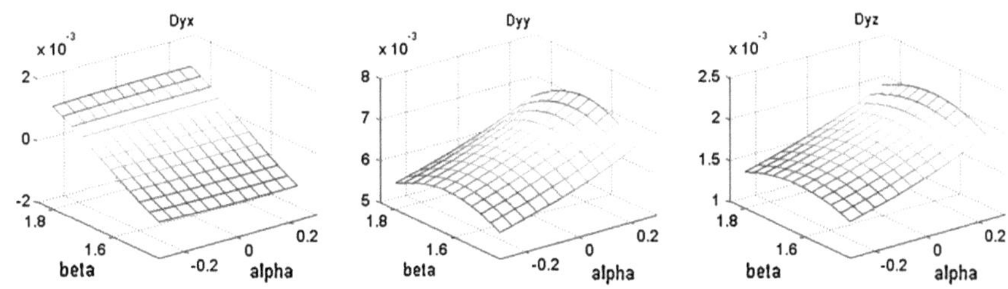

Fig. 8 Displacements of O_e for a 1 kN effort in $\boldsymbol{y}$ direction and $z_e = 1$ m for PAW-II

Note that δ_{ij} represents displacement of point O_e along axis j for an effort applied in i direction. Indeed, as PAW and PAW-II hold closed structures, an effort applied in a direction generates deflection in all the directions. As shown in Figs. 5 and 6, with an effort of 1 kN exerted in x direction, the biggest displacement are at the

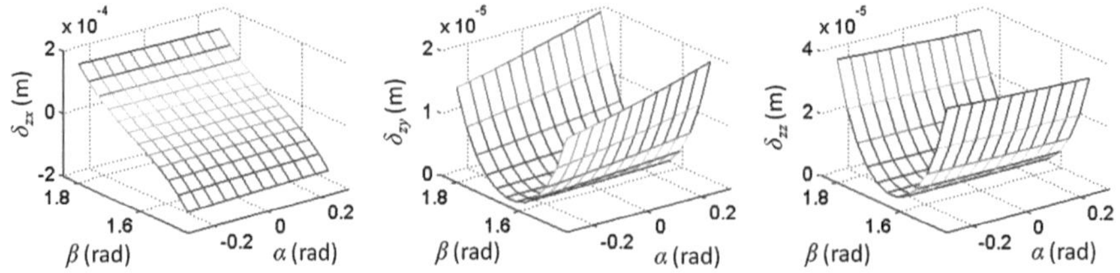

Fig. 9 Displacements of O_e for a 1 kN effort in z direction and $z_e = 1\,\mathrm{m}$ for PAW PM

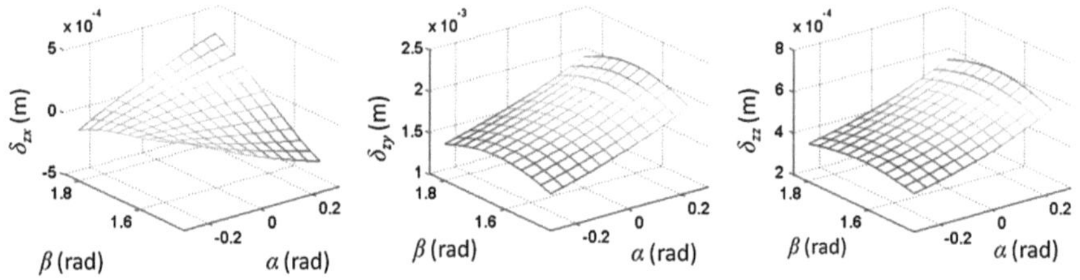

Fig. 10 Displacements of O_e for a 1 kN effort in z direction and $z_e = 1\,\mathrm{m}$ for PAW II PM

edge of workspace for both PAW and PAW-II. And the displacement of PAW is far smaller than that of PAW-II, which means that PAW is stiffer than PAW-II in x direction. Note that displacements are not symmetrical due to orientation of the load in positive direction of x-axis.

For the effort applied in y direction (see Figs. 7 and 8), both PAW and PAW-II show symmetrical distributions about $\beta = 90°$. For PAW, the maximum displacement is in the middle of the workspace while for PAW-II, it shows up at the workspace edge. Note that displacement of PAW along y direction is slightly smaller, the stiffness of PAW is better than PAW-II.

In z direction (see Figs. 9 and 10), the biggest displacement of PAW is at the edges where β reaches its limits. PAW-II gets its peak value when α is maximum. Comparing the two PKMs, displacements of PAW along z direction is smaller than PAW-II. Still, PAW has better rigidity than PAW-II in z direction.

Based on the above analysis, conclusion is drawn that PAW has overall better rigidity than PAW-II. This is mainly because the stiffness transmission from linear actuator to the moving platform is well transferred through an R joint for PAW, but not so well transferred through a spherical joint for PAW-II.

5 Conclusion

This paper first time conducts a comparative study on kinematic and static performance of two Exechon variants, PAW and PAW-II which represent two important families of 1T2R PKMs. The two PKMs show disparity merely in joint arrangement, where PAW has a S$\underline{P}$R topology for leg_2 while PAW-II has a R$\underline{P}$S topology for leg_2. It is found that the location of the joint arrangement has a dramatic effect on the workspace and stiffness performance of these PKMs. As a result, PAW has smaller workspace but better stiffness performance comparing with PAW-II. Both PAW and PAW-II have unique inverse kinematic solution, which will make ease for control. From this study, it is concluded that it is better to place a spherical joint on the base for achieving a better stiffness on the moving platform. For achieving better dexterity (larger orientation workspace), the spherical joint is better placed on the moving platform. Both two two PKMs have great potential for machine tool applications, depending on the specific operation requirements.

Acknowledgments The first author likes to acknowledge the funding support from Chinese Scholarship Council. The third author likes to acknowledge the funding support from the Engineering and Physical Science Research Council in United Kingdom under project no. EP/K004964/1.

References

1. Bi, Z.M., Jin, Y.: Kinematic modeling of Exechon parallel kinematic machine. Robot. Computer-Integrated Manuf. **27**, 186–193 (2011)
2. Neumann, K.E.: System and method for controlling a robot. US Patent, No. 6301525 (2001)
3. Xi, F., Zhang, D., Xu, Z., Mechefske, C.M.: A comparative study on tripod units for machine tools. Int. J. Mech.Tools Manuf. **43**(7), 721–730 (2003)
4. Chen, X., Xie, F.G., Liu, X.J.: A comparison study on motion/force transmissibility of two typical 3-DoF parallel manipulators: the sprint Z3 and A3 tool heads. Int. J. Adv. Robot. Syst. **11**(5). doi: 10.5772/57458 (2014)
5. Sun, T., Song, Y.M., Li, Y.G., Liu, L.S.: Dimensional synthesis of a 3-DOF parallel manipulator based on dimensionally homogeneous Jacobian matrix. Sci. China Ser. E: Technol. Sci. **53**(1), 168–174 (2010)
6. Refaat, S., Herve, J.M., Nahavandi, S., Trinh, H.: Two-mode overconstrained three-DOFs rotational-translational linearmotor-based parallel-kinematics mechanism for machine tool applications. Robotica **25**(4), 461–466 (2007)
7. Li, Q.C., Herve, J.M.: 1T2R parallel mechanisms without parasitic motion. IEEE Trans. Rob. **26**(3), 401–410 (2010)
8. Hunt, K.: Structural kinematics of in-parallel-actuated robot arms. ASME J. Mech. Trans. Autom. Des. **105**, 705–712 (1983)
9. Li, Y., Xu, Q.: Kinematic analysis of a 3-PRS parallel manipulator. Robot. Computer-Integrated Manuf. **23**(4), 395–408 (2007)
10. Li, Y., Liu, H.T., Zhao, X.M., Huang, T., Chetwynd, D.G.: Design of a 3-dof PKM module for large structural component machining. Mech. Mach. Theory **45**(6), 941–954 (2010)
11. Rao, N.M., Rao, K.M.: Dimensional synthesis of a spatial 3-RPS parallel manipulator for a prescribed range of motion of spherical joints. Mech. Mach. Theory **44**(2), 477–486 (2009)
12. Sokolov, A., Xirouchakis, P.: Dynamic analysis of a 3-D of parallel manipulator with R-P-S joint structure. Mech. Mach. Theory **42**(5), 541–557 (2007)
13. Farhat, N., Mata, V., Page, A., Valero, F.: Identification of dynamic parameters of a 3-DOF RPS parallel manipulator. Mech. Mach. Theory **43**(1), 1–17 (2008)
14. Sun, T., Song, Y.M., Xu, L.: Separation of comprehensive geometrical errors of a 3-DoF parallel manipulator based on Jacobian matrix and its sensitivity analysis with Monte-Carlo method **24**(3): 1–8 (2011)

15. Enferadi, J., Tootoonchi, A.A.: Accuracy and stiffness analysis of a 3-RPS spherical parallel manipulator. Robotica **29**, 193–209 (2010)
16. Neumann, K.E.: The key to aerospace automation. In: SAE Aerospace Manufacturing and Automatic Fastening Conference and Exhibition, Michigan, USA, 2006-01-3144 (2006)
17. Jin, Y., Bi, Z., Gibson, R., Mctoal, P., Morgan, M., Mcclory, C., Higgins, C.: Kinematic analysis of a new over-constraint parallel kinematic machine. In: Proceedings of IFToMM 2011 World Congress, Guanajuato, Mexico, A7-282 (2011)
18. Jin, Y., Kong, X.W., Higgins, C., Price, M.: Kinematic design of a new parallel kinematic machine for aircraft wing assembly. In: 2012 10th IEEE International Conference on Industrial Information, Beijing, China, pp. 669–674 (2012)
19. Bonnemains, T., Chanal, H., Bouzgarrou, B.C., Ray, P.: Stiffness computation and identification of parallel kinematic machine tools. J. Manuf. Sci. Eng. **131**, 041013-1-041013-7 (2009)
20. Bonnemains, T., Chanal, H., Bouzgarrou, B.C., Ray, P.: Definition of a new static model of parallel kinematic machines: highlighting of overconstraint influence. In: 2008 IEEE/RSJ International Conference on Intelligent Robots and Systems Acropolis Convention Center, Nice, France, pp. 2416–2421 (2008)

Part VI
Parallel Manipulators

Forward Kinematic Model of a New Spherical Parallel Manipulator Used as a Master Device

H. Saafi, M.A. Laribi, M. Arsicault and S. Zeghloul

Abstract The paper discusses the Forward Kinematic Model (FKM) of a special Spherical Parallel Manipulator (SPM). The special SPM is obtained by modifying one leg of a classic SPM. This new architecture eliminates the singularity from the workspace. The SPM is used as master device for medical tele-operation system. The FKM of the new SPM is calculated using the equation of spherical four-bar mechanism. A method to improve the FKM calculation using extra sensor is proposed in this paper.

Keywords Spherical parallel manipulator · Master device · Forward kinematic model · Singularity

1 Introduction

Nowadays, parallel manipulators are widely popular. Thanks to their high load capacity, their stiffness, their low weight and their precision, parallel manipulators are used in many fields such as medicine, where, many master devices have been developed with parallel architecture [1, 2].

In previous works [3], new master device was developed to control a surgical robot. This device has a spherical parallel architecture. The master architecture was

H. Saafi (✉) · M.A. Laribi · M. Arsicault · S. Zeghloul
Department of GMSC, Pprime Institute, CNRS—University of Poitiers, ENSMA UPR 3346, Poitiers, France
e-mail: houssem.saafi@univ-poitiers.fr

M.A. Laribi
e-mail: med.amine.laribi@univ-poitiers.fr

M. Arsicault
e-mail: marc.arsicault@univ-poitiers.fr

S. Zeghloul
e-mail: said.zeghloul@univ-poitiers.fr

S. Bai and M. Ceccarelli (eds.), *Recent Advances in Mechanism Design for Robotics*, Mechanisms and Machine Science 33,
DOI 10.1007/978-3-319-18126-4_38

chosen to have a mechanism able to provide three pure rotations around a fixed point. The geometric parameters of this structure were optimized to meet a prescribed workspace. However, the main problem of the optimized structure is the presence of parallel singularity in the workspace. To cope with the presence of the singularity, in [4], we proposed to change the architecture of one leg. Then, geometric parameters of the new SPM were optimized in order to have a singular free workspace.

The FKM of the master device is needed to control the surgical robot. The FKM calculates the orientation of the moving platform of the SPM. In this paper, the FKM of the new SPM is presented. The calculation time of the FKM is reduced using an extra sensor. This makes the FKM ready to work in real time.

2 Master Device for a Medical Tele-Operation System

The developed master device is shown in Fig. 1. This device controls a surgical robot. The target application is Minimally Invasive Surgery (MIS). In MIS, the instruments enter to the patient body throw tiny incisions. This limits the motion of the instrument to three rotations around the incision and one translation along the instrument axis. For this reason, the spherical parallel architecture was chosen for the master device since it is able to produce the similar motions.

Fig. 1 Master device of a tele-operation system

The translation motion is not taken into account in this paper because it is uncoupled in the model of the SPM.

The classical SPM has three identical legs, each leg is made of two links and three revolute joints (Fig. 2a). All axes of the revolute joints are intersecting in one common point, called CoR (Center of Rotation). Each link is characterized by the angle between its two revolute joints, as shown in Fig. 2b. This angle is constant and it represents the dimension of the link. The actuated joint axes are located along an orthogonal frame. The orientation of the SPM is described by the ZXZ Euler angles (ψ, θ, and φ).

The Jacobian matrix of the SPM can be written as follows:

$$\mathbf{J}=\mathbf{A}^{-1}\mathbf{B} \tag{1}$$

where,

$$\mathbf{A}=\left[Z_{2A}\times Z_{3A}\quad Z_{2B}\times Z_{3B}\quad Z_{2C}\times Z_{3C}\right]^{\mathrm{T}} \tag{2}$$

$$\mathbf{B}=\begin{bmatrix} Z_{1A}(Z_{2A}\times Z_{3A}) & 0 & 0 \\ 0 & Z_{1B}(Z_{2B}\times Z_{3B}) & 0 \\ 0 & 0 & Z_{1C}(Z_{2C}\times Z_{3C}) \end{bmatrix} \tag{3}$$

Dexterity is used to identify the presence of singularity in the workspace of the SPM. Dexterity $\eta(\mathbf{J})$ is given by the inverse of the condition number of the Jacobian matrix as follows:

$$\eta(\mathbf{J})=1/\kappa(\mathbf{J})\ \text{with}\ \kappa(\mathbf{J})=\|\mathbf{J}\|\,\|\mathbf{J}^{-1}\| \tag{4}$$

Figure 3 shows the distribution of the dexterity for a self-rotation $\varphi = 50°$. It shows the presence of parallel singularity inside the workspace (dotted line).

To cope with the problem of the singularity, we proposed in [4] to change the architecture of the leg A. The next Section presents the architecture of the new SPM.

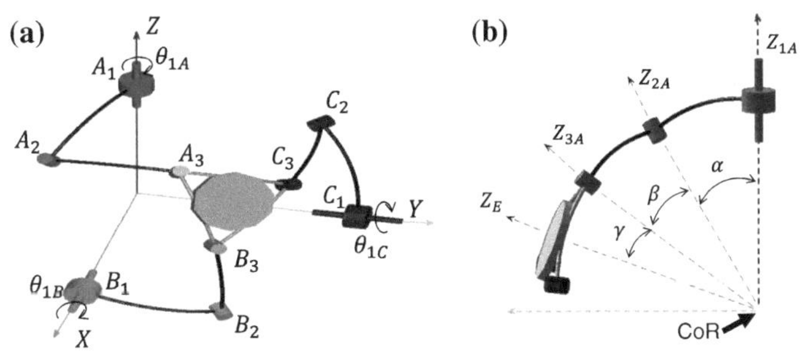

Fig. 2 **a** Classic spherical parallel manipulator. **b** Kinematic of leg A

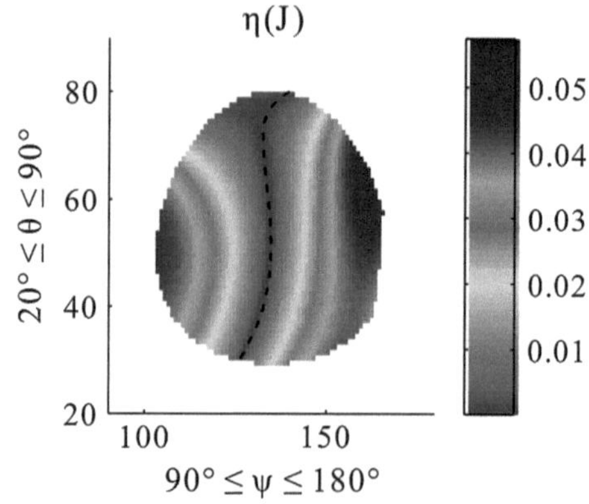

Fig. 3 The dexterity distribution for φ = 50°

3 New Architecture of the Master Device

The new structure of the SPM is shown in Fig. 4a The RRR kinematic structure of leg A was replaced by URU structure as illustrated in Fig. 4b (R for Revolute joint and U for Universal joint). The two links of the new leg have the same length, *L*. The legs B and C are not changed. This new leg generates similar motions to those obtained by the RRR leg.

The matrices **A** and **B** of the new SPM become as follows:

$$\mathbf{A}= \begin{bmatrix} Z_{4A} \times Z_{5A} & Z_{2B} \times Z_{3B} & Z_{2C} \times Z_{3C} \end{bmatrix}^{T} \tag{5}$$

$$\mathbf{B} = \begin{bmatrix} Z_{1A}(Z_{4A} \times Z_{5A}) & 0 & 0 \\ 0 & Z_{1B}(Z_{2B} \times Z_{3B}) & 0 \\ 0 & 0 & Z_{1C}(Z_{2C} \times Z_{3C}) \end{bmatrix} \tag{6}$$

Figure 5 shows the distribution of the dexterity for the self-rotation φ = 50°. It shows that parallel singularities are outside the useful area of the workspace (dotted lines).

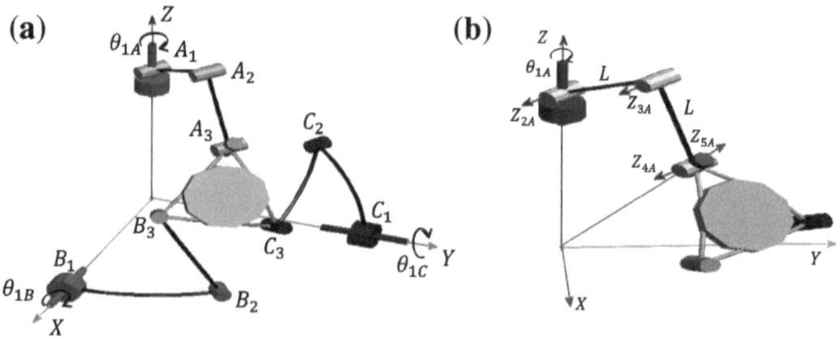

Fig. 4 **a** Kinematic structure of the new SPM. **b** New kinematic of leg A

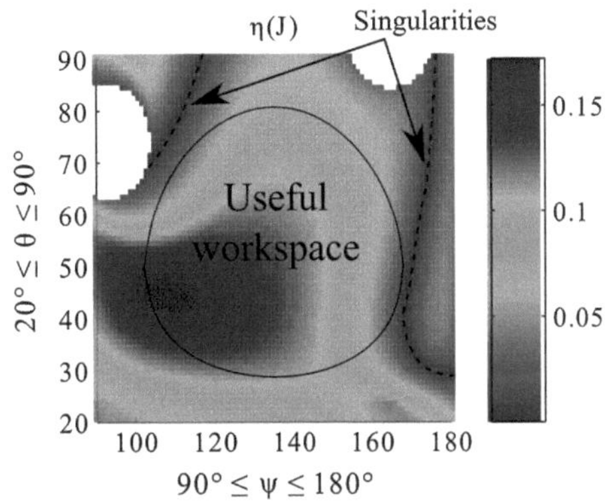

Fig. 5 The dexterity distribution for φ = 50°

4 Forward Kinematic Model

The FKM calculates the orientation (ψ, θ, and φ) of the SPM in a function of active joint angles (θ_{1A}, θ_{1B}, and θ_{1C}). The FKM is solved using a method presented in [5]. This method is based on the input/output equations of spherical four-bar mechanisms. Figure 6 shows a spherical four-bar mechanism. the input and output angles are σ and ξ, respectively.

The input/output equation of the spherical four-bar mechanism is as follows:

$$L(\xi)\cos\sigma + M(\xi)\sin\sigma + N(\xi) = 0 \tag{7}$$

With L, M and N are variables that depend on cos(ξ) and sin(ξ). Their expressions are detailed in [5].

For the SPM, two four-bar mechanisms are considered $A_2A_3B_3B_2$ and $A_2A_3C_3C_2$ (Fig. 7).

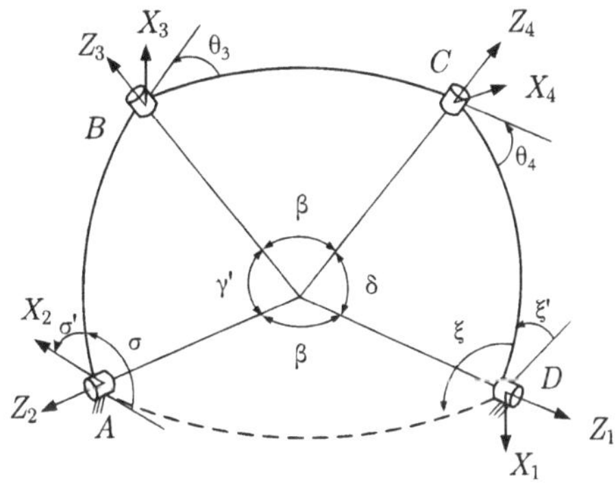

Fig. 6 Spherical four-bar mechanism

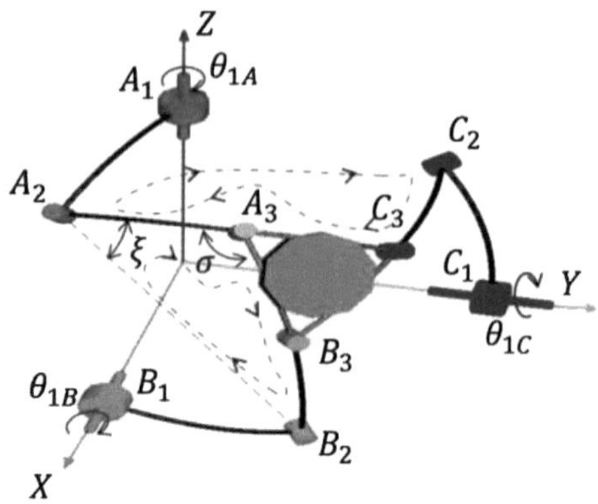

Fig. 7 The two considered spherical four-bar mechanisms

The two input/output equations are as follows:

$$\begin{cases} L_1(\xi)\cos\sigma + M_1(\xi)\sin\sigma + N_1(\xi) = 0 \\ L_2(\xi)\cos\sigma + M_2(\xi)\sin\sigma + N_2(\xi) = 0 \end{cases} \tag{8}$$

cos(σ) and sin(σ) can be expressed as follows:

$$\cos\sigma = \frac{M_1N_2 - M_2N_1}{L_1M_2 - L_2M_1}, \qquad \sin\sigma = \frac{L_1N_2 - L_2N_1}{L_1M_2 - L_2M_1} \tag{9}$$

All the possible solutions of angle, ξ, can be found by solving the Eq. (10), obtained by calculating the square sum of cos(σ) and sin(σ), defined bellow:

$$\begin{aligned} N_2^2L_2^2 + 2L_1M_1L_2M_2 - 2L_1N_1L_2N_2 + N_2^2M_1^2 - L_2^2M_1^2 \\ -2M_1N_2M_2N_2 - M_2^2L_1^2 + N_1^2L_2^2 + N_1^2M_2^2 = 0 \end{aligned} \tag{10}$$

The obtained solutions for angle, ξ, are then used to calculate the values of the angle, σ, using Eq. (9). The platform orientation (ψ, θ, and φ) can be calculated using the forward kinematic of leg A.

For the new SPM, only one four-bar mechanism can be considered (Fig. 8). The input/output equation is as follows:

$$L_1(\xi)\cos\sigma + M_1(\xi)\sin\sigma + N_1(\xi) = 0 \tag{11}$$

The second equation is obtained by expressing Z_{5A} in function of ξ and σ by using the forward kinematic of leg B as follows:

$$\begin{aligned} Z_{5A} = R_{0B}R_Z(\theta_{1A})R_X(\alpha)R_Z(\xi + \xi')R_X(\beta)R_Z(\sigma - \mu) \\ R_X(\gamma)R_Z(\frac{2\pi}{3})R_X(-\gamma)\,Z_{1A} \end{aligned} \tag{12}$$

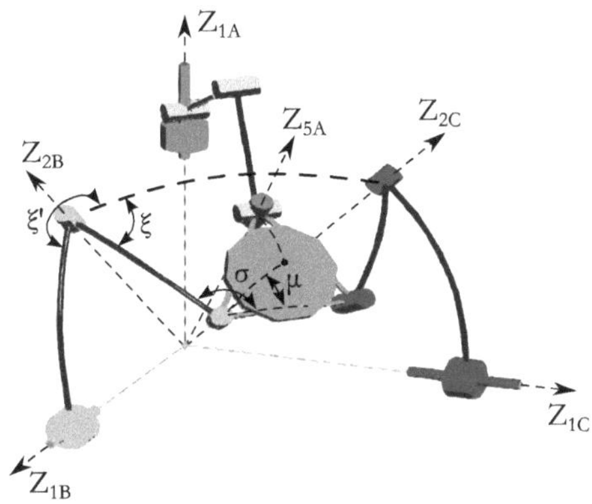

Fig. 8 The considered spherical four-bar mechanism

Y_{1A} is perpendicular to Z_{5A}, this condition leads to the following equation after the arrangement:

$$L_2(\xi)\cos\sigma + M_2(\xi)\sin\sigma + N_2(\xi) = 0 \tag{13}$$

The system composed of Eqs. (11) and (13) can be solved using the same procedure as the classic SPM. This method gives accurate solutions of the FKM. However, due to the complexity of Eq. (10), the calculation time is more than 100 ms on a PC computer with a processor running at 2.2 GHz. An improvement of the forward kinematic model is presented in the next section.

5 Improvement of the Forward Kinematic Model

In [6], we proposed a solution to improve the FKM calculation. This solution is based on the use of an extra sensor. The extra sensor improves the FKM accuracy and reduces the complexity of the model and the calculation time to less than 50 µs. The same solution is used here. An extra sensor is added in the joint with axis Z_{3B}. This extra sensor leads to a system with only one unknown, since the angle σ is now given directly by the fourth sensor.

Equations (11) and (13) can be arranged as follows:

$$\begin{cases} \overline{L}_1\cos\xi + \overline{M}_1\sin\xi + \overline{N}_1 = 0 \\ \overline{L}_2\cos\xi + \overline{M}_2\sin\xi + \overline{N}_2 = 0 \end{cases} \tag{14}$$

With, $\overline{L}_i$, $\overline{M}_i$ and $\overline{N}_i (i = 1, 2)$ are variables that depend on cos (σ) and sin (σ). The unique solution of ξ is calculated as follows:

$$\xi = a\tan 2(\sin\xi, \cos\xi) \tag{15}$$

With,

$$\cos\xi = \frac{\overline{M}_1\overline{N}_2 - \overline{M}_2\overline{N}_1}{\overline{L}_1\overline{M}_2 - \overline{L}_2\overline{M}_1}, \qquad \sin\xi = \frac{\overline{L}_1\overline{N}_2 - \overline{L}_2\overline{N}_1}{\overline{L}_1\overline{M}_2 - \overline{L}_2\overline{M}_{11}} \tag{16}$$

The Euler angles ψ, θ and φ can be obtained using the same approach presented in Sect. 3 using the forward kinematic of leg B.

The use of an extra sensor makes the FKM a linear problem with a unique solution. The calculation time of the improved FKM is less than 50 μs on a PC computer with a processor running at 2.2 GHz.

6 Conclusions

In this paper, a new master device for a tele-operation system has been presented. This device has a special spherical parallel architecture. The forward kinematic model of the SPM has been studied and then has been improved to work in real time.

Acknowledgments This research is supported by ROBOTEX, the French national network of robotics platforms (N° ANR-10-EQPX-44-01).

References

1. Tobergte, A. et al.: The sigma.7 haptic interface for mirosurge: a new bi-manual surgical console. In: Proceedings of the IEEE/RSJ International Conference on Intelligent Robots and Systems, pp. 3023–3030, 25–30 Sept 2011
2. Ma, A.; Payandeh, S.: Analysis and experimentation of a 4-DOF haptic device. In: Haptic interfaces for virtual environment and teleoperator systems, symposium on haptics, pp. 351–356, 13–14 March 2008
3. Chaker, A., Mlika, A., Laribi, M.A., Romdhane, L., Zeghoul, S.: Synthesis of a spherical parallel manipulator for a dexterous medical task. Front. Mech. Eng. **7**(2), 150–162 (2012)
4. Saafi, H., Laribi, M.A., Zeghloul, S.: Optimal design of a new Spherical Parallel Manipulator. In: 23rd International Conference on Robotics in Alpe-Adria-Danube region, Smolenice castle, Slovakia, 3–5 Sept 2013
5. Bai, S., Hansen, M.R., Angeles, J.: A robust forward-displacement analysis of spherical parallel robots. Mech. Mach. Th. **44**(12), 2204–2216 (2009)
6. Saafi, H., Laribi, M. A., Zeghloul, S.: Improvement of the direct kinematic model of a haptic device for medical application in real time using an extra sensor. In: IEEE/RSJ International Conference on Intelligent Robots and Systems, pp. 1697–1702, 14–18 Sept 2014

Kinematics of a 6 DOFs Manipulator with Interchangeable Translation and Rotation Motions

V. Glazunov, N. Nosova and M. Ceccarelli

Abstract In this paper, a parallel mechanism with six degrees of freedom is presented with capability for decoupled translational and rotational movements as a significant advantage for use in lifting and transport operations for measuring, assembly, manufacturing operations, working within a limited workspace. Direct and inverse kinematics is formulated for the proposed mechanism with the capability to provide rotational and translational in different operation modes.

Keywords Parallel kinematic machines · Mechanism kinematics · Workspace · Operation analysis

1 Introduction

Parallel kinematics machines are used for several technological and technical operations, depending on task and control features. There are two main types of mechanisms, namely closed-loop mechanisms and open-chain mechanisms [1–4]. Both types of mechanisms have their advantages and disadvantages. In general the open kinematics mechanisms are easier to manage, show decoupled motion by the actuators and are simple enough to operate, but they are with less load-carrying capacity. The parallel manipulators (as closed-loop mechanisms) may offer higher

V. Glazunov (✉)
Mechanical Engineering Research Institute of RAS, Moscow, Russia
e-mail: vaglznv@mail.ru

N. Nosova
Moscow State University of Design and Technology, Moscow, Russia
e-mail: natahys@mail.ru

M. Ceccarelli
University of Cassino and South Latium, Cassino, Italy
e-mail: ceccarelli@unicas.it

S. Bai and M. Ceccarelli (eds.), *Recent Advances in Mechanism Design for Robotics*, Mechanisms and Machine Science 33,
DOI 10.1007/978-3-319-18126-4_39

payload capacity, higher speed and acceleration, higher static and dynamic accuracy and higher stiffness than what is possible with of serial structures of robots [4, 5].

One of the most famous parallel manipulators is DELTA robot that was conceived by Clavel in 1998, [6]. The output link of this mechanism performs only translational motion, since despite the fact that the manipulator is actuated by the rotational motors, the output platform does not change its orientation. Because intermediate links of mechanisms have small weight, the mechanism is quick in motion response and has very high speed.

Other relevant manipulator is the mechanism ORTHOGLIDE that has also a translational motion [7, 8]. The mechanism consists of three kinematic chains, with three translational engines and three parallelogram joints that provide a constant orientation of the output body. This design is in many respects similar to the architecture of the DELTA robot, but rotational engines are replaced by translational engines.

Parallel mechanisms have also disadvantages for limited workspace, presence of singularities, and complexity in mechanism synthesis. Recently development of analytical and numerical methods for the analysis and synthesis of spatial parallel kinematic machines has greatly expanded the area of their applications. The possibility of expanding the workspace through coherent synthesis parameters of the mechanism and control law is considered in [9–11].

One of the most important tasks of analysis, synthesis and control of parallel manipulators is kinematic modeling to determine the relationship between generalized and absolute coordinates. Absolute coordinates are coordinates in a Cartesian system, which determine the position of the output link. Generalized (independent) coordinates are called coordinates describing the change in the position of the input links. Inverse and forward kinematics can determine the workspace of the manipulator, which is the reachable space by a reference point on the output link. Manipulator workspace is defined as the region of reachable points by a reference point on the extremity of manipulator chain and it is usually referenced as a basic characteristic for a successful use and design of robot manipulator [5, 12, 13]. In general, the determination of the manipulator workspace is due to the solution of the direct kinematic problem.

A current task of great interest is to implement a kinematic decoupling, which means to divide the generalized coordinates for handling the translational movements and handling the rotational movements. Many investigators have solved this problem [2–4]. The "POLMAN" manipulator became one of the most effective solutions [14] where the author K. Mianowski used the drive shaft in each of three kinematic chains. In order to provide a kinematic decoupling may be used an apparatus of closed groups of screws as proposed in [15], through which is possible to obtain not only a circuit with decoupled kinematic motion, but the property of isomorphism, i.e. constancy of the ratio between movement of the actuator and output link.

In this paper the problem has been attached as referring to a new kinematic design of a 6 DOF manipulator and solution is proposed as based on modernization of the ORTHOGLIDE manipulator aimed at getting an interchange between translational and rotational motions. The advantage of this mechanism is to increase functionality without adding additional links.

We know the robot ORTHOGLIDE. The motion of the output link provides translational engines. This problem is solved through the use of in each the three kinematic chains hinged parallelogram, just as is done in the DELTA robot. Drives the robot ORTHOGLIDE located along the axes of the Cartesian coordinate system. For providing additional rotation of the work body in robot ORTHOGLIDE can be introduced additional rotary engines that are transmitted rotation via the additional shaft [8, 16]. However, it does not provide sufficient rigidity and complicates design. The basic idea of the new mechanism proposed in this paper are to use parallelograms not only to transmit the translational motion and for transmitting rotary motion. This should simplify the structure and enhance the rigidity. This article shows a scheme of the robot, as well as we consider the solution of kinematics problems, which are important in the analysis, synthesis and management.

This paper discusses the robot which is obtained by converting the ORTHOGLIDE robot. This transformation is done in order to increase functionality and to ensure rotation of working body. This problem is solved by adding in each of kinematic chains of the rotary engine. Parallelograms were put not to only for translational motion, but also for rotation. Therefore this manipulator is really new.

2 Kinematics for Mechanisms with Translational Engines

Consider a parallel mechanism with six degrees of freedom (Fig. 1) [17]. The spatial mechanism includes base 1, output link 2, working body 3 and three kinematic chains. Each kinematic chain contains a translational engine 4, 4′, 4″; initial rotary kinematic pair 5, 5′, 5″; initial link of the parallel crank mechanism 6, 6′, 6″; rotary kinematic pairs of the parallel crank mechanism 7, 7′, 7″; terminal rotary kinematic pair 8, 8′, 8″; terminal link of the parallel crank mechanism 9, 9′, 9″; intermediate links of the parallelogram and terminal link of the kinematic chain 10, 10′, 10″. Rotary engines 11, 11′, 11″ and output rotary kinematic pairs 12, 12′, 12″ are located in three kinematic chains. Terminal links 10, 10′, 10″ of the kinematic chains are interfaced with the working body 3 by the intermediate rotary kinematic pairs 13, 13′, 13″.

Besides, they have two turning kinematic pairs with axes perpendicular to axes of the parallelogram i.e., each kinematic chain contains at least two universal (two degrees of freedom) joints. This fact may be used for transmitting rotation. It is known [18] that the kinematic chain, including two degrees of freedom joints with parallel axes, may be equipped with one or two additional chains. Given that, the kinematic chain is able to transfer rotation with a transmission ratio that equals one and keeps the mutual orientation of the input and output chains.

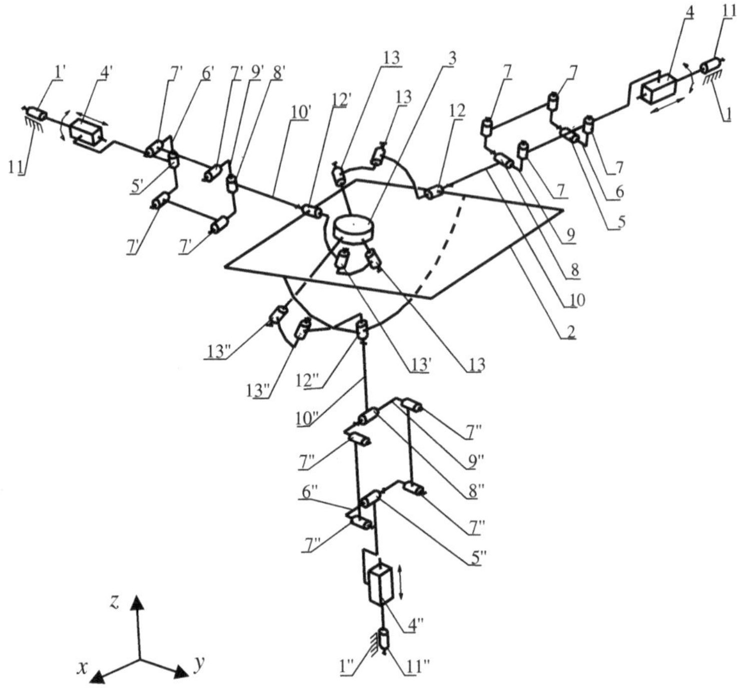

Fig. 1 Parallel mechanism with six degrees of freedom

Therefore, if the kinematic chain is equipped with a rotary actuator (11, 11′, 11″) and a rotary kinematic pair (12, 12′, 12″) which connects the kinematic chain with the output link (3), then rotation from the engine to the rotary kinematic pair will be transmitted with the constant transmission ratio. With that, the position of the output element will stay unchanged.

The spherical mechanism may be located on the output element; rotation will be transferred from the additional rotary actuators. If there are three rotary actuators, then the spherical mechanism may have three degrees of freedom. Thereafter, with fewer rotary actuators, the number of degrees of freedom of the spherical mechanism may be decreased.

Let us note that the parallel crank mechanism is the necessary element. In the case where every kinematic chain has two degrees of freedom joints, as in the POLMAN manipulator, then the number of imposed constraints would not be enough and the output element would have uncontrolled motion at the brake off actuator.

Consider the mechanism with 6 DOF and solve the inverse problem in the part which determines the translational motion. The translational motion is caused by the fact that each chain has a parallelogram, two rotational kinematic pairs, and a

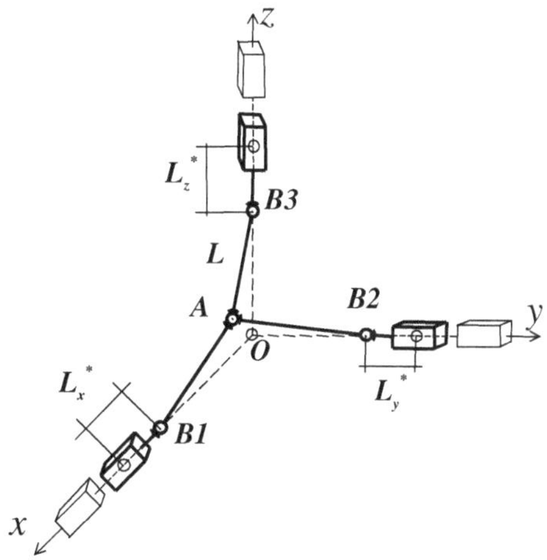

Fig. 2 The so-called pyramid mechanism

linear motor. This structure is replaced by the mechanism of the so-called pyramid (Fig. 2), which can be used for schemes such as ORTHOGLIDE.

Mechanism “pyramid” is derived from the source as follows. All the kinematic chains shifted to the center of the output link. Generalized coordinates are sections describing the positions of the points B_1, B_2, B_3. Absolute coordinates of the output link (in other words, the output coordinates) are the coordinates of the point A.

The inverse kinematic modeling is useful to determine the values of input variables (x_{B1}, y_{B2}, z_{B3}) from the known values of the output variables (x_A; y_A; z_A).

In the coordinate system x-y-z have point A, the position does not coincide with the center point O. It is necessary to determine the coordinates x_{B1}, y_{B2}, z_{B3}, that is on which a distance L_x^*, L_y^*, L_x^* will move input links of mechanism. Where L_x^*, L_y^*, L_x^* are the distance at which point move x_{B1}, y_{B2}, z_{B3}, i.e. input link mechanism relative to the origin O and an arbitrary point A.

Point A has coordinates (x_A; y_A; z_A). We can write the following equations to find out how far input links of mechanism moved relative to the center

$$\begin{aligned}(x_A - x_{B1})^2 + (y_A - y_{B1})^2 + (z_A - z_{B1})^2 = L^2\\(x_A - x_{B2})^2 + (y_A - y_{B2})^2 + (z_A - z_{B2})^2 = L^2\\(x_A - x_{B3})^2 + (y_A - y_{B3})^2 + (z_A - z_{B3})^2 = L^2\end{aligned} \quad (1)$$

where L—is the length of intermediary link for each kinematic chain.

Coordinates y_{B1}^2, z_{B1}^2, x_{B2}^2, z_{B2}^2, x_{B3}^2 and y_{B3}^2 turn to zero as the corresponding input links are located along the coordinate axes.

Thus, solving the system of independent quadratic equations (1), we obtain the coordinates x_{B1}, y_{B2}, z_{B3}.

For forward kinematic modeling values input links of mechanism (x_{B1}, y_{B2}, z_{B3}) are considered known, while the output parameters (x_A, y_A, z_A) are unknown. This task admits several solutions corresponding to different assemblies of mechanism.

From these three equations in Eq. (1), it is necessary to find three unknowns, namely x_A, y_A, z_A, and these equations are dependent.

3 Kinematics for Spherical Mechanism

On the output body of 3 DOF translation mechanism (Fig. 1a) a spherical mechanism is disposed. All the rotations are transmitted from the rotary engines. Note that the manipulator under consideration is decoupled. That is why the kinematic problems for translation and rotation motions can be solved separately. For the translation mechanism the kinematic problem is solved above. Here we solve the inverse and forward kinematic problems for the spherical parallel mechanism with rotational motors located along the axes of the Cartesian coordinate system.

Let us consider a spherical manipulator (Fig. 3). It consists of three kinematic chains with intersecting axes at an angle of 90°. Each input link is associated with the rotational engine. The output body consists of two intersecting hemisphere and the end-effector revolving around the three axes x, y, z with the intersection at the point O.

Output coordinates are angles of rotation platform: α, β, γ—rotations around the axes x, y, z, respectively. Generalized coordinates are angles φ_{11}, φ_{21}, φ_{3}—respectively, the rotation angle of the input links of the first, second and third kinematic chains. Where φ_{ij}: φ is angle of rotation of the input link; i is 1, 2, 3—the first, second, third kinematic chain; j is 1, 2, 3—the first, second, third rotational kinematic pair.

To determine the velocity and the singularity of the mechanism, it is necessary to solve the kinematic models to find a relationship between input and output coordinates. Let us calculate based on the analytical method.

Fig. 3 Spherical manipulator

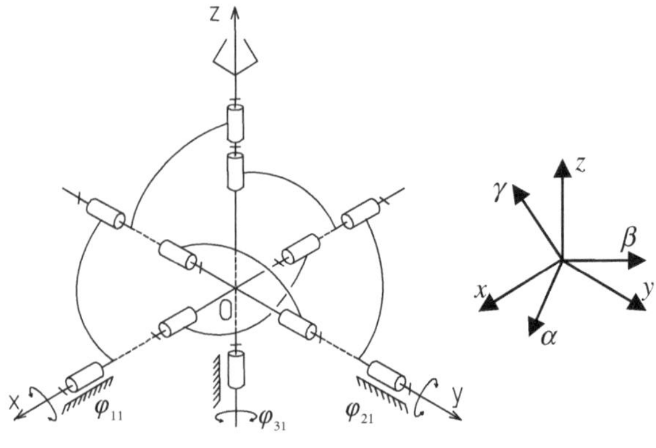

The output body executes a sequential rotation around the axes x, y, z. Transition matrix of the moving coordinate system to the fixed coordinate system, describing the rotation around the axes x, y, z, respectively, have the form:

$$\boldsymbol{A}_1 = \begin{pmatrix} 1 & 0 & 0 \\ 0 & \cos\alpha & -\sin\alpha \\ 0 & \sin\alpha & \cos\alpha \end{pmatrix}, \boldsymbol{A}_2 = \begin{pmatrix} \cos\beta & 0 & \sin\beta \\ 0 & 1 & 0 \\ -\sin\beta & 1 & \cos\beta \end{pmatrix},$$

$$\boldsymbol{A}_3 = \begin{pmatrix} \cos\gamma & -\sin\gamma & 0 \\ \sin\gamma & \cos\gamma & 0 \\ 0 & 0 & 1 \end{pmatrix}$$

Matrix describing the transition of the output body from the moving coordinate system to the fixed system will be $A = A_3 \cdot A_2 \cdot A_1$.

The input link of the first kinematic chain rotates in the following sequence: axes x, y, z. Corresponding rotation matrices are of the form:

$$\boldsymbol{B}_1 = \begin{pmatrix} 1 & 0 & 0 \\ 0 & \cos\varphi_{11} & -\sin\varphi_{11} \\ 0 & \sin\varphi_{11} & \cos\varphi_{11} \end{pmatrix}, \boldsymbol{B}_2 = \begin{pmatrix} \cos\varphi_{21} & 0 & \sin\varphi_{21} \\ 0 & 1 & 0 \\ -\sin\varphi_{21} & 1 & \cos\varphi_{21} \end{pmatrix},$$

$$\boldsymbol{B}_3 = \begin{pmatrix} \cos\varphi_{31} & -\sin\varphi_{31} & 0 \\ \sin\varphi_{31} & \cos\varphi_{31} & 0 \\ 0 & 0 & 1 \end{pmatrix}$$

where φ_{11}—the rotation angle of the input link; φ_{21}—the rotation angle of the second link; φ_{31}—the rotation angle of the third link.

Matrix describing the transition of the input link from the moving coordinate system to the fixed system will be $\mathrm{B}' = \mathrm{B}_1' \cdot \mathrm{B}_2' \cdot \mathrm{B}_3'$.

The vector of kinematic pair of the first chain conjugate with the output body has the coordinates (0, 0, 1) (Fig. 3). Substitute into the coupling equation the values found for the matrices A and B', obtain equation from which can be expressed the angle φ_{11} through the angles α, β and γ:

$$A \cdot \begin{pmatrix} 0 \\ 0 \\ 1 \end{pmatrix} = B' \cdot \begin{pmatrix} 0 \\ 0 \\ 1 \end{pmatrix}, \begin{pmatrix} \sin\beta \\ -\cos\beta \cdot \sin\alpha \\ \cos\alpha \cdot \cos\beta \end{pmatrix} = \begin{pmatrix} \sin\varphi_{12} \\ -\cos\varphi_{12} \cdot \sin\varphi_{11} \\ \cos\varphi_{11} \cdot \cos\varphi_{12} \end{pmatrix},$$

$$tg\varphi_{11} - \frac{\cos\beta \cdot \sin\alpha}{\cos\alpha \cdot \sin\alpha} = 0$$

Matrix describing the transition of the second chain from the moving coordinate system to the fixed system can be represented as $\mathrm{B}'' = \mathrm{B}_2' \cdot \mathrm{B}_1' \cdot \mathrm{B}_3'$. The vector of kinematic pair of the second chain conjugate with the output body has the coordinates (0, 0, 1). Substitute into the coupling equation the values found for the matrices A and B''. We obtain, from which can be expressed the angle φ_{21} through the angles α, β and γ:

$$A\begin{pmatrix}0\\0\\1\end{pmatrix} = B''\cdot\begin{pmatrix}0\\0\\1\end{pmatrix}, \begin{pmatrix}\sin\beta\\-\cos\beta\cdot\sin\alpha\\\cos\alpha\cdot\cos\beta\end{pmatrix} = \begin{pmatrix}\cos\varphi_{22}\cdot\sin\varphi_{21}\\-\sin\varphi_{22}\\\cos\varphi_{21}\cdot\cos\varphi_{22}\end{pmatrix}$$

$$tg\varphi_{21} - \frac{\sin\beta}{\cos\alpha\cdot\cos\beta} = 0$$

Matrix describing the transition of the third chain from the moving coordinate system to the fixed system can be represented as $B''' = B_3'\cdot B_1'\cdot B_2'$. The vector of kinematic pair of the third chain conjugate with the output body has the coordinates (0, 1, 0). Substitute into the coupling equation the values found for the matrices A and B'''. We obtain, from which can be expressed the angle φ_{31} through the angles α, β and γ:

$$A\cdot\begin{pmatrix}0\\1\\0\end{pmatrix} = B'''\begin{pmatrix}0\\1\\0\end{pmatrix}, \begin{pmatrix}-\cos\beta\cdot\sin\gamma\\\cos\alpha\cdot\cos\gamma-\sin\alpha\cdot\sin\beta\cdot\sin\gamma\\\cos\gamma\cdot\sin\alpha+\cos\alpha\cdot\sin\beta\cdot\sin\gamma\end{pmatrix} = \begin{pmatrix}-\cos\varphi_{32}\cdot\sin\varphi_{31}\\\cos\varphi_{31}\cdot\cos\varphi_{32}\\\sin\varphi_{32}\end{pmatrix}$$

$$tg\varphi_{31} - \frac{\cos\beta\cdot\sin\gamma}{\cos\alpha\cdot\cos\gamma-\sin\alpha\cdot\sin\beta\cdot\sin\gamma} = 0$$

The kinematic models are necessary to determine the relationship between the coordinates of the output body and generalized coordinates and describe a function of position. The function of the position of the mechanism implicitly expressed by the equation

$$F(\alpha,\beta,\gamma,\varphi_{11},\varphi_{21},\varphi_{31}) = 0 \tag{2}$$

The angles φ_{11}, φ_{21}, φ_{31} expressed through the angles α, β, γ, we substitute the coupling equation:

$$\begin{cases} F_1 = tg\varphi_{11} - \dfrac{\cos\beta\cdot\sin\alpha}{\cos\alpha\cdot\sin\alpha} = 0 \\ F_2 = tg\varphi_{21} - \dfrac{\sin\beta}{\cos\alpha\cdot\cos\beta} = 0 \\ F_3 = tg\varphi_{31} - \dfrac{\cos\beta\cdot\sin\gamma}{\cos\alpha\cdot\cos\gamma-\sin\alpha\cdot\sin\beta\cdot\sin\gamma} = 0 \end{cases} \tag{3}$$

Thus, the coupling equations constituted relationship between the rotation angle of the input links and the rotation angle of the output body.

4 Conclusions

Included in each kinematic chain the hinged parallelogram allowed to create new structural scheme of the mechanism on the basis of the mechanism ORTHOGLIDE. It became possible to transmit the rotation of the same kinematic chain that allows translational motion without introducing the additional transmission elements (shafts). The main advantage presented by a parallel manipulator with 6 DOFs is the presence of kinematic decoupling between the translational and rotational motions. Due to the fact that the rotation is transmitted by the same kinematic chain, which provides translational movement, it is possible to solve the inverse and forward kinematics separately for translational parts and separately for the rotational part of the manipulator. Simplified kinematic pyramid mechanism has helped to solve the forward kinematic modeling for spatial mechanism with three degrees of freedom explicitly.

References

1. Glazunov, V.A., Koliskor, A.S., Krainev, A.F.: Spatial mechanisms of parallel structure. (Пространственные механизмы параллельной структуры) (1991)
2. Gogu, G.: Structural Synthesis of Parallel Robots. Part 2: Translational Topologies with Two and Three Degrees of Freedom. Springer, Dordrecht (2009)
3. Kong, X., Gosselin, C.: Type Synthesis of Parallel Mechanisms. Springer, Dordrecht (2007)
4. Merlet, J.P.: Parallel Robots. Kluwer Academic Publishers, Dordrecht (2000)
5. Ceccarelli M.: Fundamentals of Mechanics of Robotic Manipulation. Kluwer, Springer, Dordrecht (2004). ISBN 1-4020-1810-X
6. Clavel, R.: Delta, a fast robot with parallel geometry. In: 18th International Symposium on Industrial Robot, Lausanne, pp. 91–100 (1988)
7. Chablat, D., Wenger, P., Staicu, S.: Dynamics of the orthoglide parallel robot. UPB Sci. Bull. Ser. D: Mech. Eng. **71**(3), 3–16 (2009)
8. Chablat, D., Wenger, P.: United States Patent Application Publication. Pub. No.: US 2007/0062321 Al. Pub. Date: Mar. 22, 2007. Device for the movement and orientation of an object in space and use thereof in rapid machining. Sainte Luce Sur Loire (FR)
9. Kreinin, G.V., Misyurin, S.Y.: A systematic approach to synthesis of a drive system. J. Mach. Manuf. Reliab. **40**(6), 507–511 (2011)
10. Kreinin, G.V., Misyurin, S.Y.: Influence of the gear ratio on the dynamics of a control-system drive. J. Mach. Manuf. Reliab. **41**(6), 486–491 (2012)
11. Kreinin, G.V., Misyurin, S.Y.: On some general laws in drive dynamics. J. Mach. Manuf. Reliab. **37**(6), 546–551 (2008)
12. Ceccarelli, M.: A Formulation for the workspace boundary of general N-revolute manipulators. IFToMM J. Mech. Mach. Theory **31**, 637–646 (1996)
13. Ceccarelli, M.: A new 3 D.O.F. parallel spatial mechanism. IFToMM J. Mech. Mach. Theory, **32**(8):895–902 (1997)
14. Mianowski, K.: Singularity analysis of parallel manipulator POLMAN 3 × 2 with six degrees of freedom. In: Proceedings 12th IFToMM World Congress, Besanцn (2007)
15. Glazunov, V.: Design of decoupled parallel manipulators by means of the theory of screws. Mech. Mach. Theory **45**, 239–250 (2010)

16. Ur-Rehman, R., Caro, S., Chablat, D., Wenger, P.: Kinematic and dynamic analysis of the 2 DOF spherical wrist of Orthoglide 5-axis. In: Proceedings 3rd International Congress Design and Modelling of Mechanical Systems CMSM, Hammamet (2009)
17. Nosova, NY., Glazunov, V.A., Palochkin, S.V., Terekhova, A.N.: Synthesis of mechanisms of parallel structure with kinematic interchange. J. Mach. Manuf. Reliab. **43**(5), 378–383 (2014)
18. Tyves, L., Glazunov, V., Danilin, P., Thanh, N.M.: Decoupled parallel manipulator with universal joints and additional constraints. In: Proceedings 18th CISM-IFToMM Symposium ROMANSY 18. Robot Design, Dynamics and Control, Vienna, New York, pp. 65–72. Springer, Dordrecht (2010)

Operation Modes and Self-motions of a 2-RUU Parallel Manipulator

Latifah Nurahmi, Stéphane Caro and Philippe Wenger

Abstract This paper deals with the characterization of the operation modes of the 2-RUU parallel manipulator with an algebraic approach, namely the Study kinematic mapping of the Euclidean group **SE**(3). The manipulator is described by a set of eight constraint equations and the primary decomposition reveals that the mechanism has three operation modes. The singularity conditions are obtained by deriving the determinant of the Jacobian matrix of the constraint equations with respect to the Study parameters. It is shown that there exist singular configurations in which the 2-RUU manipulator may switch from one operation mode to another operation mode. All the singular configurations are mapped onto the joint space and are geometrically interpreted. Finally, the mechanism may switch from the 1st Schönflies mode to the 2nd Schönflies mode through the additional mode that contains self-motions.

Keywords Schönflies motion · Operation mode · Study parameters · 2-RUU manipulator · Self-motion

1 Introduction

Lower-mobility parallel manipulators are suitable for wide range of applications that require fewer than six degree-of-freedom end-effector motion (6-*dof*), for example Schönflies Motion Generators (SMGs). The SMGs are manipulators which can exhibit three independent translations and one pure rotation about an axis of

L. Nurahmi · S. Caro (✉) · P. Wenger
Institut de Recherche En Communications et Cybernétique de Nantes, Nantes, France
e-mail: stephane.caro@irccyn.ec-nantes.fr

L. Nurahmi
e-mail: latifah.nurahmi@irccyn.ec-nantes.fr

P. Wenger
e-mail: philippe.wenger@irccyn.ec-nantes.fr

S. Bai and M. Ceccarelli (eds.), *Recent Advances in Mechanism Design for Robotics*, Mechanisms and Machine Science 33,
DOI 10.1007/978-3-319-18126-4_40

fixed direction, for example the 2-RUU parallel manipulator. This mechanism is composed of two RUU limbs in which two joints are actuated in each limb.

By using an algebraic description of the manipulator and the Study parametrization based on [1, 2], the operation modes of the 2-RUU manipulator are discussed in more details in this paper. The constraint equations are initially derived. Then the primary decomposition is computed over the set of constraint equations to reveal the existence of three operations modes. The singularities are examined by deriving the determinant of the Jacobian matrix of the constraint equations with respect to the Study parameters. It is shown that the mechanism is able to change from one mode to another mode by passing through the configurations that belong to both modes. The singularity conditions are mapped onto the joint space. Eventually, the changes of operation modes for the 2-RUU parallel manipulator are illustrated.

2 Manipulator Architecture

The 2-RUU parallel manipulator shown in Fig. 1, is composed of a base, a moving platform, and two identical limbs. Each limb is composed of five revolute joints such that the second and the third ones, as well as the forth and the fifth ones, are built with intersecting and perpendicular axes. Thus they are assimilated to U-joint.

The first revolute joint is attached to the base and is actuated. Its rotation angle is defined by θ_{1i} $(i = 1, 2)$. The axes of the first, the second, and the fifth joints are directed along **z**-axis. The second axis and the fifth axis are denoted by $\mathbf{v}_i$ and $\mathbf{n}_i$ $(i = 1, 2)$, respectively. The second revolute joint is also actuated and its rotation angle is defined by θ_{2i} $(i = 1, 2)$, in Fig. 2.

The axes of the third and the forth joints are parallel. The axis of the third joint is denoted by $\mathbf{s}_i$ $(i = 1, 2)$ and it changes instantaneously as a function of θ_{2i} as shown in Fig. 2, i.e.: $\mathbf{s}_i = [0, \cos(\theta_{2i}), \sin(\theta_{2i}), 0]^T$ $(i = 1, 2)$.

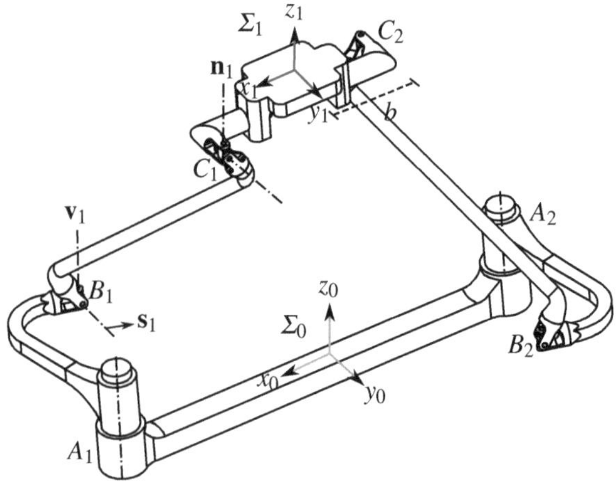

Fig. 1 The 2-RUU parallel manipulator

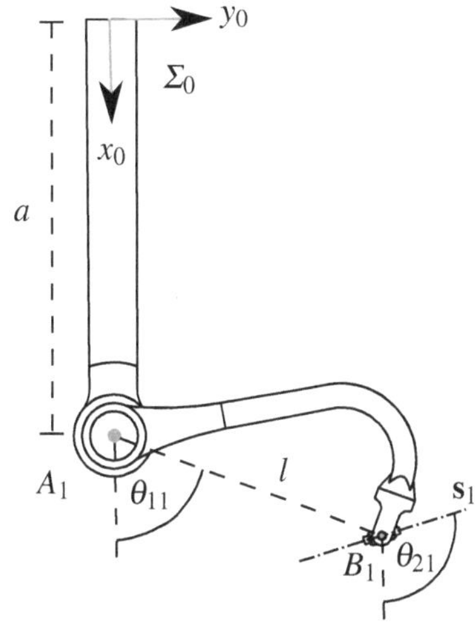

Fig. 2 The axes of R-joints from top view

The fixed frame Σ_0 is located at the center of the base. The first revolute joint of the i-th limb is located at point A_i with distance a from the origin of Σ_0. The first U-joint is denoted by point B_i with distance l from point A_i. Link A_iB_i always moves in a plane normal to $\mathbf{v}_i$. Hence the coordinates of points A_i and B_i expressed in the fixed frame Σ_0 are:

$$\begin{aligned} \mathbf{r}^0_{A_1} &= [1, a, 0, 0]^T \quad \mathbf{r}^0_{B_1} = [1, l\cos(\theta_{11}) + a, l\sin(\theta_{11}), 0]^T \\ \mathbf{r}^0_{A_2} &= [1, -a, 0, 0]^T \quad \mathbf{r}^0_{B_2} = [1, l\cos(\theta_{12}) - a, l\sin(\theta_{12}), 0]^T \end{aligned} \tag{1}$$

The moving frame Σ_1 is located at the center of the moving platform. The moving platform is connected to the limbs by two U-joints, of which the intersection point of the revolute joint axes is denoted by C_i. The length of the moving platform from the origin of Σ_1 to point C_i is defined by b. The length of link B_iC_i is defined by r. The coordinates of point C_i expressed in the moving frame Σ_1 are:

$$\mathbf{r}^1_{C_1} = [1, b, 0, 0]^T \qquad \mathbf{r}^1_{C_2} = [1, -b, 0, 0]^T \tag{2}$$

3 Constraint Equations

In this section, the constraint equations are derived whose solutions illustrate the possible poses of the moving platform (coordinate frame Σ_1) with respect to Σ_0. To obtain the coordinates of points C_i and vectors $\mathbf{n}_i$ expressed in Σ_0, the Study parametrization of a spatial Euclidean transformation matrix $\mathbf{M}$ based on [3] is used.

The parameters $x_0, x_1, x_2, x_3, y_0, y_1, y_2, y_3$, which appear in matrix $\mathbf{M}$, are called Study parameters. They are useful in the representation of a spatial Euclidean displacement and they should satisfy [3] $x_0^2 + x_1^2 + x_2^2 + x_3^2 \neq 0$. This condition will be used in the following computations to simplify the algebraic expressions. First of all, the half-tangent substitutions for θ_{ij} $(i,j = 1,2)$ are performed to remove the trigonometric functions:

$$\cos(\theta_{ij}) = \frac{1 - t_{ij}^2}{1 + t_{ij}^2}, \quad \sin(\theta_{ij}) = \frac{2t_{ij}^2}{1 + t_{ij}^2}, \quad i,j = 1,2 \tag{3}$$

where $t_{ij} = \tan(\frac{\theta_{ij}}{2})$. The coordinates of points C_i and vectors $\mathbf{n}_i$ expressed in Σ_0 are obtained by:

$$\mathbf{r}_{C_i}^0 = \mathbf{M}\,\mathbf{r}_{C_i}^1, \quad \mathbf{n}_i^0 = \mathbf{M}\,\mathbf{n}_i^1, \quad i = 1,2 \tag{4}$$

As the coordinates of all points are given in terms of Study parameters and the design parameters, the constraint equations can be obtained by examining the design of the RUU limb. The link connecting points B_i and C_i is coplanar to the vectors $\mathbf{v}_i$ and $\mathbf{n}_i^0$. Accordingly, the scalar triple product of vectors $\left(\mathbf{r}_{C_i}^0 - \mathbf{r}_{B_i}^0\right)$, $\mathbf{v}_i$ and $\mathbf{n}_i^0$ vanishes, namely:

$$(\mathbf{r}_{C_i}^0 - \mathbf{r}_{B_i}^0)^T . \quad (\mathbf{v}_i \times \mathbf{n}_i^0) = 0, \quad i = 1,2 \tag{5}$$

After computing the corresponding scalar triple product, and removing the common denominators, the following constraint equations come out:

$$\begin{aligned}
g_1\colon \quad & (at_{11}^2 - bt_{11}^2 - lt_{11}^2 + a - b + l)x_0x_1 + 2lt_{11}x_0x_2 - (2t_{11}^2 + 2)x_0y_0 + 2lt_{11}x_3x_1 \\
& + (-at_{11}^2 - bt_{11}^2 + lt_{11}^2 - a - b - l)x_3x_2 + (-2t_{11}^2 - 2)y_3x_3 = 0 \\
g_2\colon \quad & (at_{12}^2 - bt_{12}^2 + lt_{12}^2 + a - b - l)x_0x_1 - 2lt_{12}x_0x_2 + (2t_{12}^2 + 2)x_0y_0 - 2lt_{12}x_1x_3 \\
& + (-at_{12}^2 - bt_{12}^2 - lt_{12}^2 - a - b + l)x_2x_3 + (2t_{12}^2 + 2)x_3y_3 = 0
\end{aligned} \tag{6}$$

To derive the constraint equations corresponding to the link length r of link B_iC_i, the distance equation can be formulated as: $\|(\mathbf{r}_{C_i}^0 - \mathbf{r}_{B_i}^0)\|^2 = r^2$. As a consequence, the following two equations are obtained:

$$\begin{aligned}
g_3 : \quad & (a^2t_{11}^2 - 2abt_{11}^2 - 2alt_{11}^2 + b^2t_{11}^2 + 2blt_{11}^2 + l^2t_{11}^2 - r^2t_{11}^2 + a^2 - 2ab + 2al + b^2 \\
& - 2bl + l^2 - r^2)x_0^2 - 8blt_{11}x_0x_3 + (4at_{11}^2 - 4bt_{11}^2 - 4lt_{11}^2 + 4a - 4b + 4l)x_0y_1 \\
g_4 : \quad & + \cdots (a^2t_{12}^2 - 2abt_{12}^2 + 2alt_{12}^2 + b^2t_{12}^2 - 2blt_{12}^2 + l^2t_{12}^2 - r^2t_{12}^2 + a^2 - 2ab - 2al + b^2 \\
& + 2bl + l^2 - r^2)x_0^2 + 8blt_{12}x_0x_3 + (-4at_{12}^2 + 4bt_{12}^2 - 4lt_{12}^2 - 4a + 4b + 4l)x_0y_1 + \cdots
\end{aligned} \tag{7}$$

To derive the constraint equations corresponding to the actuation of the second joint of each limb, the scalar product of vector $(\mathbf{r}^0_{C_i} - \mathbf{r}^0_{B_i})$ and vector s_i is expressed as: $(\mathbf{r}^0_{C_i} - \mathbf{r}^0_{B_i})^T \mathbf{s}_i = 0$. Hence, the following constraint equations are obtained:

$$
\begin{aligned}
g_5: \quad & (-at_{11}^2t_{21}^2 + bt_{11}^2t_{21}^2 + lt_{11}^2t_{21}^2 + at_{11}^2 - at_{21}^2 - bt_{11}^2 + bt_{21}^2 - lt_{11}^2 + 4lt_{11}t_{21} - lt_{21}^2 \\
& + a - b + l)x_0^2 - 4bt_{21}(t_{11}^2+1)x_0x_3 - 2(t_{21}^2-1)(t_{11}^2+1)x_0y_1 + 4t_{21}(t_{11}^2+1)x_0y_2\ldots \\
g_6: \quad & (at_{12}^2t_{22}^2 - bt_{12}^2t_{22}^2 + lt_{12}^2t_{22}^2 - at_{12}^2 + at_{22}^2 + bt_{12}^2 - bt_{22}^2 - lt_{12}^2 + 4lt_{12}t_{22} - lt_{22}^2 \\
& - a + b + l)x_0^2 + 4bt_{22}(t_{12}^2+1)x_0x_3 - 2(t_{22}^2-1)(t_{12}^2+1)x_0y_1 + 4t_{22}(t_{12}^2+1)x_0y_2\ldots
\end{aligned}
\tag{8}
$$

For detail expressions of equations g_3, g_4, g_5, g_6, the reader may refer to [4]. All solutions have to be within the Study quadric, i.e.: $g_7 : x_0y_0 + x_1y_1 + x_2y_2 + x_3y_3 = 0$. To exclude the exceptional generators ($x_0 = x_1 = x_2 = x_3 = 0$), the normalization equation is added: $g_8 : x_0^2 + x_1^2 + x_2^2 + x_3^2 - 1 = 0$.

4 Operation Modes

The design parameters are assigned as $a = 2, b = 1, l = 1, r = 2$. The set of eight constraint equations is written as a polynomial ideal with variables $\{x_0, x_1, x_2, x_3, y_0, y_1, y_2, y_3\}$ over the coefficient ring $\mathbb{C}[t_{11}, t_{12}, t_{21}, t_{22}]$, defined as: $\mathcal{I} = \langle g_1, g_2, g_3, g_4, g_5, g_6, g_7, g_8\rangle$. At this point, the following ideal is examined: $\mathcal{J} = \langle g_1, g_2, g_7\rangle$.

The primary decomposition is computed and it turns out that the ideal $\mathcal{J}$ is decomposed into three components as: $\mathcal{J} = \bigcap_{i=1}^3 \mathcal{J}_i$, with the results of primary decomposition:

$$
\begin{aligned}
\mathcal{J}_1 = &\ \langle x_0, x_3, x_1y_1 + x_2y_2\rangle \\
\mathcal{J}_2 = &\ \langle x_1, x_2, x_0y_0 + x_3y_3\rangle \\
\mathcal{J}_3 = &\ \langle (t_{11}^2t_{12}^2 + 2t_{12}^2 + 1)x_0x_1 + (-t_{11}^2t_{12} + t_{11}t_{12}^2 + t_{11} - t_{12})x_0x_2\ldots\rangle
\end{aligned}
\tag{9}
$$

For complete results of the primary decomposition, the reader may refer to [4]. Accordingly, the 2-RUU manipulator under study has three operation modes. The computation of the Hilbert dimension of ideal $\mathcal{J}_i$ with $t_{11}, t_{12}, t_{21}, t_{22}$ treated as variables shows that: $dim(\mathcal{J}_i) = 4\ i = 1..3$. To complete the analysis, the remaining equations are added by writing:

$$
\mathcal{K}_i : \mathcal{J}_i \cup \langle g_3, g_4, g_5, g_6, g_8\rangle, \quad i = 1..3 \tag{10}
$$

It follows that the 2-RUU manipulator has 4-*dof*. Each system $\mathcal{K}_i$ corresponds to a specific operation mode that will be discussed in the following.

4.1 System $\mathcal{K}_1$: 1st Schönflies Mode

In this operation mode, the moving platform is reversed about an axis parallel to the **xy**-plane by 180° from the home position [5]. The condition $x_0 = 0, x_3 = 0, x_1 y_1 + x_2 y_2 = 0$ are valid for all poses and are substituted into the transformation matrix **M**, as:

$$M_1 = \begin{pmatrix} 1 & 0 & 0 \\ 2(x_1 y_0 - x_2 y_3) & x_1^2 - x_2^2 & 2x_1 x_2 & 0 \\ 2(x_1 y_3 + x_2 y_0) & 2x_1 x_2 & -x_1^2 + x_2^2 & 0 \\ -\frac{2y_2}{x_1} & 0 & 0 & -1 \end{pmatrix} \tag{11}$$

From the transformation matrix it can be seen that the manipulator has 3-*dof* translational motions, which are parametrized by y_0, y_2, y_3 and 1-*dof* rotational motion, which is parametrized by x_1, x_2 in connection with $x_1^2 + x_2^2 - 1 = 0$ [6].

4.2 System $\mathcal{K}_2$: 2nd Schönflies Mode

In this operation mode, the condition $x_1 = 0, x_2 = 0, x_0 y_0 + x_3 y_3 = 0$ are valid for all poses. The transformation matrix in this operation mode is:

$$M_2 = \begin{pmatrix} 1 & 0 & 0 & 0 \\ -2(x_0 y_1 - x_3 y_2) & x_0^2 - x_3^2 & -2x_0 x_3 & 0 \\ -2(x_0 y_2 + x_3 y_1) & 2x_0 x_3 & x_0^2 - x_3^2 & 0 \\ -\frac{2y_3}{x_0} & 0 & 0 & 1 \end{pmatrix} \tag{12}$$

From this transformation matrix, it can be seen that the manipulator has 3-*dof* translational motions, which are parametrized by y_1, y_2, y_3 and 1-*dof* rotational motion, which is parametrized by x_0, x_3 in connection with $x_0^2 + x_3^2 - 1 = 0$ [6].

4.3 System $\mathcal{K}_3$: Third Mode

In this operation mode, the moving platform is no longer parallel to the base. The variables x_3, y_0, y_1 can be solved linearly from ideal $\mathcal{J}_3$. Accordingly, the 2-RUU parallel manipulator will perform two translational motions, which are parametrized by variables y_2, y_3 and two rotational motions, which are parametrized by variables x_0, x_1, x_2 in connection with normalization equation g_8.

Under this operation mode, the joint angles t_{21} and t_{22} can be computed from Eq. (8). It turns out that no matter the value of the first actuated joint (t_{11}, t_{12})

in each limb, these equations vanish for two real solutions, namely (1) $t_{21} = -\frac{1}{t_{22}}$ and (2) $t_{21} = t_{22}$. As a consequence, in this operation mode, the links B_iC_i $(i = 1, 2)$ from both limbs are always parallel to the same plane and the axes s_i $(i = 1, 2)$ from both limbs are always parallel too.

5 Singularity Conditions and Operation Mode Changing

The 2-RUU manipulator reaches a singularity condition when the determinant of its Jacobian matrix vanishes. The Jacobian matrix is the matrix of all first order partial derivatives of the eight constraint equations with respect to the eight Study parameters. Since the 2-RUU manipulator has more than one operation mode, the singular configurations can be classified into two different types, i.e. the configurations that belong to a single operation mode and the singularity configurations that belong to more than one operation mode. The common configurations that belong to more than one operation mode allow the 2-RUU manipulator to switch from one mode to another mode. However, the 1st Schönflies mode and the 2nd Schönflies mode do not have configurations in common, since x_0, x_1, x_2, x_3 can never vanish simultaneously. It means that the 2-RUU manipulator cannot switch from the Schönflies mode into the reversed Schönflies mode directly.

To change from the 1st Schönflies mode into the 2nd reversed Schönflies mode, the mechanism should pass through the third mode, namely system $\mathscr{K}_3$, as shown in Fig. 3. There exist some configurations in which the mechanism can change from the 1st Schönflies mode to the third mode and these configurations belongs to both

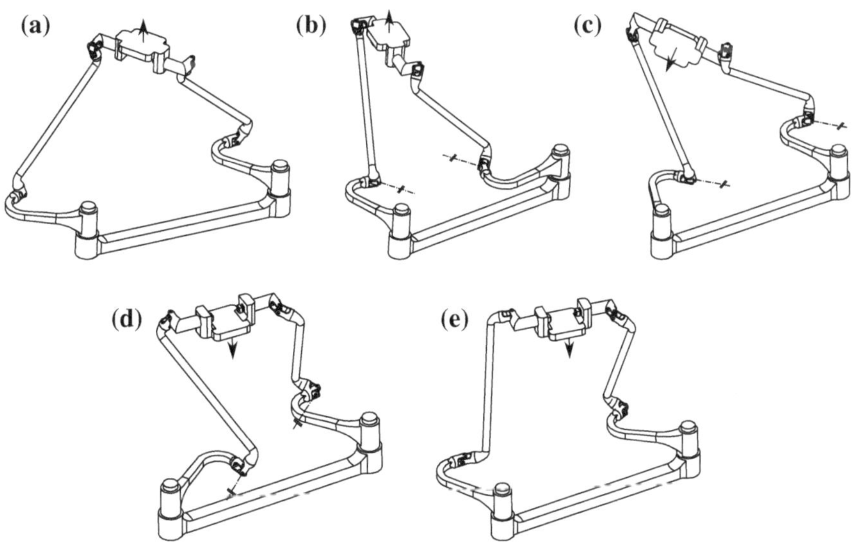

Fig. 3 Transition from the 2nd Schönflies mode to the 1st Schönflies mode via the third mode

modes. Noticeably, these configurations are also singular configurations since they lie in the intersection of two operation modes.

5.1 1st Schönflies Mode ($\mathcal{K}_1$) $\leftrightarrow$ Third Mode ($\mathcal{K}_3$)

For the 1st Schönflies mode, the determinant of Jacobian matrix S_1:$det(\mathbf{J}_1) = 0$ (for complete results of $det(\mathbf{J}_1)$, the reader may refer to [4]) has four factors. The first factor is $y_2 = 0$, when the moving platform is coplanar to the base, the mechanism is always in a singular configuration. The second factor shows the singularity configurations that lie in the intersection with $\mathcal{K}_2$. However, this factor is neglected since systems $\mathcal{K}_1$ and $\mathcal{K}_2$ do not have configurations in common.

The inspection of the third factor yields the singularity configurations that belong to the 1st Schönflies mode $\mathcal{K}_1$ and the third mode $\mathcal{K}_3$. The third factor is added to the system $\mathcal{K}_1$. Then, all Study parameters are eliminated. The elimination yields two polynomials of degree eight and degree nine in $t_{11}, t_{12}, t_{21}, t_{22}$, respectively. The factorization splits both polynomials into four factors as follows:

$$
\begin{aligned}
f_1 :&\quad (t_{21}t_{22} + 1)(t_{21} - t_{22})(t_{12} + t_{11})(3t_{11}^2 t_{12}^2 - 2t_{11}t_{12} + 8t_{12}^2 + 3)\\
f_2 :&\quad (t_{21}t_{22} + 1)(t_{21} - t_{22})(t_{12}^2 + 1)(3t_{11}^2 t_{12}^2 - 2t_{11}t_{12} + 8t_{12}^2 + 3)
\end{aligned}
\tag{13}
$$

Polynomials f_1,f_2 vanish simultaneously when: (1) $t_{21} = -\frac{1}{t_{22}}$ and (2) $t_{21} = t_{22}$. It means that when the second link B_iC_i $(i = 1, 2)$ from both limbs are parallel to the same plane and the moving platform is twisted about an axis parallel to the **xy**-plane by 180°, the 2-RUU manipulator is in the intersection of the 1st Schönflies mode and the third mode.

Eventually, the last factor of the determinant of Jacobian matrix $S_1 : det(\mathbf{J}_1) = 0$ is analysed. This factor is added to the system $\mathcal{K}_1$ and all Study parameters are eliminated. Due to the heavy elimination process, the joint angles are assigned as $t_{11} = 1, t_{12} = 0, t_{21} = -1$ and the result of elimination is a univariate polynomial of degree 12 in t_{22}.

5.2 2nd Schönflies Mode ($\mathcal{K}_2$) $\leftrightarrow$ Third Mode ($\mathcal{K}_3$)

For the 2nd Schönflies mode, the determinant of Jacobian matrix $S_2 : det(\mathbf{J}_2) = 0$ (for complete results of $det(\mathbf{J}_2)$, the reader may refer to [4]) has four factors too. The first factor is $y_3 = 0$ in which the moving platform is coplanar to the base, hence the mechanism is always in a singular configuration. The second factor gives the condition in which the mechanism is in the intersection of systems $\mathcal{K}_1$ and $\mathcal{K}_2$. As explained in Sect. 5.1, this factor is removed.

The analysis of the third factor yields the singularity configurations that belong to the 2nd Schönflies mode $\mathscr{K}_2$ and the third mode $\mathscr{K}_3$. The third factor is added to the system $\mathscr{K}_2$. Then all Study parameters are eliminated and yields:

$$\begin{aligned} f_1 &: \quad (t_{21}t_{22}+1)(t_{21}-t_{22})(t_{12}+t_{11})(3t_{11}^2t_{12}^2-2t_{11}t_{12}+8t_{12}^2+3) \\ f_2 &: \quad (t_{21}t_{22}+1)(t_{21}-t_{22})(t_{12}^2+1)(3t_{11}^2t_{12}^2-2t_{11}t_{12}+8t_{12}^2+3) \end{aligned} \tag{14}$$

Notably, the elimination results have the same mathematical expressions as in Eq. (13) and it leads to the same solutions, namely: (1) $t_{21}=-\frac{1}{t_{22}}$ and (2) $t_{21}=t_{22}$. It means that when the second link B_iC_i $(i=1,2)$ from both limbs are parallel to the same plane and the moving platform is parallel to the base, the 2-RUU manipulator is in the intersection of the 2nd Schönflies mode and the third mode.

Finally, the last factor of the determinant of Jacobian matrix $S_2{:}det(\mathbf{J}_2)=0$ is analysed. Due to the heavy elimination process, the joint angles are assigned as $t_{11}=1, t_{12}=0, t_{21}=-1$ and the result of elimination is also a univariate polynomial of degree 12 in t_{22}.

6 Self-motions in System $\mathscr{K}_3$

The determinant of the Jacobian matrix is computed in the system $\mathscr{K}_3$, which consists of five constraint equations over five variables. Hence the 5×5 Jacobian matrix can be obtained. The determinant of this Jacobian matrix $S_3:det(\mathbf{J}_3)=0$ (for complete results of $det(\mathbf{J}_3)$, the reader may refer to [4]) consists of 10 factors, two of them are: $(t_{21}t_{22}+1)$ and $(t_{21}-t_{22})$. It can be seen from Sect. 4.3 that these factors are the necessary conditions for the manipulator to be in the system $\mathscr{K}_3$, namely (1) $t_{21}=-\frac{1}{t_{22}}$ and (2) $t_{21}=t_{22}$. It means that every configuration in the system $\mathscr{K}_3$ amounts to a self-motion. Furthermore, the image of the transition configurations in the joint space from Eq. (13), is contained in $S_3:det(\mathbf{J}_3)=0$. As a consequence, the transition between systems $\mathscr{K}_1$ and $\mathscr{K}_2$ occurs through the third system $\mathscr{K}_3$ that contains self-motions. These transition configurations are illustrated in Fig. 3a–e.

7 Conclusions

In this paper, the method of algebraic geometry was applied to characterize the type of operation modes of the 2-RUU parallel manipulator. The set of eight constraint equations are derived and the primary decomposition is computed. It is shown that the 2-RUU parallel manipulator has three operation modes and the interpretation of each operation mode was given. The singularity conditions were computed and represented in the joint space. It turns out that the mechanism is able to switch from the 1st Schönflies mode to the 2nd Schönflies mode or vice versa, by passing through the third mode that contains self-motions.

References

1. Nurahmi, L., Schadlbauer, J., Caro, Stephane., Husty, M., Wenger, P.: Kinematic analysis of the 3-RPS cube parallel manipulator. J. Mech. Robot. **7**(1), 011008–1011008-10 (2015)
2. Schadlbauer, J., Walter, D.R., Husty, M.: The 3-RPS parallel manipulator from an algebraic viewpoint. Mech. Mach. Theory **75**, 161–176 (2014)
3. Husty, M.L., Pfurner, M., Schröcker, H.-P., Brunnthaler, K.: Algebraic methods in mechanism analysis and synthesis. Robotica **25**(6), 661–675 (2007)
4. http://www.irccyn.ec-nantes.fr/%7enurahmi/MEDER2015/Appendix.pdf
5. Kong, X.: Reconfiguration analysis of a 3-DOF parallel mechanism using Euler Parameter Quaternions and algebraic geometry method. Mech. Mach. Theory **74**, 188–201 (2014)
6. Schadlbauer, J., Nurahmi, L., Husty, M., Wenger, P., Caro, S.: Operation modes in lower-mobility parallel manipulators, second conference on interdisciplinary applications of kinematics, Lima, Peru, 9–11 September (2013)

Singularity Analysis of 2R1P Spherical Parallel Mechanisms

Yaohong Zhao, Ruiqin Li, Shaoping Bai and Lei Sui

Abstract This paper studies the possible configurations and singularities of 3-DOF spherical parallel mechanisms with revolute and prismatic pairs. In particular, the 2R1P spherical parallel mechanisms are studied, for which the kinematic models are established and Jacobian matrices are derived. Three kinds of singularities, namely the boundary, configuration and structure singularities, are identified and analyzed for RRP, PRR and RPR type of spherical parallel mechanisms. The results can be used for further modeling and analysis on workspace and trajectory planning of 3-DOF spherical parallel mechanisms towards the practical application of the kind of mechanisms.

Keywords 2R1P spherical parallel mechanism · Singularity analysis · Jacobian matrix

1 Introduction

The spherical parallel mechanism is a special type of parallel mechanisms, for which all the axes of revolute pairs intersect at the center of the sphere and the end-effector can freely rotate around the center of the sphere. With their advantages, such as simple structure, large workspace, not easy to interference, etc., spherical parallel mechanisms can be found in applications, for example, ultra-precision machining, solar tracking devices, radio telescope, the orientation of the satellite antenna tracking system, and a variety of aircraft ground automatic tracking

Y. Zhao · R. Li (✉) · L. Sui
School of Mechanical and Power Engineering, North University of China,
Taiyuan 030051, China
e-mail: liruiqin@nuc.edu.cn

S. Bai
Department of Mechanical and Manufacturing Engineering, Aalborg University,
Aalborg, Denmark
e-mail: shb@m-tech.aau.dk

S. Bai and M. Ceccarelli (eds.), *Recent Advances in Mechanism Design for Robotics*, Mechanisms and Machine Science 33,
DOI 10.1007/978-3-319-18126-4_41

devices. It can also be widely used in the robot's shoulder, wrist, waist, hip and medical rehabilitation equipment.

Intensive research on spherical parallel mechanisms has been conducted. Many results on 3-RRR spherical parallel mechanisms have been reported. Guo [1] mainly studied kinematics performance of 3-DOF spherical parallel mechanism. Kong et al. [2] discussed type synthesis of 3-DOF spherical parallel manipulators based on screw theory. Tao et al. [3] discussed and analyzed the method of path planning for 3-RRR spherical mechanism. Gosselin et al. [4] developed 3-RRR spherical parallel mechanism for an electronic smart eye. Zeng et al. [5] developed a 3-RRR NC swivel worktable based on spherical parallel mechanism. Li et al. [6] discussed symmetric 3-DOF 3-PRR spherical parallel manipulator. Bai [7] presented a method for performance optimization over a prescribed workspace.

Compared with 3-RRR spherical parallel mechanism, there are fewer reports about research on other type spherical parallel mechanism, in particular, the spherical parallel mechanisms including prismatic pairs. Bonev et al. [8] presented the computation and representation of the Type-2 singularity loci of symmetric spherical parallel mechanisms based on a not-well-known intuitive orientation representation.

This paper is focus on a type of spherical parallel robots, namely, the 2R1P type spherical parallel mechanisms. Its kinematics and singularity are addressed. Kinematic models are established by taking into the particular spatial relationships among limbs. Then, the Jacobian matrix is further derived. Singularity types of 2R1P spherical parallel mechanisms are identified and analyzed.

2 Configurations of 3-DOF 2R1P Spherical Parallel Mechanisms

The 3-DOF spherical parallel mechanism has three limbs. Each limb contains three kinematic pairs. If the kinematic pairs are limited to revolute pair R and prismatic pair P, eight kinds of limbs with different spatial arrangements can be obtained according to the principle of permutations and combinations, i.e. RRR, PPP, PRR, RPP, PRP, RPR, RRP, PPR.

The 2R1P type spherical parallel mechanisms contain three kinds of configurations, i.e. RRP, PRR, RPR types, as shown in Fig. 1. All axes of revolute pairs in each configuration intersect at one point, i.e. the center of rotation [9].

Let $\mathbf{w}_k$, $\mathbf{r}_k$ and $\mathbf{t}_k (k = 1, 2, 3)$ denote the unit vectors of driving pairs, passive kinematic pairs, the third joint kinematic pair, respectively.

Figure 1a shows a 3-RRP spherical parallel mechanism. The mechanism is composed by a triangular base, a spherical moving platform and three limbs. Three limbs have the same structure and connect the base to the moving platform. Each limb contains two revolute pairs and one prismatic pair with arc-shaped guide [10]. The driving revolute pairs are evenly arranged on the base. The axes of the driving

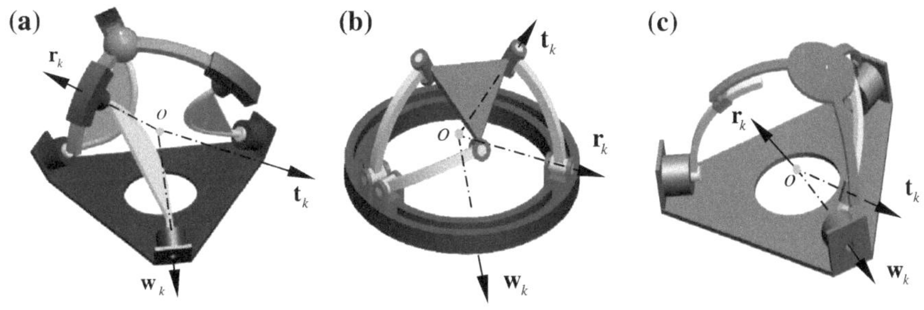

Fig. 1 Configurations of 2R1P type spherical parallel mechanism. **a** RRP. **b** PRR. **c** RPR

pairs and the horizontal base form an angle of 60°, respectively. The passive revolute pairs are coplanar with the arc-shaped prismatic pairs. Moreover, the axis of the passive revolute pair is normal to the circular guide of the prismatic pair. The vector directions of the kinematic pairs in one limb when the mechanism is in a position are marked as shown in Fig. 1a.

Figure 1b shows a 3-PRR spherical parallel mechanism. The prismatic pairs provide circular guides for the revolute pairs of second joint; the revolute pairs of third joint are connected with the mobile platform. These six axes of revolute pairs in all three limbs intersect at the center of rotation.

Figure 1c is a 3-RPR spherical parallel mechanism. The directions of the unit vector $\mathbf{w}_k$ of driving revolute pairs remain horizontal. The prismatic pairs with arc-guide and its direction vector $\mathbf{r}_k$ passes through the center of rotation and always perpendicular to the plane of the arc-shaped guide. The direction vectors $\mathbf{t}_k$ of the revolute pairs connecting with the moving platform are always perpendicular to the arc-shaped guide and pass through the center of the rotation. All three direction vectors of kinematic pairs intersect one point, i.e. the center of rotation.

3 Jacobians of the Spherical Parallel Manipulators

The Jacobian matrix of the robot can be readily obtained [11], which links the input angular velocity of driving pairs and the output angular velocity of the moving platform.

$$\omega = \mathbf{J}\dot{\gamma} \tag{1}$$

or

$$\omega = \mathbf{J}_2^{-1}\mathbf{J}_1\dot{\gamma} \tag{2}$$

where, $\dot{\gamma}$ is the array of angular velocities of active joints, and ω is the angular velocity of the moving platform. $\mathbf{J}_1$ and $\mathbf{J}_2$ are the forward and backward kinematic Jacobian matrices, respectively,which satisfy

$$\mathbf{J}_1\dot{\gamma} - \mathbf{J}_2\omega = 0 \tag{3}$$

The angular velocity of the moving platform in the fixed coordinate system is

$$\dot{\gamma}_k\mathbf{w}_k + \dot{\mu}_k\mathbf{r}_k + \dot{\beta}_k\mathbf{t}_k = \omega \tag{4}$$

where, $\dot{\gamma}_k$ is the driving angular velocity of the kth limb, $\dot{\mu}_k$ and $\dot{\beta}_k$ are the angular velocities of the second and third kinematic pairs.

In Eq. (4), multiplying both sides by $-\mathbf{r}_k^T$ and $-\mathbf{t}_k^T$ separately yields:

$$\dot{\mu}_k = -\mathbf{r}_k^T\omega,\ \dot{\beta}_k = \dot{\gamma}_k\mathbf{t}_k^T\mathbf{w}_k - \mathbf{t}_k^T\omega$$

Similarly, multiplying both sides of Eq. (4) by $(\mathbf{r}_k \times \mathbf{t}_k)^T$ leads to:

$$(\mathbf{r}_k \times \mathbf{t}_k)^T\mathbf{w}_k\dot{\gamma}_k - (\mathbf{r}_k \times \mathbf{t}_k)^T\omega = 0 \tag{5}$$

From Eq. (3), $\mathbf{J}_1$ and $\mathbf{J}_2$ can be found as:

$$\mathbf{J}_1 = \begin{bmatrix} c_1 & 0 & 0 \\ 0 & c_2 & 0 \\ 0 & 0 & c_3 \end{bmatrix},\ \mathbf{J}_2 = \begin{bmatrix} (\mathbf{r}_1 \times \mathbf{t}_1)^T \\ (\mathbf{r}_2 \times \mathbf{t}_2)^T \\ (\mathbf{r}_3 \times \mathbf{t}_3)^T \end{bmatrix}$$

where, $c_i = (\mathbf{r}_k \times \mathbf{t}_k)^T\mathbf{w}_k$, $i = 1, 2, 3$.

4 Singularity Analysis of the 2R1P Spherical Parallel Mechanisms

When the mechanism is in singular configuration, the moving platform obtains or loses instantaneous extra degrees of freedom which reduces the stiffness of mechanism and makes the mechanism uncontrollable, even damages the mechanism. In general situation, the singular configuration of mechanism often appears in the boundary or internal of theoretical workspace. In this case, its Jacobian matrix becomes singular.

4.1 Boundary Singularity

When $\det(\mathbf{J}_1) = 0$ and $\det(\mathbf{J}_2) \neq 0$, there exists a non-zero vector making ω become a zero vector, it leads to singularity. That is to say, there exists at least one set of vectors making $(\mathbf{r}_k \times \mathbf{t}_k) \cdot \mathbf{w}_k = 0$. This situation belongs to reverse kinematics singularity, i.e. boundary singularity. That generally refers to the configuration generated at the time when the mechanism is in the boundary of workspace or the number of reverse kinematics change.

(1) In 3-RRP spherical parallel mechanism, $\mathbf{r}_k$ and $\mathbf{t}_k$ are always vertical which constitute a plane. $\mathbf{r}_k$ and $\mathbf{w}_k$ are always coplanar. $\mathbf{w}_k$ must be vertical to $(\mathbf{r}_k \times \mathbf{t}_k)$ if making $(\mathbf{r}_k \times \mathbf{t}_k) \cdot \mathbf{w}_k = 0$. Thus, there must exist at least one set of coplanar vectors $\mathbf{r}_k$, $\mathbf{w}_k$ and $\mathbf{t}_k$ in three limbs. This situation does not occur for 3-RRP spherical parallel mechanism.
(2) In 3-PRR spherical parallel mechanism, $\mathbf{w}_k$ must be vertical $(\mathbf{r}_k \times \mathbf{t}_k)$ to satisfy $(\mathbf{r}_k \times \mathbf{t}_k) \cdot \mathbf{w}_k = 0$. Because the direction of $\mathbf{w}_k$ is the axial direction of the circular prismatic pair and is vertical to the fixed base, when the plane constituted by two revolute joints is vertical to the plane defined by the fixed base, there exist at least one group of coplanar vectors $\mathbf{r}_k$,$\mathbf{w}_k$ and $\mathbf{t}_k$ making it lead to singular configuration, as shown in Fig. 2.
(3) In 3-RPR spherical parallel mechanism, to satisfy $(\mathbf{r}_k \times \mathbf{t}_k) \cdot \mathbf{w}_k = 0$, three vectors must be coplanar. Because $\mathbf{w}_k$ and $\mathbf{t}_k$ are continuously coplanar with the circular guide when the mechanism is operating, and the direction of $\mathbf{r}_k$ is continuously vertical to the circular surface of the rail, as is shown in Fig. 3a, b. Therefore, the situation that $\mathbf{w}_k$,$\mathbf{t}_k$ and $\mathbf{r}_k$ are coplanar does not exist, there are not singular configuration.

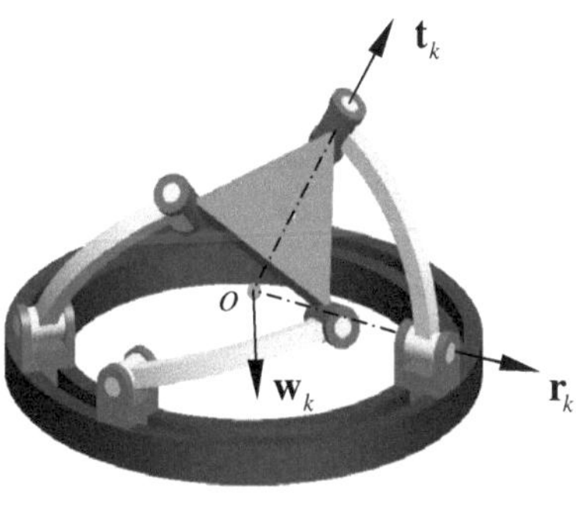

Fig. 2 Singular configuration of 3-PRR spherical parallel mechanism

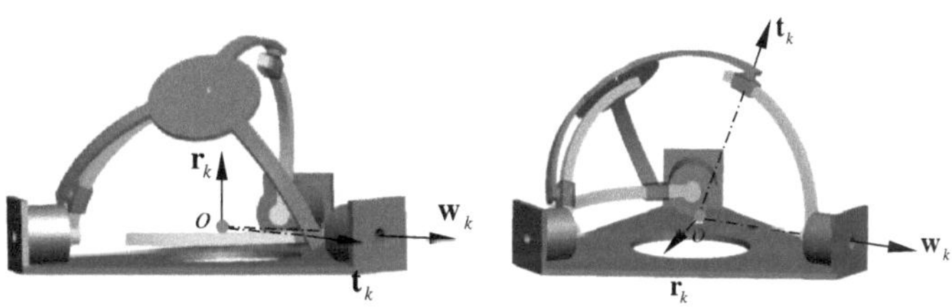

Fig. 3 Vector directions of the 3-RPR spherical parallel mechanism

4.2 Configuration Singularity

When $\det(\mathbf{J}_2) = 0$ and $\det(\mathbf{J}_1) \neq 0$, the mechanism will add one degree of freedom. Even if the drive is locked, an instantaneous motion of the movable platform can still be possible. This situation belongs to positive kinematics singularity, i.e. a configuration singularity. With driving parts fixed, the moving platform is still movable, thus the mechanism will lose stability. At this time $(\mathbf{r}_k \times \mathbf{t}_k)$ obtains three vectors.

For the 3-RRP spherical parallel mechanism, this type of singularity occurs with the following two conditions:

① Two rows of matrix $\mathbf{J}_2$ are linear dependent. When two limbs become coplanar, this singularity will occur. According to the structure of the moving platform which three limbs are 120° distributions, so this situation does not exist.

② A row of matrix $\mathbf{J}_2$ is the linear combination of the other two rows. In this situation, three planes identified by vector $(\mathbf{r}_k \times \mathbf{t}_k)$ intersect in a straight line, and singular configuration of the mechanism occurs, as shown in Fig. 4.

For the 3-PRR spherical parallel mechanism, the singularity occurs with the following two conditions.

① Three planes identified by vector $(\mathbf{r}_k \times \mathbf{t}_k)$ are coincident. According to the structure of the moving platform, this situation does not exist.

② Three planes identified by vector $(\mathbf{r}_k \times \mathbf{t}_k)$ intersect at a common axis. Because the active joints are evenly distributed in 120° and the three limbs are vertical to the fixed base, three planes identified by vector $(\mathbf{r}_k \times \mathbf{t}_k)$ intersect at a common axis. The second singularity of the mechanism occurs, as shown in Fig. 5.

Finally, two conditions of this type of singularity can be identified for 3-RPR spherical parallel mechanism:

① Three planes with normal vectors $(\mathbf{r}_k \times \mathbf{t}_k)$ are coincident. According to the structure of the movable platform of the spherical parallel mechanism, this case does not exist.

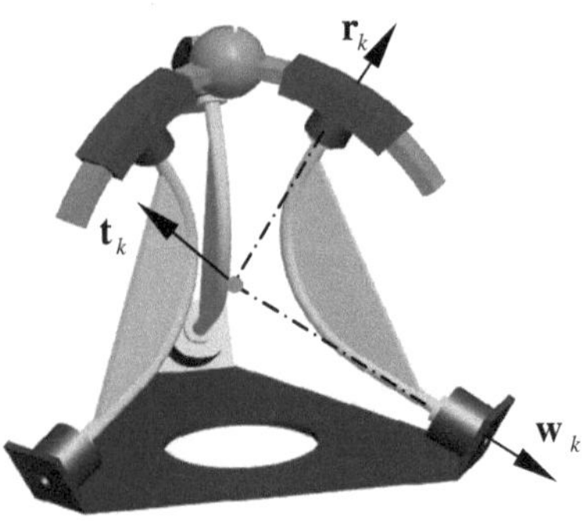

Fig. 4 Configuration singularity of 3-RRP spherical parallel mechanism

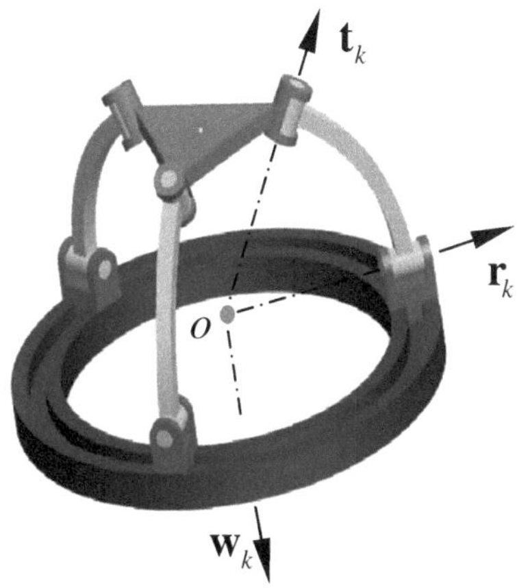

Fig. 5 Configuration singularity of 3-PRR spherical parallel mechanism

② Three planes with normal vectors $(\mathbf{r}_k \times \mathbf{t}_k)$ intersect at a common axis. Due to the construction of the mechanism, three planes identified by vector $(\mathbf{r}_k \times \mathbf{t}_k)$ are vertical to the three curved rails, and cannot intersect at a common axis, but intersect at one point that is the rotation center of the mechanism, so this situation does not exist.

4.3 Structure Singularity

When $\det(\mathbf{J}_1) = 0$ and $\det(\mathbf{J}_2) = 0$, the occurrence of singularity is mixed. In this case, the active joint input and the end-effector exist instantaneous and non-zero input and output independently. This type of singularity occurs when the two singularities previously defined occur simultaneously. The occurrence of the singularity has nothing with position and attitude of the mechanism, but depends entirely on the mechanism configuration, so it is called structure singularity that needs to be avoided in the design stage.

(1) For 3-RRP spherical parallel mechanism, because of the specificity of structure that these are three sub-chains uniformly distributed over the movable platform, structure singularity does not exist.
(2) For 3-PRR spherical parallel mechanism, when these are three sliding pairs evenly distributed on the circular rail, $\det(\mathbf{J}_1) = 0$ and $\det(\mathbf{J}_2) = 0$ hold simultaneously, the third singularity occurs, as shown in Fig. 5.
(3) For 3-RPR spherical parallel mechanism, because the previous two singularities do not occur, there isn't the third singularity.

According to the analysis of the Jacobian matrixes of diffident mechanisms, 3-RRP spherical parallel mechanism contains configuration singularity; 3-PRR spherical parallel mechanism contains boundary singularity, configuration singularity and structure singularity; 3-RPR spherical parallel mechanism does not contain any singularity.

5 Conclusions

The 2R1P type spherical parallel mechanisms are studied in this work, addressing the singularity analysis. Three types of singularities, i.e., the boundary, configuration and structure singularities are identified on the basis of the Jacobian matrices of the manipulator. The analysis and identification of the singularity for the 2R1P spherical parallel manipulators will be useful for the proper selection of robots in different applications. Further analyzing and quantifying the singularity for a certain design will be considered.

Acknowledgments The authors gratefully acknowledge the financial and facility support provided by the National Natural Science Foundation of China (Grant No. 51275486) and the Specialized Research Fund for the Doctoral Program of Higher Education (Grant No. 20111420110005).

References

1. Guo, Y.: Research on kinematics performance of 3-DOF spherical parallel mechanism. Dissertation, North University of China (2014)
2. Kong, X., Gosselin, C.M.: Type synthesis of 3-DOF spherical parallel manipulators based on screw theory. ASME J. Mech. Des. **126**(1), 101–108 (2004)
3. Tao, Z.J., An, Q.: Path planning method for optimum 3-RRR spherical parallel manipulator. J. East China Univ. Sci. Technol. **40**(2), 267–272 (2014)
4. Gosselin, C.M., Angeles, J.: The optimum kinematics design of a spherical 3-DOF parallel manipulator. ASME J. Mech. Trans. Autom Des. **111**(2), 202–207 (1989)
5. Zeng, X., Huang, T., Zeng, Z.: Accuracy analysis of the 3-RRR NC swivel worktable. J. Mech. Eng. **37**(11), 42–46 (2001)
6. Li, Q., Chen, H., Li, Y et al.: Kinematics analysis of a 3-DOF 3-$P_c(RR)_N$ spherical parallel manipulator. China Mech. Eng. **20**(11), 1280–1285 (2009)
7. Bai, S.: Optimum design of spherical parallel manipulators for a prescribed workspace. Mech. Mach. Theory **45**(2), 200–211 (2010)
8. Bonev, I.A., Gosselin, C.M.: Singularity loci of spherical parallel mechanisms. In: Proceedings of the 2005 IEEE International Conference on Robotics and Automation, vol. 3, pp. 2957–2962. Barcelona (2005)
9. Wu, G., Bai, S., Kepler, J.: Mobile platform center shift in spherical parallel manipulators with flexible limbs. Mech. Mach. Theory **75**, 12–26 (2014)
10. Enferadi, J., Tootoonchi, A.A.: Accuracy and stiffness analysis of a 3-RRP spherical parallel manipulator. Robot. **29**, 193–209 (2011)
11. Liu, W., Chang, S.: Study on singularity and workspace of a novel parallel manipulator. Chin. Mech. Eng. **23**(7), 786–790 (2012)

Kinematics Analysis of 5-Rod Car Suspension Mechanism with Singularities

S. Yu Misyurin and A.P. Nelyubin

Abstract In the paper kinematics of the independent 5-rod car suspension mechanism is analyzed to calculate camber and toe angles, lateral and longitudinal displacements that affect the performance of the whole vehicle guiding mechanism. The problem is solved numerically by multidimensional Newton method with regularization. The proposed regularization parameter provides acceptable problem conditioning and proximity to the original problem.

Keywords Independent multi-link suspension · Kinematics · Numerical methods

1 Introduction

Multi-link structure of car suspension can offer better compromise between handling, safety, ride comfort and space efficiency. It provides high stiffness in the lateral direction and a reasonable compliance in the longitudinal (fore/aft) direction. The first 5-rod (or 5-link) suspension was patented by Mercedes Benz in the late 80s. Since that time, multi-link rear suspension is increasingly used in modern sedans and coupes both at front and rear axles. Its construction is not strictly defined and differs from one model to another. In this paper we will consider Honda Accord's suspension which is essentially a double wishbone suspension added with the fifth control arm. Honda's 5-rod double wishbone suspension features a Watt link arrangement (layout).

Numerical simulation of vehicle suspension movement allows automating its design and reducing the number of full-scale experiments. For example in [1] a lane

S.Y. Misyurin
National Research Nuclear University «MEPhI», Moscow, Russia
e-mail: ssmmrr@mail.ru

A.P. Nelyubin (✉)
Institute of Machines Science of the Russian Academy of Sciences, Moscow, Russia
e-mail: nelubin@gmail.com

S. Bai and M. Ceccarelli (eds.), *Recent Advances in Mechanism Design for Robotics*, Mechanisms and Machine Science 33,
DOI 10.1007/978-3-319-18126-4_42

change maneuver has been simulated to optimize comfort and handling criteria. In Knapczyk and Maniowski [2, 3] optimized elastokinematic and dynamic characteristics of a 5-rod suspension. There are many models of other suspension systems. For example, in [4] an analytical planar quarter-car model of McPherson suspension was proposed which accounts both kinematics and dynamics. However, the 5-rod suspension is more complex and it will be hard to model it analytically.

In this paper we will study some important kinematical characteristics of a vehicle suspension, such as toe and camber angles, lateral and longitudinal displacements [5, 6]. The values of these characteristics and their changes on rebound and pressure strokes affect the vehicle handling and stability, ride comfort, tire wear, etc.

Kinematic analysis of suspension mechanism involves solving systems of polynomial equations. In Raghavan and Roth [7] reviewed three exact computational methods for solving such systems. These are Grobner bases (symbolical), Polynomial Continuation (numerical) and Dialytic Elimination (hybrid). Dialytic Elimination procedure is very useful only for small problems with up to 6 variables. Grobner bases method suffers from exploding computations due to the large number of intermediate polynomials generated.

Though in Uchida and McPhee [8] succeed in generating the Grobner bases to triangularize the kinematic constraint equations. Their approach has been applied to the Stewart-Gough platform and 5-link suspension systems, whose topological similarity motivates their concurrent study. Although pre-generating Grobner basis is attractive in real-time applications, it is computationally expensive and requires individual approach to each problem.

In Papegay et al. [9] used interval analysis to construct all possible trajectories corresponding to a given range of a free parameter. They have shown that it is not practicable, even for the simple MacPherson mechanism, to compute the symbolical expression of the output parameters.

In this paper we will find kinematical characteristics by following the trajectory of the suspension's mechanism from its rebound to pressure strokes. We will solve the system of polynomial equations arising in kinematic analysis of 5-rod suspension mechanism numerically by multidimensional Newton method. It is the usual approach to follow the trajectory of a mechanism, because we can use the result obtained at one step as the initial guess of the Newton method that will be used to determine the configuration for the next step.

2 Kinematical Characteristics

Given the suspensions stroke, coordinates of its joints can be calculated. Using these coordinates we can determine the kinematical characteristics under interest: toe and camber angles, lateral and longitudinal displacements.

Camber angle γ is employed to enhance lateral bite of wheels with road surface during the turn maneuver. The best plot of the camber angle versus wheel stroke is as follows: γ is minimal under compression s_1 and maximal under rebound s_2 [5]:

$$K_1 = \gamma(s_2) - \gamma(s_1) \rightarrow \max.$$

If toe angle δ is positive, then wheels undergo stabilizing lateral forces. So to achieve better vehicle stability the desirable change of the toe angle should be as minimal as possible [5]:

$$K_2 = \Delta\delta \rightarrow \min.$$

Hereafter the symbol Δ is defined as the difference between maximal and minimal values of the corresponding parameter on the whole suspension's stroke.

Axle base L and wheel span b affect maneuverability, roadholding ability, passability [5, 6]. Minimal changes of axle base (longitudinal displacement), wheel span (lateral displacement) and toe angle further not only better handling, but also less tire wear:

$$K_3 = \Delta b \rightarrow \min, \qquad K_4 = \Delta L \rightarrow \min.$$

All these performance indexes can be used as criteria of optimality to adjust design parameters in a synthesis procedure.

Kinematics of the 5-rod suspension mechanism is presented of the Fig. 1. The reference system $Oxyz$ is associated with the car body and its origin is placed in the wheel center point in constructive position (zero wheel stroke). So the coordinates of joints A_0, B_0, C_0, D_0 and F_0 of the suspension's construction with the car body

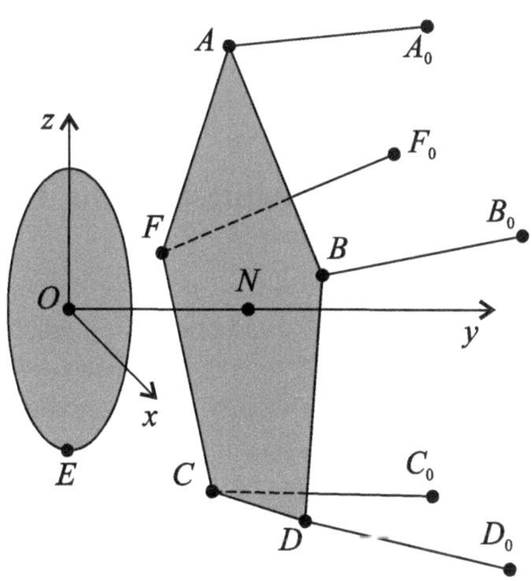

Fig. 1 Kinematics of the 5-rod suspension

Table 1 Initial parameters

Point	x, mm	y, mm	z, mm
A_0	304	370.5	180
B_0	159.6	497.8	20
C_0	−353.4	418	−30
D_0	−53.2	418	−45
F_0	−60.8	380	195
A	22.8	87.4	200
B	159.6	163.4	20
C	−110.2	106.4	−50
D	−53.2	76	−50
F	−60.8	121.6	210

are fixed. All 5 rods have constant lengths. The wheel spin axis *ON* and the wheel carrier (moving platform) *ABCDF* are considered as one rigid body. The point *E* is the contact point of the wheel with the road surface. The system has 1-DOF given by vertical position $z_O = s$ (the wheel stroke) of the wheel center *O*.

The required kinematical characteristics can be expressed as follows:

$$\sin\gamma = (z_N - z_O)/|ON|; \mathrm{tg}\delta = (x_N - x_O)/(y_N - y_O); \Delta b = \Delta y_E; \Delta L = \Delta x_E \quad (1)$$

To find their changes on the wheel stroke we will follow the trajectory of the mechanism from its initial position ($s = 0$) to its pressure stroke ($s_1 = 100$ mm) and rebound stroke ($s_2 = -100$ mm) with the step 1 mm. Approximate initial coordinates (in mm) of joints of the considered suspension's mechanism with Watt link arrangement [10] are presented in the Table 1:

These parameters in constructive position determine design of the mechanism and its trajectory. This design can be used as a prototype in a synthesis or optimization procedure.

3 Numerical Solution of the Kinematics

Following the trajectory requires solving forward kinematic problem: by a given *s* find position of the suspension's mechanism, i.e. coordinates of the points *O*, *E*, *N*, *A*, *B*, *C*, *D* and *F*. It is sufficient to find coordinates of the points *O*, *A*, *B*, *C*, *D*, *F* and then independently calculate coordinates of the points *E* and *N*. So we have to find 17 unknown coordinates by solving the system of 17 nonlinear constraints on the constant lengths of A_0A, *AB*, B_0B, *AC*, *BC*, C_0C, *AD*, *BD*, *CD*, D_0D, *AF*, *BF*, *CF*, F_0F, *AO*, *BO*, *CO*. The constraints are enumerated correspondingly:

$$
\begin{aligned}
f_1 &= {}^1\!/_2\left((x_A - x_{A0})^2 + (y_A - y_{A0})^2 + (z_A - z_{A0})^2 - |A_0A|\right) = 0;\\
f_2 &= {}^1\!/_2\left((x_B - x_A)^2 + (y_B - y_A)^2 + (z_B - z_A)^2 - |AB|\right) = 0;\\
f_3 &= {}^1\!/_2\left((x_{B0} - x_B)^2 + (y_{B0} - y_B)^2 + (z_{B0} - z_B)^2 - |B_0B|\right) = 0;\\
f_4 &= {}^1\!/_2\left((x_C - x_A)^2 + (y_C - y_A)^2 + (z_C - z_A)^2 - |AC|\right) = 0;\\
f_5 &= {}^1\!/_2\left((x_C - x_B)^2 + (y_C - y_B)^2 + (z_C - z_B)^2 - |BC|\right) = 0;\\
f_6 &= {}^1\!/_2\left((x_C - x_{C0})^2 + (y_C - y_{C0})^2 + (z_C - z_{C0})^2 - |C_0C|\right) = 0;\\
f_7 &= {}^1\!/_2\left((x_D - x_A)^2 + (y_D - y_A)^2 + (z_D - z_A)^2 - |AD|\right) = 0;\\
f_8 &= {}^1\!/_2\left((x_D - x_B)^2 + (y_D - y_B)^2 + (z_D - z_B)^2 - |BD|\right) = 0;\\
f_9 &= {}^1\!/_2\left((x_D - x_C)^2 + (y_D - y_C)^2 + (z_D - z_C)^2 - |CD|\right) = 0;\\
f_{10} &= {}^1\!/_2\left((x_D - x_{D0})^2 + (y_D - y_{D0})^2 + (z_D - z_{D0})^2 - |D_0D|\right) = 0;\\
f_{11} &= {}^1\!/_2\left((x_F - x_A)^2 + (y_F - y_A)^2 + (z_F - z_A)^2 - |AF|\right) = 0;\\
f_{12} &= {}^1\!/_2\left((x_F - x_B)^2 + (y_F - y_B)^2 + (z_F - z_B)^2 - |BF|\right) = 0;\\
f_{13} &= {}^1\!/_2\left((x_F - x_C)^2 + (y_F - y_C)^2 + (z_F - z_C)^2 - |CF|\right) = 0;\\
f_{14} &= {}^1\!/_2\left((x_F - x_{F0})^2 + (y_F - y_{F0})^2 + (z_F - z_{F0})^2 - |F_0F|\right) = 0;\\
f_{15} &= {}^1\!/_2\left((x_O - x_A)^2 + (y_O - y_A)^2 + (z_O - z_A)^2 - |AO|\right) = 0;\\
f_{16} &= {}^1\!/_2\left((x_O - x_B)^2 + (y_O - y_B)^2 + (z_O - z_B)^2 - |BO|\right) = 0;\\
f_{17} &= {}^1\!/_2\left((x_O - x_C)^2 + (y_O - y_C)^2 + (z_O - z_C)^2 - |CO|\right) = 0.
\end{aligned}
\tag{2}
$$

The factor ½ was introduced to simplify the following calculations. Each nonlinear function f_i, $i = 1, \ldots, 14$ depends on 3 or 6 unknown coordinates (the last three functions depend on 5 unknowns as $z_O = s$ is given).

We solve the system (2) numerically by multidimensional Newton method [11, 12]. This is an iterative procedure which calculates progressive approximations of the solution. It is essential to use previous trajectory position of the mechanism as the initial approximation $x^0 = (x_1^0, \ldots, x_{17}^0)$. Each next approximation x^{k+1} is calculated by solving the system of linear algebraic equations (SLAE):

$$
\begin{cases}
\sum_{i=1}^{17} J_{1i}(\vec{x}^k) z_i = -f_1(\vec{x}^k)\\
\cdots\\
\sum_{i=1}^{17} J_{17i}(\vec{x}^k) z_i = -f_{17}(\vec{x}^k)
\end{cases},
\tag{3}
$$

where

$$J_{ji}(\vec{x}^k) = \frac{\partial f_j(\vec{x}^k)}{\partial x_i^k},$$

$i, j = 1, \ldots, 17$, are elements of the Jacobian matrix of the system (2). Solution z of the SLAE (3) is the incremental vector, so that:

$$\vec{x}^{k+1} = \vec{x}^k + \vec{z}.$$

In practice to verify the method's convergence a simplified rule is employed. The iterations stop when a norm of the incremental vector is less than a small positive parameter ε. We use the following rule:

$$\max_{i=1,\ldots,17}(|z_i|) < 0.05 \text{ mm}. \tag{4}$$

In case the method doesn't converge a limit number K of iterations should be provided. This may take place near zones of singularity [9, 13, 14], where two or more trajectories come close to each other. In such zones the Jacobian matrix J is ill-conditioned or even degenerate (singular). In [9] the authors used Kantorovich theorem to guarantee that the Newton method will converge to the solution that belongs to the same branch. In this paper for this purpose we will use regularization technique. Unlike [9], we don't need to find all branches of trajectory, but only that corresponding to a given design of the mechanism. And if two or more brunches collapse then such design should be discarded on synthesis stage.

To construct the Jacobian matrix of the system (2) it is sufficient to calculate only three partial derivatives of the functions f_i, $i = 1, \ldots, 17$, because for example ($i = 2$):

$$\partial f_2/\partial x_A = x_A - x_B = -\partial f_2/\partial x_B, \quad \partial f_2/\partial y_A = -\partial f_2/\partial y_B,$$
$$\partial f_2/\partial z_A = -\partial f_2/\partial z_B$$

Denote by J'_{i1}, J'_{i2}, J'_{i3}, $i = 1, \ldots 17$ partial derivatives of the function f_i by coordinates of the second point in the corresponding segment. For example ($i = 2$):

$$J'_{21} = \partial f_2/\partial x_B = x_B - x_A, \quad J'_{22} = \partial f_2/\partial y_B = y_B - y_A,$$
$$J'_{23} = \partial f_2/\partial z_B = z_B - z_A.$$

Hence the 17 × 17-matrix J of the SLAE (3) is determined by 17 × 3 numbers. We will write just part of it:

$$\begin{bmatrix} J'_{11} & J'_{12} & J'_{13} & 0 & 0 & 0 & 0 & 0 & 0 & \cdots \\ -J'_{21} & -J'_{22} & -J'_{23} & J'_{21} & J'_{22} & J'_{23} & 0 & 0 & 0 & \cdots \\ 0 & 0 & 0 & J'_{31} & J'_{32} & J'_{33} & 0 & 0 & 0 & \cdots \\ -J'_{41} & -J'_{42} & -J'_{43} & 0 & 0 & 0 & J'_{41} & J'_{42} & J'_{43} & \cdots \\ 0 & 0 & 0 & -J'_{51} & -J'_{52} & -J'_{53} & J'_{51} & J'_{52} & J'_{53} & \cdots \\ 0 & 0 & 0 & 0 & 0 & 0 & J'_{61} & J'_{62} & J'_{63} & \cdots \\ \vdots & \vdots & \vdots & \vdots & \vdots & \vdots & \vdots & \vdots & \vdots & \ddots \end{bmatrix} \quad (5)$$

This is a quite sparse matrix: each row consists of only 3–6 non-zero elements. And even these numbers can be equal or close to zero if any two points will be on the same axis. However, due to large-scale of the matrix it is practically impossible to solve the SLAE symbolically (for example, using such programs as MAPLE, MATHCAD, etc.). The same conclusion was made in [9]. Therefore it is necessary to use numerical methods.

Approximate numerical methods of solving SLAE are usually used when the use of exact methods is difficult or impossible, such as when the order of the system is thousands of variables. In the case where the number of variables doesn't exceed 10^2–10^3, it is reasonable to use exact methods [12]. Among exact methods we chose Gauss elimination method because of its efficiency from a computational point of view [11]. To ensure numerical stability of the algorithm we use partial pivoting, which selects the entry with the largest absolute value from the column of the matrix that is currently being considered as the pivot (leading coefficient) element.

Gauss elimination method also finds the rank of the matrix J, so we can check consistency of the SLAE during the solution process. By the Rouché–Capelli (Kronecker–Capelli) theorem, the SLAE (3) has a solution if and only if the rank of the matrix J is equal to the rank of its augmented matrix $[J|{-f}]$. The matrix J is square; hence in the non-degenerate case ($\det J \neq 0$) there exists a unique solution of the SLAE (3).

This was the case on the whole trajectory of the prototype design with initial coordinates given in the Table 1. All computational process of following the trajectory took 0.5 s on Core Duo 1.6 GHz processor. Each of the 200 steps needed not more than 2 iterations of the Newton method to stop according to the rule (4). The maximum deviation of the links lengths was not more than 10^{-9} mm, which can be accounted for round-off errors. The resulting values of kinematical characteristics are

$$\gamma(s_2) - \gamma(s_1) = 0.16^\circ,\ \Delta\delta = 0.0075^\circ,\ \Delta b = 46.9\ \text{mm},\ \Delta L = 12.8\ \text{mm}.$$

Testing the algorithm on other designs of the mechanism sometimes we occurred with the problem, when in some positions the matrix J was ill-conditioned ($|\det J| << 1$) or even degenerate ($\det J = 0$). The SLAE (3) was always consistent, i.e. underdetermined. In the degenerate case there was infinite number of solutions. Although the ill-conditioned SLAE had one solution, it was pointless to find it from practical point of view. In these problem positions there are zones of singularity of the mechanism. One of the ways to solve this problem is to discard such designs at all. Suppose that for ill-conditioned SLAE we want to find the solution anyway. For this purpose we adopted Tikhonov regularization [15].

Since the SLAE is underdetermined Tikhonov regularization gives some preference to a particular solution z_0 with desirable properties. In our problem it is reasonable to minimize the norm of the incremental vector z. Hence, we choose $z_0 = 0$. The regularization term is included in the residual minimization functional:

$$\Omega(\vec{z}, \lambda) = \left\| J\vec{z} + \vec{f} \right\|^2 + \lambda \|\vec{z} - \vec{z}_0\|^2.$$

This regularization improves the conditioning of the problem, thus enabling a numerical solution. The new SLAE can be written as follows:

$$(J^T J + \lambda I)\vec{z} = -J^T \vec{f} + \lambda \vec{z}_0.$$

For $\lambda \rightarrow 0$ it turns into the original ill-conditioned systems, and for large λ, being a well-conditioned, has a solution z_0. Obviously, the best would be some intermediate value of λ, establishing a trade-off between acceptable conditioning and proximity to the original problem. In our experiments acceptable results were obtained for λ from the interval 0.001–0.002.

4 Conclusions

In the paper kinematics of the independent 5-rod vehicle suspension mechanism was analyzed to calculate the following kinematical characteristics: toe and camber angles, lateral and longitudinal displacements. For this purpose we solved the system of 17 nonlinear equations with 17 unknowns numerically by multidimensional Newton method with regularization. This approach is universal and reasonably efficient for such problems.

The changes of the chosen kinematical characteristics on the wheel stroke affect the performance of the whole vehicle guiding mechanism. So the proposed algorithm can be adopted to optimize the suspension parameters with regard to these kinematical characteristics.

To perform calculation that is more accurate it is advisable to take into account the dynamic effects of hydro (pneumatic) shock absorbers in the mathematical model. However, it will make the whole model more sophisticated. Considered in [16–19] is the method of iteratively growth of complexity of the mathematical model of a dynamical system mechanism—drive (pneumatic, hydro). This method is also applicable to the vehicle suspension.

References

1. Samin, J.C., Brüls, O., Collard, J.F., Sass, L., Fisette, P.: Multiphysics modeling and optimization of mechatronic multibody systems. Multibody Syst. Dyn. **18**(3), 345–373 (2007)
2. Knapczyk, J., Maniowski, M.: Elastokinematic modeling and study of five-rod suspension with subframe. Mech. Mach. Theory **41**, 1031–1047 (2006)
3. Knapczyk, J., Maniowski, M.: Optimization of 5-rod car suspension for elastokinematic and dynamic characteristics. Arch. Mech. Eng. **42**(2), 133–147 (2010)
4. Hurel, J., Mandow, A., García-Cerezo, A.: Kinematic and dynamic analysis of the McPherson suspension with a planar quarter-car model. Veh. Syst. Dyn. **51**(9), 1422–1437 (2013)
5. Reimpell, J.: Fahrwerktechnik: Radaufhängungen. Vogel-Buchverlag, Würzburg (1986)
6. Litvinov, A.S., Farobin, Y.E.: Automobile. The theory of operation properties. Mashinostroenie, Moscow (1989)
7. Raghavan, M., Roth, B.: Solving polynomial systems for the kinematic analysis and synthesis of mechanisms and robot manipulators. J. Mech. Des. **117**, 71–79 (1995)
8. Uchida, T., McPhee, J.: Efficient solution of kinematics for multi-loop mechanisms using Grobner bases. In: Proceedings of the 13th World Congress in Mechanism and Machine Science, Guanajuato, Mexico, 19–25 June 2011
9. Papegay, Y.A., Merlet, J.-P., Daney, D.: Exact kinematics analysis of car's suspension mechanisms using symbolic computation and interval analysis. Mech. Mach. Theory **40**, 395–413 (2005)
10. Honda site: http://world.honda.com/news/1997/t970702b.html
11. Samarskiy, A.A., Gulin, A.V.: Numerical Methods. Nauka, Moscow (1989)
12. Kalitkin, N.N.: Numerical Methods. Nauka, Moscow (1978)
13. Lunev, V.V., Misyurin, S.Y.: Osobye mnogoobraziya ploskih i prostranstvennyh mechanizmov s neskolkimi stepenyami svobody. Problemy mashinostroeniya i nadezshnosti mashin **1**, 102–109 (1993)
14. Kreinin, G.V., Misyurin, S.Y.: Selection of the scheme for incorporating a drive into the structure of a mechanism in solving problems of kinematic synthesis. J. Mach. Manuf. Reliab. **37**(1), 1–5 (2008)
15. Tikhonov, A.N., Goncharskii, A.V., Stepanov, V.V., Yagola, A.G.: Numerical methods of solving ill-posed problems. Nauka, Moscow (1990)
16. Kreinin, G.V., Misyurin, S.Y.: On some general laws in drive dynamics. J. Mach. Manuf. Reliab. **37**(6), 546–551 (2008)
17. Kreinin, G.V., Misyurin, S.Y., Yashina, M.A.: Synthesis of a positioning system with a hydrodrive, a motion transmission mechanism with variable transfer ratio, and a combined load. J. Mach. Manuf. Reliab. **5**, 1–7 (2004)
18. Kreinin, G.V., Misyurin, SY.: Influence of the gear ratio on the dynamics of a control-system drive. J. Mach. Manuf. Reliab. **41**(6), 486–491 (2012)
19. Kreinin, G.V., Misyurin, S.Y.: A systematic approach to synthesis of a drive system. J. Mach. Manuf. Reliab. **40**(6), 507–511 (2011)

MIX
Papier aus verantwortungsvollen Quellen
Paper from responsible sources
FSC® C105338

If you have any concerns about our products, you can contact us on
ProductSafety@springernature.com

In case Publisher is established outside the EU, the EU authorized representative is:
Springer Nature Customer Service Center GmbH
Europaplatz 3, 69115 Heidelberg, Germany

Printed by Libri Plureos GmbH
in Hamburg, Germany